Game Theory
with Engineering Applications

Advances in Design and Control

SIAM's Advances in Design and Control series consists of texts and monographs dealing with all areas of design and control and their applications. Topics of interest include shape optimization, multidisciplinary design, trajectory optimization, feedback, and optimal control. The series focuses on the mathematical and computational aspects of engineering design and control that are usable in a wide variety of scientific and engineering disciplines.

Editor-in-Chief

Series Volumes

Game Theory

with Engineering Applications

DARIO BAUSO

The University of Sheffield
Sheffield, UK

Università di Palermo
Palermo, Italy

Society for Industrial and Applied Mathematics
Philadelphia

Publisher	David Marshall
Acquisitions Editor	Elizabeth Greenspan
Developmental Editor	Gina Rinelli
Managing Editor	Kelly Thomas
Production Editor	David Riegelhaupt
Copy Editor	Bruce Owens
Production Manager	Donna Witzleben
Production Coordinator	Cally Shrader
Compositor	Techsetters, Inc.
Graphic Designer	Lois Sellers

Library of Congress Cataloging-in-Publication Data

Names: Bauso, Dario.
Title: Game theory with engineering applications / Dario Bauso, The
   University of Sheffield, Sheffield, UK, and Università di Palermo,
   Palermo, Italy.
Description: Philadelphia : Society for Industrial and Applied Mathematics,
   [2016] | Series: Advances in design and control ; 30 | Includes
   bibliographical references and index.
Identifiers: LCCN 2015045970 (print) | LCCN 2016004150 (ebook) | ISBN
   9781611974270 | ISBN 9781611974287 ()
Subjects: LCSH: Engineering–Social aspects. | Game theory.
Classification: LCC TA157 .B3447 2016 (print) | LCC TA157 (ebook) | DDC
   620.001/5193–dc23
LC record available at http://lccn.loc.gov/2015045970

To the loving memory of my parents.

Contents

List of Figures

List of Tables

List of Algorithms

Preface

Why this book now?

A key direction for research in systems and control involves *engineering systems*. These are highly distributed collective systems with humans in the loop. *Highly distributed* means that decisions, information, and objectives are distributed throughout the system. *Humans in the loop* implies that the *players* have bounded rationality and limited computation capabilities. In addition, decisions may also be influenced by societal and cultural habits. Engineering systems emphasize the potential of control and games beyond traditional applications.

The reason why I chose to write this book now is that, within the realm of engineering systems, a key point is the use of *game theory* to design incentives to obtain *socially desirable behaviors* on the part of the players. As an example, in *demand side management*, an increase of the electricity price on the part of the network operator may induce a change in the consumption patterns on the part the prosumers (producers-consumers). In *opinion dynamics*, sophisticated marketing campaigns may influence the market share, assuming that the customers are susceptible players sharing opinions with their neighbors. In *pedestrian flow*, informing the pedestrians on the congestion at different locations may lead to a better redistribution of the traffic. These are only some of the applications discussed in this book.

In this context, *game theory* offers a rich set of model elements, solution concepts, and evolutionary notions. The model elements are the players, the action sets, and the payoffs; the solution concepts include the Nash equilibrium, the Stackelberg equilibrium, and Pareto and social optimality; evolutionary notions shed light on the fact that equilibria are relevant only if the players can converge to such solutions in a dynamic setting. Evolutionary notions essentially turn the game into a kind of dynamic feedback system.

However, *a game theory model is more than just a dynamic feedback system*, as each player learns the environment, which in turn learns the player, and so forth. Such a coupled learning introduces a higher level of difficulty to the feedback structure.

A large portion of this book is dedicated to *games with a large number of players*. Here each player uses an aggregate description of the environment based on a distribution function on actions or states, which is the main idea in a *mean-field game*. Thus, in most examples the game is a mapping from distributions (congestion levels) to payoffs (think of the replicator dynamics).

If a game is a mapping from congestion levels to payoffs, the evolution model is a dynamic model that operates in the opposite direction: it maps flows of payoffs to flows of congestion levels. Here, *systems and control theory* provides a set of sophisticated stabilizability tools to design self-organizing and resilient systems characterized by cooperation and competition. This book will mainly use the Lyapunov approach both in a deterministic and a stochastic setting.

Goal of this book

This book's goal is to bring together game theory and systems and control theory in the unconventional framework of engineering systems. The goal of Part I is to cover the foundations of the theory of noncooperative and cooperative games, both static and dynamic. Part I also highlights new trends in cooperative differential games, learning, approachability (games with vector payoffs), and mean-field games (large number of homogeneous players). The treatment emphasizes theoretical foundations, mathematical tools, modeling, and equilibrium notions in different environments.

The goal of Part II is to illustrate stylized models of engineered and societal situations. These models aim at providing fundamental insights on several aspects, including the individuals' strategic behaviors, scalability and stability of the collective behavior, and the influence of heterogeneity and local interactions. Other relevant issues discussed throughout the book are uncertainty and model misspecification. Remarkably, the framework of robust mean-field games is developed with an eye to grand engineering challenges such as *resilience* and *big data*.

What this book is not

This book is not an encyclopedia of game theory, and the material covered reflects my personal taste. More importantly, this book is not a collection of takeaway models and solutions to specific applications. These models need not be interpreted literally but are guidelines towards a better understanding and an efficient design of collective systems.

Structure of this book

This book is organized in two parts. Part I follows [24] and goes from Chapter 1 to 12. Chapters 1 to 4 review the foundations of noncooperative games. Chapters 5 to 6 deal with cooperative games. Chapter 7 surveys evolutionary games. Chapter 8 analyzes the replicator dynamics and provides a brief overview of learning in games. Chapter 9 deals with differential games. Chapter 10 discusses stochastic games. Chapter 11 pinpoints basics and trends in games with vector payoffs, such as *approachability* and *attainability*. Chapter 12 provides an overview of mean-field games.

Part II builds upon articles of the author and goes from Chapters 13 to 21. In particular, under the umbrella of power systems, Chapters 14 to 15 analyze demand side management and synchronization of power generators, respectively. Within the realm of sociophysical systems, Chapter 13 discusses consensus in multi-agent systems, and Chapters 16 to 18 illustrate, in order, opinion dynamics, bargaining, and pedestrian flow applications. Within the context of production/distribution systems, Chapters 19 to 21 deal with supply chain, population of producers, and cyber-physical systems.

At the end of each chapter a section entitled "Notes and references" acknowledges the work on which the chapter is based and related works.

Audience

The primary audience is students, practitioners, and researchers in different areas of Engineering such as Industrial, Aeronautical, Manufacturing, Civil, Mechanical, and Electrical Engineering. However, the topic also interests scientists in Computer Science, Economics, Physics, and Biology. Young researchers may benefit from reading Part II. The comprehensive reference list enables further research. The book is self-contained and makes the path from undergraduate students to young researchers short.

Using this book in courses

This book can be used as a textbook, especially Part I. This part covers material that can be taught in first-year graduate courses. I use a tutorial style to illustrate the major points so that the reader can quickly grasp the basics of each concept.

Part I assembles the material of three graduate courses given at the Department of Mathematics of the University of Trento, at the Department of Engineering Science of the University of Oxford, and at the Department of Electrical and Electronic Engineering of Imperial College in 2013. The material has also been used for the short course given at the Bertinoro International Spring School 2015 held in Bertinoro, Forlì, Italy.

The book can also be used for an undergraduate course. To this purpose, the book is complemented with Appendix sections on mathematical review, optimization, Lyapunov stability, basics of probability theory, and stochastic stability theory. Part II shows a number of simulation algorithms and numerical examples that may help improve the coding skills of the students. The software used for the simulations is MATLAB®. *Prior knowledge* includes the material discussed in the Appendix sections.

Acknowledgments

A large part of this book is based on my research over the past 10 years. I was honored to have a number of brilliant co-authors, and I would like to mention those with whom I have worked extensively. The collaboration with Tamer Başar and Hamidou Tembine is the origin of many ideas in robust mean-field games. The collaboration with Ehud Leher, Eilon Solan, and Xavier Venel has inspired research on attainability (cf. Chapter 11). The joint work with Franco Blanchini is the source of several ideas on robust stabilizability of network flows appearing throughout the book. Raffaele Pesenti and Laura Giarré have helped me develop the ideas discussed in the multi-agent consensus application in Chapter 13. The bargaining model in Chapter 17 has been developed in a joint work with Angelia Nedić. The supply-chain model in Chapter 18 has been studied in a collaboration with Judith Timmer. The collaboration with Fabio Bagagiolo has inspired the design of objective functions in differential and mean-field games.

More recently, the approximation technique based on state space extension to compute mean-field equilibria has resulted from the fruitful interactions with Alessandro Astolfi and Thulasi Mylvaganam during my sabbatical at Imperial College London in 2013. The collaboration with Antonis Papachristodoulou and Xuan Zhang has inspired the pedestrian flow model in Chapter 18. My special thanks to Xuan, who has contributed the simulations in Chapter 18. I really enjoyed sharing thoughts with Mark Cannon, and the resulting ideas combining games and receding horizon are discussed in Chapter 16 in the context of opinion dynamics. The collaborations with Antonis, Xuan, and Mark started during my sabbatical period in Oxford in 2013.

Many thanks are due to the several PhD students, postdocs, and fellows who have attended the courses and have contributed to the improvement of the material with their comments and questions.

I deeply thank my sister Helga, who has been generous with love and encouragement despite the distance between us. Finally, I would like to thank Claudia for her enormous support and for sharing the ups and downs with me.

I hope you will enjoy reading the book as much as I did writing it!

List of Notation

We use the following abbreviations and symbols throughout the book.

$\mathbb{R}$	set of real numbers
$\mathbb{R}^n$	n-dimensional vector space over $\mathbb{R}$
$\mathbb{R}^+$	set of nonnegative real numbers
x^T	transpose of a vector x
A^T	transpose of a matrix A
x_i or $[x]_i$	ith coordinate component of a vector x
a_{ij} or $[A]_{ij}$ or a^i_j	ijth entry of a given matrix A
$x < y$ $(x \le y)$	$x_i < y_i$ $(x_i \le y_i)$ for all coordinate indices i of two vectors x and y
$[\xi]_+$	positive part of real $\xi \in \mathbb{R}$
$\|x\|$	Euclidean norm of a vector x
$\|x\|^2_A$	weighted two-norm $x^T A x$ of given vector $x \in \mathbb{R}^n$ and matrix $a \in \mathbb{R}^{n \times n}$
Δ^n	simplex in $\mathbb{R}^n$
$\Pi_X[x]$	projection of a vector x on a set X, i.e., $\Pi_X[x] = \arg\min_{y \in X} \|x - y\|$
$\text{dist}(x, X)$	distance from vector x to set X, i.e., $\text{dist}(x, X) = \|x - \Pi_X[x]\|$
$U \subset S$	U is a proper subset of S
$|S|$	cardinality of a given finite set S
∂_x	first partial derivative with respect to x or gradient with respect to x
∇_x or ∇	gradient
∂^2_{xx}	second derivative with respect to x
∇^2	Hessian matrix
$\mathbb{E}$	expectation
$\mathbb{P}$	probability
$\bar{m}(.)$	mean of a given density function $m(.)$
$std(m(.))$	standard deviation of a given density function $m(.)$

Part I

Theory

Chapter 1

Introduction to Games

1.1 ▪ Introduction

This chapter is an introduction to the foundations of game theory, i.e., the *theory of strategic thinking*.

Game theory intersects several disciplines. From Table 1.1, we understand that game theory conventionally involves multiple players and multiple payoffs. From this perspective, game theory is a generalization of *optimization* theory, which deals with one player, the optimizer, and one payoff, called objective function. Game theory also differs from *multi-objective optimization*, which deals with one player and multiple payoffs. Likewise, game theory differs from *team theory*, which considers multiple decision makers and one payoff.

Table 1.1. *Connections of game theory with other disciplines.*

	1 payoff	n payoffs
1 player	Optimization	Multi-objective optimization
n players	Team theory	Game theory

In Section 1.2, we browse applications. In Section 1.3, we introduce different types of games, such as simultaneous and sequential games. Different types of games admit different game formulations as, for instance, the strategic or normal representation and the extensive or tree representation. We also distinguish between *cooperative* and *noncooperative games*. We continue with the introduction of basic concepts such as pure and mixed strategy, Nash equilibrium, and dominant strategy, in Section 1.4. Furthermore, we streamline seminal results on the existence of equilibria. Section 1.5 discusses the iterated dominance algorithm. We also introduce the *Cournot duopoly* as an example of an infinite game and illustrate the aforementioned algorithm on it. We conclude with the presentation of some stylized finite games, such as the *Coordination game*, the *Hawk and Dove game*, or the *Stag-Hunt game*, in Section 1.6. Finally, in Section 1.7 we provide notes and references for this chapter.

1.2 ▪ Applications

Game-theoretic models arise in several application domains, and below is a partial list of them.

3

Field and board games. In field games such as football or rugby or board games like chess or draughts, we can review the players' decisions as elements of a given set, called a decisions set, and the probability of a win is the payoff that every player seeks to maximize. Thus, such games admit a mathematical description via game theory. It is known, for instance, that certain tactics in rugby are successful only under the assumption that the opponent is playing a certain tactic. A common perspective is the one of assimilating the tactic choice to a play of the *Rock-Paper-Scissors game. Algorithmic game theory* provides theoretical foundations for this. Algorithmic game theory overlaps algorithm design, game theory, and artificial intelligence.

Commercial and business operations. When firms operate in the same market they usually are competitors. Survival is sometimes related to their ability to predict the impact of a new product on the market and how such a new product will change the competitor's strategic operations. This involves a strategic analysis of the current market demand and of the reactions of the potential competitors in consequence of the introduction of the new product.

Politics. Game theory is used in politics to produce indices to measure the power of parties involved in a governing coalition. Game theory is also useful for analyzing voting methods. In social policymaking, game theory provides tools to governmental agencies for a better understanding of the impact of specific social policy choices, such as pension rules, education, or labor reforms.

Military and civil defense. It is in defense that game theory first contributed the notion of *strategic thinking*. This is about putting ourselves in the place of the opponent before making a decision. Such a role-playing game is a milestone of the theory. In military applications, as, for instance, the one involving missile pursuing fighter airplanes, missions are usually developed on game-theoretic models.

Engineering applications, robotics, and multi-agent systems. Within the broad area of engineering applications, game theory develops models for the movement of automated robot vehicles with distributed task assignment. Robotic manipulation and path planning in the presence of moving obstacles is also a classical game theory application.

Networks. In social networks, one deals with the analysis of the spread of innovation, or the propagation of opinions. Here, game theory provides fundamental insights on why certain behaviors or opinions emerge. In communication networks, game theory is commonly used to design band allocation policies, improve security, and reduce threats.

1.3 ▪ Overview on different types of games

In this section, we survey different types of games. For each type we have a corresponding representation. We start by providing a formal description of a game in generic terms. Then, we introduce two distinct classes: *cooperative* and *noncooperative* games. After doing this, a second distinction we highlight is between *simultaneous* and *sequential* games. Finally, we illustrate the *strategic* or *normal* representation used for simultaneous games and the *extensive* or *tree* representation used for sequential games.

1.3.1 ▪ Ingredients for a game

A game in strategic form involves a tuple $\langle N, (\mathscr{A}_i)_{i \in N}, (u_i)_{i \in N} \rangle$, where

- the set $N = \{1, 2, \ldots, n\}$ is the *set of players*, which we assume to be maximizers if not specified differently;

- the set $\mathscr{A}_i$ is the *set of actions of player i* for all $i \in N$;

- the set $A := \{a | a = (a_i)_{i \in N}, a_i \in \mathscr{A}_i, \forall i \in N\}$ is the *set of action profiles*, where an *action profile* (also called *outcome*) is an n-tuple of actions;

- the function $u_i : A \to \mathbb{R}$ is the *payoff function of player i*, i.e.,

$$(a_1, \ldots, a_n) \mapsto u_i(a_1, \ldots, a_n).$$

Here the payoff u_i is conventionally assumed to be a profit, and therefore it has to be maximized. In other circumstances it can also be a cost to be minimized.

Action profiles can be equivalently written in order to isolate player i's action as

$$(a_j)_{j \in N} = (a_1, \ldots, a_n) = (a_i, a_{-i}),$$

where $a_{-i} = (a_j)_{j \in N, j \neq i}$ is the action profile of all players except i.

1.3.2 ▪ A first distinction: Noncooperative and cooperative games

The theory of noncooperative games differs substantially from the one of cooperative games. The two theories have given rise to two independent literatures. To understand the main differences, in noncooperative games

(i) every player maximizes his own payoff by choosing his *best response* on the basis of what he knows about others' actions;

(ii) the players have no binding agreements on joint actions that are optimal for the group;

(iii) the players are not involved in any pre-play communication stage.

Cooperative games admit two different formulations. One formulation is about *games with transferable utility (TU games)*, whereas the second formulation deals with games with *nontransferable utility (NTU games)*. Differently from noncooperative games, in cooperative games

(i) the players look for joint actions which may turn optimal for the group, which is the case in NTU games, or seek reasonable cost or reward sharing rules that make the coalitions stable, as in TU games;

(ii) the players may be involved in a pre-play communication stage;

(iii) the players are allowed to use side payments in order to stabilize the coalition as in TU games.

It is well known that noncooperative game theory plays a dominant role in most textbooks on game theory. Also, noncooperative game theory is to some extent more common than cooperative game theory among scientists of disciplines other than economics. However, there are controversial opinions about the fact that cooperative game theory may have a broader range of applications.

The next example is a milestone of the theory of noncooperative games.

Example 1.1 (*Prisoner's dilemma*). The *Prisoner's dilemma* represents one of the most common stylized noncooperative games. The underlying story involves two criminals who are arrested under the suspicion of having committed a crime. The maximal sentence for the crime is four years. Each fellow has two possibilities: *to cooperate* with the other fellow, denoted by C, or *to defect* the other fellow, denoted by D. If the fellows play (D,D), the sentence is mitigated to three years, which corresponds to saying that each one gets one year of freedom. If the fellows play (C,C), both are released after one year due to lack of evidence, which corresponds to each one getting three years of freedom. If only one fellow *cooperates*, namely (C,D) or (D,C), the fellow who plays D is released immediately, which corresponds to four years of freedom, whereas the other fellow is sentenced to the maximal punishment; that is, he gets zero years of freedom. Such a strategic scenario can be represented in bimatrix form, as displayed in Fig. 1.1.

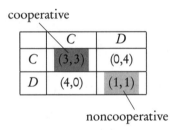

Figure 1.1. Prisoner's dilemma: *cooperative vs. noncooperative solutions.*

The above scenario describes a noncooperative context, where every player, having no guarantee on the opponent's play, is better off by playing D. The corresponding outcome is then (D,D) depicted in light gray in Fig. 1.1. The above scenario can be reviewed also in a cooperative context. Here we assume that both players can collude and negotiate on joint actions. In this case, a better solution for the group requires that both cooperate, i.e., (C,C), and this is depicted in dark gray in Fig. 1.1. ∎

1.3.3 ▪ A second distinction: Simultaneous and sequential games

A second distinction is between *simultaneous* and *sequential games*. Simultaneous games are characterized as follows:

   (i) players take actions or make decisions once and for all and at the same time;

   (ii) just because the game is played in one shot, there is no variable that summarizes the *state* of the game, and consequently the actions cannot depend on states, which implies that there is no need to build strategies (this can be better understood after we introduce the notion of strategy as a map from states to decisions in sequential games);

   (iii) there exists a common *representation in normal form*, which is also called *strategic form* or, in the case of two players, *bimatrix form*.

The *Prisoner's dilemma* introduced earlier is an example of simultaneous game in strategic form; see Fig. 1.1. In the bimatrix representation the rows and the columns scan the actions or decisions of the two players, also called row and column players, and each entry of the bimatrix involves two elements which are the payoffs of the players. Note that this kind of representation does not carry any inbuilt information structure.

Differently from simultaneous games, in *sequential games* we find the following aspects:

(i) there exists a specific order of events which establishes "who makes decisions when";

(ii) at each stage a *state* variable collects information on earlier decisions;

(iii) at each stage the player who is in turn to make a decision may have perfect or imperfect information on the actual state (earlier decisions);

(iv) the players make decisions based on the state; in other words, they build a map from states to decisions, and such a map is called *strategy*;

(v) the sequential nature of the game is conveniently represented in *extensive* or *tree form*, as displayed in Fig. 1.2.

In an extensive or tree representation, the nodes indicate the states and the corresponding labels identify the players who are to act. The edges are the actions that can be taken in a given state. The end of the game is represented by leaf nodes which indicate the payoffs resulting from the whole history of actions taken. Such a representation has an inbuilt information structure.

The example in Fig. 1.2 shows a two-player extensive game where in stage 1, player 1 can play *left*, L, or *right*, R. Then, in stage 2, player 2 can play *left*, l, or *right*, r. In stage 2, player 2 is in one of the two states, indicated as state 1 (light gray node) or state 2 (dark gray node), which yields four possible actions: l_1, r_1, l_2, r_2 (the index indicates the state; for instance, r_1 means *right* in state 1, and l_2 means *left* in state 2).

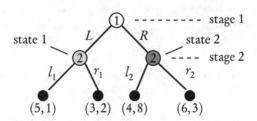

Figure 1.2. *Example of extensive/tree form representation: (stage 1) player 1 can play L or R; (stage 2) player 2 can play l_1 or r_1 in state 1 (light gray node), and l_2 or r_2 in state 2 (dark gray node). Reprinted with permission from Hindustan Book Agency* [239].

It is worth noting that it is possible to have a strategic form representation also for an extensive game. To do this, we need to substitute decisions with strategies for some of the players. This is illustrated later in Example 1.5 and Fig. 1.5. There we have four strategies for player 2, i.e., $l_1 l_2$ (always *left*), $l_1 r_2$ (*left* only in state 1, that is, when player 1 plays L), $l_2 r_1$ (*left* only in state 2, that is, when player 1 plays R), and $r_1 r_2$ (always *right*). Thus, the set of "actions" for player 2 is $\mathscr{A}_2 = \{l_1 l_2, l_1 r_2, r_1 l_2, r_1 r_2\}$, while the one for player 1 is simply $\mathscr{A}_1 = \{L, R\}$.

When simultaneous games are played repeatedly in time, the game is called *repeated game*. A natural representation for these games is in extensive form. Fig. 1.3 shows this for the *Prisoner's dilemma* example. Here we consider streams of instantaneous payoffs, which are summed up over the rounds of a finite horizon or an infinite horizon. This can be done by considering a discounted sum or a long-term average over the horizon window.

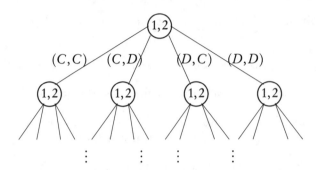

Figure 1.3. *Extensive or tree form representation of the* Prisoner's dilemma. *Reprinted with permission from Hindustan Book Agency* [239].

1.4 ▪ Nash equilibrium and dominant strategy

The concepts of Nash equilibrium and dominant strategy are among the foundations of game theory. In this section we review both concepts, which will accompany the reader through the whole book.

1.4.1 ▪ Nash equilibrium

A Nash equilibrium is an action profile such that no player can be better off by deviating from it, assuming that the other players do not change their actions. This leads to the notion of *unilateral deviations*, namely, situations where only one player changes his own decision while the others stick to their current choices, actions, or decisions. Thus we can say shortly that, in a Nash equilibrium, unilateral deviations do not benefit any of the players.

Definition 1.2 (Nash equilibrium). *The action profile/outcome* $(a_1^*, a_2^*, \ldots, a_n^*)$ *is a Nash equilibrium if none of the players by deviating from it can gain anything, i.e.,*

$$u_i(a_i^*, a_{-i}^*) \geq u_i(a_i, a_{-i}^*) \quad \forall a_i \in \mathscr{A}_i, \forall i \in N.$$

A Nash equilibrium can be equivalently defined using *best-response* sets.

Definition 1.3 (Best-response set). *The best-response set for player i is the set*

$$\mathscr{B}_i(a_{-i}) := \{a_i^* \in \mathscr{A}_i \,|\, u_i(a_i^*, a_{-i}) = \max_{a_i \in \mathscr{A}_i} u_i(a_i, a_{-i})\}.$$

Then in a Nash equilibrium all players play a best response, namely

$$a_i^* \in \mathscr{B}_i(a_{-i}^*) \quad \forall i \in N.$$

Example 1.4. In the *Prisoner's dilemma* (see Fig. 1.4) the solution (D, D) is a Nash equilibrium, as player 1 by deviating from it would get 0 years of freedom rather than 1 (stick to 2nd column and move vertically to 1st row) and therefore would be worse off. Likewise for player 2. Note that the Nash equilibrium corresponds to the noncooperative solution introduced earlier (see Fig. 1.1) for the same example. ∎

	C	D
C	(3,3)	(0,4)
D	(4,0)	(1,1)

Figure 1.4. Prisoner's dilemma: (D,D) is a Nash equilibrium.

Example 1.5. In the extensive game of Fig. 1.5, player 2 has four strategies, i.e.,

- $l_1 l_2$ (always *left*),

- $l_1 r_2$ (*left* only in state 1, that is, when player 1 plays L),

- $l_2 r_1$ (*left* only in state 2, that is, when player 1 plays R), and

- $r_1 r_2$ (always *right*).

Thus, the set of "actions" for player 2 is $\mathscr{A}_2 = \{l_1 l_2, l_1 r_2, r_1 l_2, r_1 r_2\}$, while the one for player 1 is $\mathscr{A}_1 = \{L, R\}$. There exists one Nash equilibrium $(R, r_1 l_2)$. This can be computed via dynamic programming backwards. To see this, in state 1 (light gray node), player 2's rational choice is r_1 (dashed line), as he gets 2, while by playing l_1 he would get 1. In state 2 (dark gray node), player 2 could play l_2 and get 8 or play r_2 and get 3; then his rational choice is l_2 (dashed line). In stage 1 (top node), player 1 gets 4 by playing R, and 3 by playing L, so his best response is R (dashed line). The equilibrium payoffs are then $(4,8)$. ∎

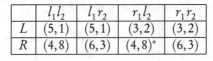

	$l_1 l_2$	$l_1 r_2$	$r_1 l_2$	$r_1 r_2$
L	(5,1)	(5,1)	(3,2)	(3,2)
R	(4,8)	(6,3)	(4,8)*	(6,3)

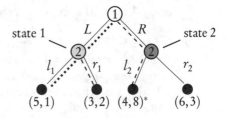

Figure 1.5. *Example of normal representation of a sequential game. Reprinted with permission from Hindustan Book Agency* [239].

The representation in normal form of the game (left) shows another solution, $(R, l_1 l_2)$, returning the same payoffs as the equilibrium payoffs. Note that this solution is not a Nash equilibrium, as player 2 would benefit from changing from l_1 to r_1. So, in principle there may exist solutions that are not equilibria and which are equivalent to equilibria in terms of payoffs.

The literature provides a weaker equilibrium solution, called ϵ-*Nash equilibrium*.

Definition 1.6 (ϵ-Nash equilibrium). *For a given $\epsilon \geq 0$, the action profile $(a_1^\epsilon, a_2^\epsilon, \ldots, a_n^\epsilon)$ is an ϵ-Nash equilibrium if no player by deviating from it can gain more than ϵ, i.e.,*

$$u_i(a_i^\epsilon, a_{-i}^\epsilon) \geq u_i(a_i, a_{-i}^\epsilon) - \epsilon \quad \forall a_i \in \mathscr{A}_i, \forall i \in N.$$

The ϵ-Nash equilibrium is a generalization of the Nash equilibrium. Actually, the ϵ-Nash equilibrium coincides with the Nash equilibrium for $\epsilon = 0$.

1.4.2 ▪ On the existence of equilibria in mixed strategies

A milestone in the theory of games is the *equilibrium point theorem* by Nash (1950) [183]. The theorem establishes the existence of equilibrium solutions for nonzero-sum games.

The *equilibrium point theorem* makes use of the notion of *mixed strategy*.

Definition 1.7 (Mixed strategy). *A mixed strategy is a strategy defined by a probability distribution over the finite set of the feasible strategies.*

In the parlance of game theory, one uses the term *pure strategy* in contrast to mixed strategy. In plain words, a mixed strategy is a randomization on pure strategies.

The *equilibrium point theorem* builds on *Kakutani's (fixed point) theorem* (1941). *Kakutani's theorem* provides sufficient conditions for a set-valued function, defined on a convex and compact subset of a Euclidean space, to have a *fixed point*, i.e., a point which is mapped to a set containing it.

Theorem 1.8 (*Kakutani's (fixed point) theorem* (1941)). *Let K be a nonempty subset of a finite-dimensional Euclidean space. Let $f : K \to K$ be a correspondence, with $x \in K \mapsto f(x) \subseteq K$, satisfying the following conditions:*

- *K is a compact and convex set;*

- *$f(x)$ is nonempty for all $x \in K$;*

- *$f(x)$ is a convex-valued correspondence: for all $x \in K$, $f(x)$ is a convex set;*

- *$f(x)$ has a closed graph; that is, if $\{x_n, y_n\} \to \{x, y\}$ with $y_n \in f(x_n)$, then $y \in f(x)$.*

Then, f has a fixed point; that is, there exists some $x \in K$ such that $x \in f(x)$.

Kakutani's theorem is a generalization of *Brouwer's (fixed point) theorem*. This fixed point theorem says that if S is a compact and convex subset of $\mathbb{R}^n$ and f is a continuous function mapping S into itself, then there exists at least one $x \in S$ such that $f(x) = x$.

Rather than the formal proof of *Kakutani's theorem*, we provide a graphical illustration for a simple scalar case of the main ideas used in the proof. Let x be plotted in the horizontal axis and $f(x)$ in the vertical axis, as in Fig. 1.6. Fixed points, if they exist, must solve $f(x) = x$ and therefore can be found at the intersection between the function $f(x)$ and the dotted line. On the left, the function $f(x)$ is not convex valued, and therefore it does not admit a fixed point. On the right, the function $f(x)$ does not have a closed graph, which again implies that there exists no fixed point.

Nash used *Kakutani's theorem* to prove the existence of a Nash equilibrium for nonzero-sum games. In plain words, Nash's equilibrium theorem establishes the existence of at least one Nash equilibrium under the following conditions:

(i) the set of actions $\mathscr{A}_i$ consists of compact and convex subsets of $\mathbb{R}^n$, as it occurs in continuous (infinite) games, or games in mixed extension (we will expand more on it later);

(ii) the payoffs $u_i(a_i, a_{-i})$ are continuous and concave in a_i for fixed strategy a_{-i} of the opponents.

In the following we provide a formal statement of the *equilibrium point theorem*.

Theorem 1.9 (*Equilibrium point theorem*). *Each finite bimatrix game has a Nash equilibrium in the mixed strategies.*

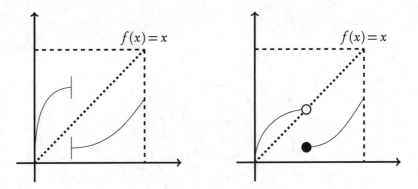

Figure 1.6. *Graphical illustration of* Kakutani's theorem. *Function $f(x)$ is not convex valued (left), and $f(x)$ has no closed graph (right). Reprinted with permission from Asuman Ozdaglar and Morgan and Claypool* [197, 178].

***Proof* (Sketch).** We here provide only a sketch of the proof. Let us introduce the best-response set,

$$\mathcal{B}_i(a_{-i}) := \{a_i^* \in \mathcal{A}_i \mid u_i(a_i^*, a_{-i}) = \max_{a_i \in \mathcal{A}_i} u_i(a_i, a_{-i})\}.$$

We can then apply *Kakutani's (fixed point) theorem* to the best-response correspondence $\mathcal{B} : \Delta \rightrightarrows \Delta$, $\Delta = \prod_{i \in N} \Delta_i$ (Δ_i is the simplex in the $\mathbb{R}^{|\mathcal{A}_i|}$):

$$\mathcal{B}(a) = \left(\mathcal{B}_i(a_{-i}) \right)_{i \in N}.$$

In other words the theorem is proved once we show that the correspondence $\mathcal{B}(a)$ satisfies the conditions of Kakutani's theorem. $\quad\square$

A relevant characteristic of Nash equilibria in mixed strategies is that every action in the support of any player's equilibrium mixed strategy is a best response and yields that player the same payoff. Such a property is commonly referred to as the *Indifference Principle* and will be used extensively in the remainder of this book.

In Chapter 3 we will see that the computation of Nash equilibrium solutions for nonzero-sum games can be performed via linear complementarity programming (see also [239, Chap. 7]).

Example 1.10. This example describes a two-player continuous infinite game. We say that the game is an *infinite game*, as the set of actions consists of segments in $\mathbb{R}$ (see horizontal and vertical axes in Fig. 1.7), and therefore each player has an infinite number of available actions. From looking at the level curves, we see that the maxima are attained at points P and Q for players 1 and 2, respectively. The Nash equilibrium is point R, which has horizontal and vertical tangents to the level curves of players 1 and 2 passing through it. Indeed, from belonging to a horizontal tangent, the horizontal coordinate of point R is the best response of player 1 to player 2. Similarly, from belonging to a vertical tangent, the vertical coordinate of R is the best response of player 2 to player 1. ∎

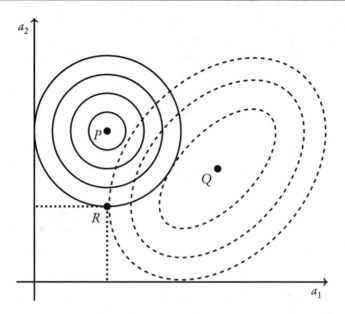

Figure 1.7. *Two-player continuous infinite game. Level curves of player 1 (solid) and player 2 (dashed); action spaces of player 1 (horizontal axis) and player 2 (vertical axis). Global maximum is P for player 1 and Q for player 2, while the Nash equilibrium is point R. Reprinted with permission from A. Bressan and Springer Science+Business Media* [65, 66].

1.4.3 ▪ Dominant strategy

We saw that the concept of equilibrium refers to action profiles. In this section we introduce the concept of *dominant strategy*, which is a characteristic related to a single action.

Dominance is a strong property, in that we will see that an action profile made by dominant strategies is a Nash equilibrium but that the converse is not true, i.e., we can have a Nash equilibrium that does not involve dominant strategies. We state in the following the definition of weak dominance.

Definition 1.11 (Weak dominance). *Given two strategies, $a_i^*, a_i \in \mathcal{A}_i$, we say that a_i^* weakly dominates a_i if it is at least as good as a_i for all choices of the other players $a_{-i} \in \mathcal{A}_{-i}$,*

$$u_i(a_i^*, a_{-i}) \geq u_i(a_i, a_{-i}) \quad \forall a_{-i} \in \mathcal{A}_{-i}.$$

If the above inequality holds strictly, then we say that a_i^* *(strictly) dominates* a_i.

Example 1.12. In the *Prisoner's dilemma*, illustrated in Fig. 1.8, strategy D is a dominant strategy.   ∎

	C	D
C	(3,3)	(0,4)
D	(4,0)	(1,1)

Figure 1.8. *D is a dominant strategy in the* Prisoner's dilemma.

We say that a strategy is (weakly) dominant if it (weakly) dominates all other strategies. It turns out that a profile of dominant strategies is a Nash equilibrium. However, the converse is not true.

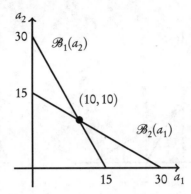

Figure 1.9. *Best-response curves for the Cournot duopoly.*

The property of dominance is used in the renowned *iterated dominance algorithm*. The algorithm, at each iteration, eliminates subsets of dominated solutions. This can be visualized in an exploration tree by "pruning" the corresponding node. This is a technique used in combinatorial optimization. Here the search for the optimum is graphically represented by an exploration tree. The nodes describe families of solutions. If, based on estimates, a family does not contain the optimum, the corresponding node is pruned. The iterated dominance algorithm builds on the property that an undominated strategy survives to the algorithm pruning. We illustrate this in the context of the Cournot duopoly.

1.5 ▪ Cournot duopoly and iterated dominance algorithm

Two manufacturers $i = 1, 2$ compete on a same market and must decide their production quantities. The production quantity of manufacturer i is q_i. From the *law of demand*, the sale price of manufacturer i decreases with the total production quantity $(q_1 + q_2)$. In particular, for the sale price we take the expression

$$c_i = 30 - (q_1 + q_2).$$

As a consequence, the income (payoff) of manufacturer i obtained by selling the produced quantity q_i at the price c_i is given by

$$u_i(q_i, q_j) = c_i q_i = 30 q_i - q_i^2 - q_i q_j.$$

As the payoff is concave in q_i, the maximum is obtained by imposing the derivative of the payoff with respect q_i equal to zero, namely, $\frac{\partial u_i}{q_i} = 0$. This yields the best-response curve of player i:

$$q_i^* = \mathcal{B}_i(q_j) = 15 - q_j/2.$$

The Nash equilibrium is at the intersection between the best-response curves of both manufacturers, which is point $(10, 10)$ (see Fig. 1.9).

1.5.1 ▪ Iterated dominance algorithm

We apply the iterated dominance algorithm to a normalized version of the Cournot duopoly model to find the Nash equilibrium. The first two steps of the algorithm are illustrated in Fig. 1.10. Every iteration involves one round of elimination of dominated strategies.

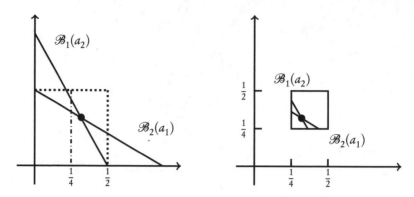

Figure 1.10. *The iterated dominance algorithm illustrated on the Cournot duopoly. Reprinted with permission from Asuman Ozdaglar and Morgan and Claypool* [197, 178].

Let S_i^j be the set of actions of player i that have survived to the elimination rounds up to iteration j. Then, the first round of elimination returns $S_1^1 = [0, \frac{1}{2}]$, $S_2^1 = [0, \frac{1}{2}]$ (left). To see this, consider that any production rate greater than $\frac{1}{2}$ is a dominated action for player 1 as $\mathscr{B}_1(a_2)$ lives in the range $[0, \frac{1}{2}]$. The same reasoning applies to player 2 as the game is symmetric. As a consequence, the equilibrium must be searched in the new domain $[0, \frac{1}{2}]$ for both players (dotted square on the left plot). In a second round of elimination, we obtain $S_1^2 = [\frac{1}{4}, \frac{1}{2}]$, $S_2^2 = [\frac{1}{4}, \frac{1}{2}]$ (right). Actually, after restricting the best responses to the dotted square (left), every player knows that his best response lives in the range $[\frac{1}{4}, \frac{1}{2}]$. In other words any production rate less than $\frac{1}{4}$ is a dominated strategy. Thus the search concentrates to the new domain $[\frac{1}{4}, \frac{1}{2}]$ for both players. This corresponds to the square (solid line) on the right plot. By replicating the same iterative procedure, the algorithm is shown to converge to the Nash equilibrium.

1.6 ▪ Stylized strategic models

In the last part of this chapter, we present some stereotypical models of strategic games, such as the *Battle of the Sexes*, the *Coordination or Typewriter game*, the *Hawk and Dove or Chicken game*, and the *Stag-Hunt game*.

Example 1.13 (*Battle of the Sexes*). The game is described by the bimatrix in Fig. 1.11. A couple decides to meet in the evening to go shopping (this action is indicated by S) or to attend a cricket match (this action is indicated by C). Preferences are different for the two players. The column player prefers to go to the cricket game, while the row player prefers to go shopping. However, both prefer to go to the same place, rather than to split and to go to different places. Here, the payoffs represent a measure of the "happiness" of the two players. In a first scenario, both go shopping, i.e., (S, S), which implies that the row player has a payoff equal to 2 while the column player has a payoff equal to 1. In a second scenario, both go to the cricket game, in which case the payoffs of the two players swap, namely 2 for the column player and 1 for the row player. In the other two scenarios, the players go to different places and both payoffs are 0. The action profiles (S, S) and (C, C) are both Nash equilibrium solutions. There exist no dominant strategies. ∎

	S	C
S	(2,1)	(0,0)
C	(0,0)	(1,2)

Figure 1.11. Battle of the Sexes: (S, S) and (C, D) are Nash equilibrium solutions; there are no dominant strategies.

Example 1.14 (*Coordination or Typewriter game*). This game is usually presented as a stylized model for the diffusion of innovation. The game is described by the bimatrix in Fig. 1.12. It provides insights on when it is convenient to adopt a new technology. The game involves a couple who agrees to meet in the evening to go to a Mozart or a Mahler concert. Both players have a small preference for Mozart, and if they both select $(Mozart, Mozart)$, then the payoffs of both players (think of the payoff as the level of happiness) are 2. The payoffs are a bit lower, say 1, if both players go to a Mahler concert, i.e., $(Mahler, Mahler)$. The other two scenarios are about going to two different concerts, which returns payoffs equal to 0 to both players. The action profiles $(Mozart, Mozart)$ and $(Mahler, Mahler)$ are both Nash equilibrium solutions. There exist no dominant strategies. ∎

	$Mozart$	$Mahler$
$Mozart$	(2,2)	(0,0)
$Mahler$	(0,0)	(1,1)

Figure 1.12. Coordination game: $(Mozart, Mozart)$ and $(Mahler, Mahler)$ are Nash equilibrium solutions; there are no dominant strategies.

Example 1.15 (*Hawk and Dove or Chicken game*). This game builds on the idea that while each player prefers not to give in to the other, the worst possible outcome occurs when both players do not yield. The game describes a situation where two drivers drive towards each other and the one who swerves at the last moment is the "chicken," as he lacks courage. The same game is also known as *Hawk and Dove game*. Here the game describes a scenario where two contestants may choose between a nonaggressive or aggressive behavior. The game was used during the Cold War to analyze the strategic scenario on the occasion of the Cuban Missile Crisis. The game is particularly meaningful if the cost of fighting exceeds the prize of victory, i.e., $C > V > 0$. The model is described by the bimatrix in Fig. 1.13. In a first scenario, both players opt for a nonaggressive behavior and share the prey. This corresponds to the outcome $(Dove, Dove)$, and each player's payoff is half of the prize of victory, $V/2$. In a second scenario, both players behave aggressively and end up fighting, which corresponds to the outcome $(Hawk, Hawk)$. In this case each player will pay a cost equal to half of the prize of victory subtracted to the cost of fighting. The other two scenarios are about one player yielding and the other player getting the entire prey. This is described by the outcomes $(Hawk, Dove)$ and $(Dove, Hawk)$. The winner gets the prize of victory, which is V, while the loser is left with zero reward. For this game, we have two Nash equilibrium solutions, $(Hawk, Dove)$ and $(Dove, Hawk)$. There exist no dominant strategies. ∎

Example 1.16 (*Stag-Hunt game*). This game is used to analyze and predict social cooperation. The model illustrates situations where two individuals can go out on a hunt and

	$Hawk$	$Dove$
$Hawk$	$\left(\frac{V-C}{2}, \frac{V-C}{2}\right)$	$(V, 0)$
$Dove$	$(0, V)$	$\left(\frac{V}{2}, \frac{V}{2}\right)$

Figure 1.13. Hawk and Dove *or* Chicken game*: (Dove, Hawk) and (Hawk, Dove) are Nash equilibrium solutions; there are no dominant strategies.*

collaborate or not collaborate. Both hunters must decide whether to hunt a stag or a hare, but without knowing a priori what the other hunter has decided. Here the challenging aspect is that hunting a stag alone is not possible. The model is described by the bimatrix in Fig. 1.14. Thus, a first scenario contemplates both players cooperating and going for a stag, which yields the outcome $(Stag, Stag)$. In this case both players share the large prey and each payoff is $\frac{3}{2}$. In a second scenario, both players hunt a hare, which corresponds to the outcome $(Hare, Hare)$. The payoff is lower for both players, say equal to 1. The other two scenarios consider both players going for different preys, that is, $(Stag, Hare)$ or $(Hare, Stag)$. Then, the hunter who goes for the smaller prey (the hare) gets the entire prey for himself, while the other hunter is left with nothing, as he cannot get a stag by himself. For this game $(Stag, Hare)$ and $(Hare, Stag)$ are two Nash equilibrium solutions. There are no dominant strategies.   ■

	$Stag$	$Hare$
$Stag$	$\left(\frac{3}{2}, \frac{3}{2}\right)$	$(0, 1)$
$Hare$	$(1, 0)$	$\left(1, 1\right)$

Figure 1.14. Stag-Hunt game*: (Stag, Hare) and (Hare, Stag) are Nash equilibrium solutions; there are no dominant strategies.*

1.7 ▪ Notes and references

Game theory has its origins in the book by the mathematician von Neumann and the economist Morgenstern [248]. The book develops ideas already available in [247]. Quoting from [15], *Morgenstern was the first economist clearly and explicitly to recognize that economic agents must take the interactive nature of economics into account when making their decisions. He and von Neumann met at Princeton in the late Thirties, and started the collaboration that culminated in the Theory of Games.*

Precursors are the French philosopher and mathematician Cournot, who first introduced the *duopoly model* in 1838, and the German economist von Stackelberg, who formulated the equilibrium concept named after him in 1934 [249].

There are different formal definitions of game theory in the literature. Maschler, Solan, and Zamir define game theory as a methodology using *mathematical tools to model and analyze situations involving several decision makers (DMs), called players* [173]. Osborne and Rubinstein say that game theory is *a bag of analytical tools designed to help us understand the phenomena that we observe when DMs interact (DMs are rational and reason strategically)* [196]. Here, by (individual) rationality and strategic reasoning one means that every DM *is aware of his alternatives, forms expectations about any unknowns, has clear preferences, and chooses his action deliberately after some process of optimization* [196]. Tijs in his book introduces game theory as *a mathematical theory dealing with models of conflict and cooperation* [239].

In the introductory paragraph we pointed out that game theory has connections with team theory, which considers multiple DMs with a common payoff. Seminal papers on team theory are [118, 172]. Team theory with binary decisions is studied in [38].

In Section 1.2 we discuss applications. Field and board games are discussed in [54]. Algorithmic game theory is the main topic in [190]. More details on the use of game theory in commercial and business operations are available in [195, 101]. Game theory applied to politics is examined in [182]. Game-theoretic approaches to military and civil defense are presented in [106]. For engineering applications, robotics, and multi-agent systems, we refer the reader to [222]. A survey on game theory and distributed control is provided in [170]. Games and networks are the main focus of [86, 209].

The *Prisoner's dilemma* was developed by Flood and Dresher of the RAND Corporation in 1950. The interpretation in terms of prison sentence and the corresponding name is due to Tucker.

A detailed analysis of cooperative game theory applied to communication networks is in [209].

The original works by John Nash where the Nash equilibrium was first formulated are [183, 184]. The example in Fig. 1.2 and Example 1.5 are borrowed from [239]. The definition of ϵ-Nash equilibrium is adapted from [23, Chap. 4.2]. *Kakutani's theorem* is a generalization of the *Brouwer fixed point theorem*, for which several proofs exist, one of the most elementary ones being given in [141]. Fig. 1.6 on *Kakutani's theorem* is courtesy of Ozdaglar, slides of the course 6.254 Game Theory with Engineering Applications, MIT OpenCourseWare (2010). The *equilibrium point theorem* by Nash is a landmark in the literature of game theory, and we refer the reader to the original work [183]. The *Indifference Principle* is discussed in [196, Lemma 33.2]. Example 1.10 and Fig. 1.7 are courtesy of Alberto Bressan, *Noncooperative Differential Games: A Tutorial* (2010) [65].

The iterated dominance algorithm illustrated on the Cournot duopoly in Fig. 1.10 is courtesy of Ozdaglar, slides of the course 6.254 Game Theory with Engineering Applications, MIT OpenCourseWare (2010).

Chapter 2

Two-Person Zero-Sum Games

2.1 • Introduction

After the general introduction to noncooperative games provided in the previous chapter, we now turn to a special class of such games: *two-person zero-sum games*. Two-person zero-sum games constitute the purest form of noncooperative games, by this meaning that there is no margin for cooperation between the players. The essence of such games is contained in the Latin expression

"Mors tua vita mea."

In brief, an improvement for one player always comes at a cost for the other player. If one player wins 1 dollar, the other loses 1 dollar and vice versa. These games are also called *minimax games*.

The relevance of two-person zero-sum games derives from the tractability of the existence conditions of Nash equilibrium solutions, now assuming the form of *saddle-points*. Furthermore, in case of multiple saddle-points, they are proven to be interchangeable and to have equal payoffs. These two properties are known as (i) *interchangeability property* and (ii) *equal payoff property*. It is in the context of two-person zero-sum games that we have the fundamental result known as the *minimax theorem*. The theorem uses the notion of a *mixed extension* of a two-person zero-sum game. Mixed extension means that the players play mixed strategies (cf. Definition 1.7); namely, they choose probabilities over their action spaces and randomize their actions based on such probabilities.

In Section 2.2 we formulate two-person zero-sum games as matrix games. Section 2.3 is the core of this chapter. Here we introduce the *minimax theorem*. Before doing this, we survey notions like *conservative strategies* and *saddle-points* and establish existence conditions. Also, we state the aforementioned properties of interchangeability equal payoff and introduce the notion of *mixed strategy*. In Section 2.4 we discuss links with robust control and in particular with H^∞-optimal control. Section 2.5 offers a few examples. Finally, Section 2.6 provides some concluding remarks and references for this chapter.

2.2 • Formalization as matrix games

Two-person zero-sum games can be formulated as matrix games. In this section we show that this is possible due to the special structure of the payoffs.

In two-person zero-sum games the sum of the payoffs resulting from any action profile is always zero, namely

$$u_1(a_1, a_2) = -u_2(a_1, a_2) \quad \forall (a_1, a_2) \in \mathscr{A}_1 \times \mathscr{A}_2,$$

where (a_1, a_2) is the action profile and $\mathscr{A}_1$ and $\mathscr{A}_2$ are the sets of actions of players 1 and 2, respectively. As one payoff is the opposite of the other payoff, we can simply use a matrix rather than a bimatrix, and we can call such games *matrix games*. Conventionally, we assume that the scalar entry represents the payoff u_2 that player 2 wishes to maximize and player 1 wishes to minimize. Note that when player 1 tries to maximize u_1, he is indeed trying to minimize u_2.

$$P_2 \text{ (max)}$$

$$
\begin{array}{c c}
& \begin{array}{|c|c|c|c|}
\hline
(1,-1) & (3,-3) & \cdots & A_{1n} \\
\hline
(5,-5) & A_{22} & \cdots & A_{2n} \\
\hline
\vdots & \vdots & \ddots & \vdots \\
\hline
A_{m1} & A_{m2} & \cdots & A_{mn} \\
\hline
\end{array}
\end{array}
$$

min
P_1 (max)

Figure 2.1. *Two-person zero-sum game: matrix game representation.*

Let us denote by A the matrix of the game, and let A_{ij} be its ijth entry. Entry A_{ij} is the payoff corresponding to player 1 playing the ith row and player 2 playing the jth column. Based on what we have said, the row player, call him P_1, is the *minimizer*, and the column player, call him P_2, is the *maximizer*. The whole procedure to turn the bimatrix into a matrix is illustrated in Fig. 2.1 for a two-person zero-sum game where P_1 has m actions and P_2 has n actions. The game is then represented by an $(m \times n)$-matrix.

2.3 ▪ From conservative strategies to saddle-points

Because of the special structure of the payoffs, in two-person zero-sum games the Nash equilibrium solutions take the form of saddle-points. Keeping in mind the matrix game formulation mentioned earlier, we now introduce *conservative strategies* and build on them existence conditions of saddle-points. Before doing this, observe that any best opponent's response yields the worst own payoff. Thus a conservative strategy is obtained by considering the best among the possible worst-case scenarios. This reasoning implies the resolution of a minimax problem for the minimizer and of a maximin problem for the maximizer. This leads to the characterization of *conservative strategies* (i^*, j^*) as follows:

$$
\begin{cases}
\bar{J}(A) := \min_i \max_j a_{ij} & \text{(loss ceiling)}, \\
\underline{J}(A) := \max_j \min_i a_{ij} & \text{(gain floor)}.
\end{cases}
\tag{2.1}
$$

The *loss ceiling* $\bar{J}(A)$—which is a different way of calling the upper bound—is obtained by maximizing over the columns thus obtaining the worst payoffs for every choice of player 1, and then taking the minimum over the rows (best worst payoff). This returns the conservative strategy i^* for player 1. Similarly, the *gain floor* $\underline{J}(A)$—which is a different way of calling the upper bound—is obtained by minimizing over the rows and then taking the maximum over the columns. This returns the conservative strategy j^* for player 2.

The following existence result for saddle-points applies to the case when both players play *pure strategies*. That the players play pure strategies means that the players' actions are in general discrete, as illustrated in the two examples provided below the theorem. Let us first state the result.

Theorem 2.1. *A saddle-point exists if the gain floor is equal to the loss ceiling, i.e.,*

$$\underline{J}(A) = a_{i^* j^*} = \overline{J}(A). \qquad (2.2)$$

Furthermore, if a saddle-point exists, this corresponds to both players playing conservative, and the equilibrium payoff is then $a_{i^* j^*}$.

Example 2.2. Fig. 2.2 depicts a matrix game for which no saddle-point exists. Indeed, the loss ceiling $\overline{J}(A) = 2$ and the gain floor $\underline{J}(A) = -1$, and therefore the existence condition (2.2) is violated. To see this more in detail, consider all of player 1's actions and the corresponding best responses of player 2. In the example, P_1 can go *Top, Center,* or *Bottom*. His strategy or action set is denoted by $A_1 = \{T, C, B\}$. The actions available to P_2 are *Left, Middle,* or *Right,* and the corresponding set of actions is denoted by $A_2 = \{L, M, R\}$.

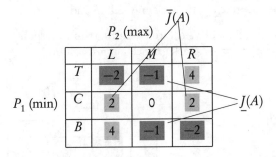

Figure 2.2. *Loss ceiling and gain floor; this game has no saddle-point.*

Now, to understand the procedure that leads to the conservative strategy for P_1, the minimizer, let us consider the following scenarios:

- If P_1 plays T (1st row), P_2 responds R (3rd column), which yields the payoff $a_{13} = 4$ (light gray cell in 1st row).

- Differently, if P_1 plays C (2nd row), then P_2 responds either L or R (1st or 3rd column), which yields the payoff $a_{21} = a_{23} = 2$ (light gray cell in 2nd row). Note that P_2 can play equivalently L or R, as the payoffs in the two cases are equal.

- Finally, if P_1 plays B (3rd row), then P_2 responds L (1st column), which produces the payoff $a_{31} = 4$ (light gray cell in 3rd row).

After comparing the three different scenarios and corresponding payoffs, P_1 selects C (2nd row), which returns the minimum payoff (both a_{21} or a_{23} are less than a_{13} and a_{31}). We can conclude that the conservative strategy of player 1 is $i^* = 2$ (2nd row). Consequently, the gain floor is $\overline{J}(A) = 2$. To understand that this represents an upper bound, note that if P_1 plays conservatively, then the payoff cannot exceed such a value for any choice of P_2 (fix the 2nd row, and compare the payoffs over the columns). Furthermore, note that the whole procedure is exactly described by the minimax expression delineated in the first line of (2.1).

By reiterating the procedure for P_2, we get that the conservative strategy for him is $j^* = 2$ (2nd column) and the gain floor is $\underline{J}(A) = -1$. It is clear that this is a lower bound, as if P_2 plays conservatively, then the payoff can never be lower than such a value whatever P_1 picks (fix the 2nd column, and span across the rows). In conclusion, the payoff corresponding to both players playing conservative strategies is $a_{i^*j^*} = a_{22} = 0$. From condition (2.2) in Theorem 2.1, the analysis culminates in the result that (i^*, j^*) is not a saddle-point as the loss ceiling and the gain floor are different. More generally, no saddle-point exists for this game. ∎

Example 2.3. This second example shows a matrix game for which a saddle-point exists; see Fig. 2.3. Briefly, for any choice of P_1 (any row), P_2 reacts by playing M (2nd column); see the light gray cells in each row. In other words, M is a dominant action for P_2. As a consequence, the conservative strategy of P_1 is $i^* = 2$ and the gain floor is $\overline{J}(A) = 10$. Let us now consider the game from the perspective of P_2.

- If P_2 plays L (1st column), then P_1 responds T (1st row), which yields the payoff $a_{11} = -40$ (dark gray cell in 1st column).

- Differently, if P_2 plays M (2nd column), then P_1 responds C (2nd row), which yields the payoff $a_{22} = 10$ (dark gray cell in 2nd column).

- Finally, if P_2 plays R (3rd column), then P_1 responds B (3rd row), which yields the payoff $a_{33} = -2$ (dark gray cell in 3rd column).

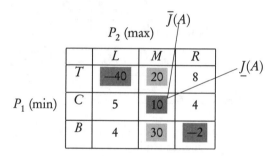

Figure 2.3. *Loss ceiling and gain floor; this game admits a saddle-point.*

After comparing the three scenarios and the corresponding payoffs, P_2 selects M (2nd row), which returns the maximum payoff a_{22} (both a_{11} or a_{33} are less than a_{22}). Thus the conservative strategy of P_2 is $j^* = 2$ (2nd column) and the gain floor is $\overline{J}(A) = 10$. Therefore the payoff obtained when both players play conservatively is $a_{i^*j^*} = a_{22} = 0$. As loss ceiling and gain floor coincide, one can see that condition (2.2) in Theorem 2.1 is satisfied, and therefore $(i^*, j^*) = (2, 2)$ is a saddle-point and $a_{i^*j^*} = a_{22}$ is an equilibrium payoff. Observe that a_{22} is the maximum over the columns and the minimum over the rows, i.e.,

$$a_{22} \geq a_{21}, a_{23}, \qquad a_{22} \leq a_{12}, a_{32}. \quad ∎$$

For a general two-person zero-sum game, there may exist multiple saddle-points. However, such points satisfy the following two properties.

Given two saddle-points (i, j) and (k, l) we have that

- (*interchangebility*) (i, l) and (k, j) are also saddle-points;

- (*equal payoff*) $a_{ij} = a_{kl} = a_{il} = a_{kj}$.

As anticipated in the introduction of this chapter, the first breakthrough in the theory of games is the *minimax theorem*, which reads as follows.

Theorem 2.4 (Minimax theorem). *Each matrix game has a saddle-point in the mixed strategies.*

We will see in the next chapter that the computation of saddle-points involves solving linear programs (see also [239, Chap. 6]). We conclude this chapter with a classical application of zero-sum games to robust H^∞-optimal control borrowed from [22].

2.4 ▪ From two-person zero-sum games to H^∞-optimal control

This section follows [22] and provides fundamental insights on the relation between two-person zero-sum games and H^∞-optimal control. The latter is the theory supporting the design of controllers in the presence of worst-case uncertainty. The name H^∞ derives from the *Hardy space*, which is the space of the operator whose ∞-norm has to be minimized. This operator is the transfer function from the disturbance to the controlled output. The classical setup of an H^∞-optimal control problem is as follows. Let the block diagram in Fig. 2.4 be given. Block G is the plant, and K is a feedback controller. Let us denote by u the controlled input (*control* in short), by w the uncontrolled disturbance (think of it as an exogenous input), by y the measured output, and by z the controlled output (this is the variable that we wish to keep as small as possible independently of the effects of the disturbance). All variables live in measurable Hilbert spaces $\mathcal{H}_u, \mathcal{H}_w, \mathcal{H}_z, \mathcal{H}_y$. A mathematical representation of the dynamics of the system is given by

$$\begin{cases} z = G_{11}(w) + G_{12}(u), \\ y = G_{21}(w) + G_{22}(u), \\ u = K(y). \end{cases} \tag{2.3}$$

The operators G_{ij} and the controller $K \in \mathcal{K}$ are assumed to be bounded, causal, and linear. Here we denote by $\mathcal{K}$ the controller space. That an operator is *causal* means that all subsystems are *nonanticipative*, i.e., that the output cannot depend on future inputs; it can rely only on past and current inputs. An operator is *bounded* if bounded inputs imply bounded outputs. Finally, an operator is *linear* if we can apply the superimposition of the effects.

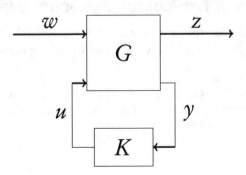

Figure 2.4. *Block diagram of plant and feedback controller. Reprinted with permission from Springer Science+Business Media* [22].

In plain words, the main goal in H^∞-optimal control is to design the controller so that the closed-loop system may absorb the energy injected by the disturbance, thus avoiding that this energy is transmitted to the controlled output. In other words, we wish that the controlled plant will be able to attenuate the effects of the disturbance. This property is referred to as *disturbance attenuation*. Such a problem is known to be describable in terms of a two-person zero-sum game between the controller and the disturbance. To see this, for every fixed $K \in \mathcal{K}$, let us introduce bounded causal linear operators $T_K : \mathcal{H}_w \to \mathcal{H}_z$:

$$T_K(w) = G_{11}(w) + G_{12}(I - KG_{22})^{-1}(KG_{21})(w).$$

Let us concentrate on the worst-case infimum of the operator norm given by

$$\begin{cases} \inf_{K \in \mathcal{K}} \langle\langle T_K \rangle\rangle =: \gamma^*, \\ \langle\langle T_K \rangle\rangle = \sup_{w \in \mathcal{H}_w} \frac{\|T_K(w)\|_z}{\|w\|_w}. \end{cases} \tag{2.4}$$

Then the problem turns into the following two-person zero-sum game, where player 1 (the minimizer) is the controller and player 2 (the maximizer) is the uncontrolled disturbance. The game takes the form

$$\overbrace{\inf_{K \in \mathcal{K}} \sup_{w \in \mathcal{H}_w} \frac{\|T_K(w)\|_z}{\|w\|_w}}^{\text{upper bound}} \geq \overbrace{\sup_{w \in \mathcal{H}_w} \inf_{K \in \mathcal{K}} \frac{\|T_K(w)\|_z}{\|w\|_w}}^{\text{lower bound}}.$$

It can be shown that the above game can be rewritten using a so-called *soft-constrained* representation as follows. Let us denote by γ^* the attenuation level, for which it holds that

$$\inf_{K \in \mathcal{K}} \sup_{w \in \mathcal{H}_w} \|T_K(w)\|_z^2 - \gamma^{*^2} \|w\|_w^2 \leq 0.$$

Let us introduce the parameter $\gamma \geq 0$ and consider the parametrized cost

$$J_\gamma(K, w) := \|T_K(w)\|_z^2 - \gamma^2 \|w\|_w^2.$$

The soft-constrained game turns into finding the smallest value of $\gamma \geq 0$ under which the upper value is bounded by zero. H^∞-optimal control formulations will be used extensively in the chapter on mean-field games and in a few applications mentioned in Part II of this book.

2.5 • Examples of two-person zero-sum games

We conclude the chapter with a series of examples. For each of them we compute the loss ceiling $\bar{J}(A)$, the gain floor $\underline{J}(A)$, and the saddle-points (i^*, j^*) in pure strategies if any exist.

Example 2.5. This example shows a game that has no saddle-point. This is due to the fact that the existence condition in Theorem 2.1 is not verified.

Consider the matrix game

		P_2	
		L	R
P_1	T	6	0
	B	-3	3

To see why the existence condition fails, note that if P_1 plays T (*top*), then P_2 responds L (*left*), and the resulting payoff is 6. Conversely, if P_1 plays B (*bottom*), then P_2 responds R (*right*), and the resulting payoff is 3. Comparing the two scenarios, the conservative strategy of P_1 is B, and the loss ceiling is $\bar{J}(A) = a_{22} = 3$.

For P_2 we have that if he plays L, then P_1 responds B. Alternatively, if he plays R, then P_1 responds T. Then, after comparison of the two payoffs, the conservative strategy of P_2 is R and the gain floor is $\underline{J}(A) = a_{12} = 0$. By noting that the loss ceiling and gain floor are different, we can conclude that there exists no saddle-point for this game. ∎

Example 2.6. As a second example, we consider a game for which the existence condition in Theorem 2.1 is verified. In particular, we will see that (B, R) is a saddle-point.

Consider the matrix game

P_2

		L	R
P_1	T	-3	8
	B	4	4

Let us start by observing that if P_1 plays T (*top*), then P_2 responds R (*right*) and the payoff is 8. Alternatively, if P_1 plays B (*bottom*), then P_2 responds equivalently L (*left*) or R (*right*) and the resulting payoff is 4. After comparison of the two payoffs, the conservative choice of P_1 is B, and therefore the loss ceiling is $\bar{J}(A) = a_{21} = a_{22} = 4$.

Repeating the analysis for P_2 we obtain that if he plays L, then P_1 responds T, whereas if he plays R, then P_1 responds B. The conservative strategy of P_2 is then R and the gain floor is $\underline{J}(A) = a_{22} = 4$. Note that here P_2 can indifferently deviate unilaterally from R to L. The loss ceiling being equal to the gain floor, the existence condition is satisfied and the game has a saddle-point, which is (B, R). ∎

Example 2.7. We show next a third example for which there is no saddle-point, as the existence condition in Theorem 2.1 is not verified.

Consider the matrix game

P_2

		L	R
P_1	T	-6	7
	B	2	1

Actually, if P_1 plays T (*top*), then P_2 plays R (*right*) and the payoff is 7. Conversely, if P_1 plays B (*bottom*), then P_2 plays L (*left*) and the payoff is 2. The conservative strategy of P_1 is B, and the loss ceiling is $\bar{J}(A) = a_{21} = 2$. Reiterating the analysis for P_2 we obtain that if he plays L, then P_1 plays T, whereas if he plays R, then P_1 plays B. By comparing the two payoffs, the conservative strategy of P_2 is R and the gain floor is $\underline{J}(A) = a_{22} = 1$. The loss ceiling and the gain floor are different, and therefore the game has no saddle-point. ∎

Example 2.8. This fourth and last example deals with a game for which the existence condition in Theorem 2.1 is matched and (B, R) is a saddle-point.

Consider the matrix game

$$P_2$$

		L	R
P_1	T	−3	8
	B	2	4

To see this, note that if P_1 plays T (*top*), then P_2 responds R (*right*) and the payoff is 8. Differently, if P_1 plays B (*bottom*), then P_2 responds again R and the payoff is 4. Note that R is a dominant strategy for P_2. As a consequence, the conservative strategy of P_1 is B and the corresponding loss ceiling is $\bar{J}(A) = a_{22} = 4$.

Replicating the analysis for P_2 we get that if he plays L (*left*), then P_1 plays T. Conversely, if he plays R, then P_1 plays B. As a result, the conservative strategy of P_2 is R and the gain floor is $\underline{J}(A) = a_{22} = 4$. As the loss ceiling and the gain floor coincide, there exists a saddle-point, which is (B, R).   ∎

2.6 ▪ Notes and references

Two-person zero-sum games occupy an important place in the history and literature of game theory. Indeed, a primary breakthrough in the theory of games is the *minimax theorem* by von Neumann (1928) [247, 248]. In this chapter, we simply state the theorem and refer the reader to the original work [248] for the proof.

Section 2.3 grew out of my notes taken during a short course on game theory given by Debasish Ghose at the University of California, Los Angeles in 2002. Section 2.4 is strongly based on the introductory chapter of [22].

Chapter 3

Computation of Saddle-Points and Nash Equilibrium Solutions

3.1 ▪ Introduction

In this chapter, we deal with the computation of saddle-points and Nash equilibrium solutions. In preparation for the general formulation of these mathematical optimization problems, we mention some examples that can be solved graphically.

In Section 3.2, we illustrate the computation approach on a simple example. In Section 3.3, we formulate the problem of computing saddle-points via linear programming. Computing Nash equilibrium solutions requires solving linear complementarity programming problems, and this is discussed in Section 3.4. Section 3.5 provides notes and references for this chapter.

3.2 ▪ Graphical resolution: An example

Let us consider the following two-person zero-sum game, for which we wish to compute the saddle-point in mixed strategies:

		P_2	
		L	R
P_1	T	6	0
	B	-3	3

P_1 can play T (*top*) or B (*bottom*). P_2 can play L (*left*) or R (*right*). Thus, the mixed strategies for both players is a probability distribution over a discrete set of two actions. In particular, for P_1 the mixed strategy is

$$y^T = [y_1 \quad y_2] \in Y, \text{ where } Y = \left\{ y \in \mathbb{R}^2 : \sum_{i=1}^{2} y_i = 1, y_i \geq 0, \forall i = 1, 2 \right\}.$$

Similarly, the mixed strategy of P_2 is

$$z^T = [z_1 \quad z_2] \in Z, \text{ where } Z = \left\{ z \in \mathbb{R}^2 : \sum_{j=1}^{2} z_j = 1, z_j \geq 0, \forall j = 1, 2 \right\}.$$

The mean payoff corresponding to P_1 playing y and P_2 playing z is given by

$$J_m(A) = \sum_i \sum_j a_{ij} y_i z_j = y^T A z.$$

The expression is a weighted sum of each payoff a_{ij} multiplied by the probabilities that P_1 plays the ith row and P_2 the jth column.

3.2.1 ▪ Conservative strategy of P_1 via minimax

To compute the conservative strategy of P_1, the first step is to consider the two pure actions of the opponent separately. This yields the following two possible mean payoffs, one obtained when P_2 plays the 1st column, namely $z_1 = 1$, and the other when P_2 plays the 2nd column, namely $z_2 = 1$:

$$\begin{cases} J_m(A) = y^T A z = 6y_1 - 3y_2, & z_1 = 1, \\ J_m(A) = y^T A z = 0y_1 + 3y_2, & z_2 = 1. \end{cases} \tag{3.1}$$

Note that the above payoffs depend only on y, as we have fixed z. This is clear in Fig. 3.1, which plots the mean payoff as a function of the mixed strategy of P_1 for the two cases of P_2 playing the 1st column ($z_1 = 1$) or 2nd column ($z_2 = 1$).

In order to identify the worst-case scenario for P_1, let us consider that the worst payoff for P_1 (the minimizer) is given by the point-wise maximum, whose expression is as follows:

$$\max_z y^T A z = \max_{z_1, z_2} z_1(6y_1 - 3y_2) + z_2(0y_1 + 3y_2).$$

The above expression corresponds to the line drawn in boldface. After comparing the worst payoffs we select the best worst case, which is obtained for

$$y^* = \arg\min_y \max_z y^T A z, \quad y^{*^T} = [0.5 \quad 0.5].$$

Actually, the above point identifies the minimum of the line drawn in boldface. In other words this is the minimum of the point-wise maximum plot. It is now clear that to find the conservative strategy y^* of P_1 (the minimizer) we need to solve a minimax problem. In the next section we replicate the same analysis for P_2 (the maximizer) in order to find his conservative strategy. This will involve the formulation and solution of a maximin optimization problem.

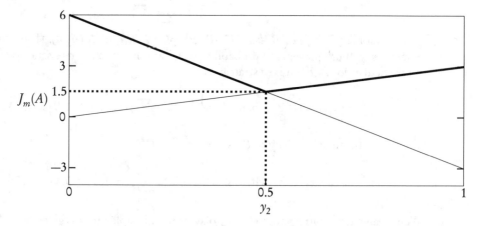

Figure 3.1. *Graphical resolution for P_1: average payoff $J_m(A)$ (vertical axis) as a function of y_2 (horizontal axis).*

3.2.2 ▪ Conservative strategy of P_2 via maximin

In order to calculate the conservative strategy of P_2, let us start by fixing the strategy for P_1 to one of the two alternative pure actions. Then, we have two possible values for the mean payoff as follows:

$$\begin{cases} J_m(A) = y^T A z = 6z_1 + 0z_2, & y_1 = 1, \\ J_m(A) = y^T A z = -3z_1 + 3z_2, & y_2 = 1. \end{cases} \tag{3.2}$$

Note that, once we fix y, the above expressions return a function of the only z. This is illustrated in Fig. 3.2, which plots the mean payoff as a function of the mixed strategy of P_2 for the two cases of P_1 playing the 1st row ($y_1 = 1$) or the 2nd row ($y_2 = 1$).

The worst payoff for P_2, which is a maximizer, is given by　point-wise minimum curve. This is the line emphasized in boldface and which is given by the expression

$$\min_y y^T A z = \min_{y_1, y_2} y_1(6z_1 + 0z_2) + y_2(-3z_1 + 3z_2).$$

We need to isolate the best among the worst cases, which corresponds to the maximum of the boldface line:

$$z^* = \arg\max_z \min_y y^T A z, \quad z^{*^T} = [0.25 \quad 0.75].$$

From the procedure illustrated in this section, it should be clear that the computation of the conservative strategy of the maximizer requires the solution of a maximin problem.

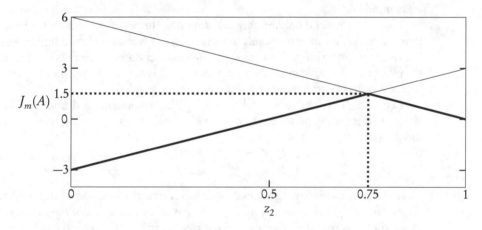

Figure 3.2. *Graphical resolution for P_2: average payoff $J_m(A)$ (vertical axis) as a function of z_2 (horizontal axis).*

Example 3.1. This example presents a two-person zero-sum game possessing two interesting characteristics. First, the conservative strategy for the minimizer is indeed a pure strategy. This is not surprising, as we can always think of a pure strategy as a special case of mixed strategy. Second, the conservative strategy for the maximizer admits infinite solutions.

Let the following game be given:

		P_2	
		L	R
P_1	T	-3	8
	B	4	4

Let us compute the saddle-point strategies of both players. In particular, let us apply the same procedure as in the previous example. For P_1, we have the following two payoff expressions related to P_2 playing $z_1 = 1$ or $z_2 = 1$:

$$\begin{cases} J_m(A) = y^T A z = -3y_1 + 4y_2, & z_1 = 1, \\ J_m(A) = y^T A z = 8y_1 + 4y_2, & z_2 = 1. \end{cases} \tag{3.3}$$

It is worth noting that the above expressions are obtained for fixed z, and therefore they express the payoff as a function of y. This is illustrated in Fig. 3.3, where we plot the mean payoff as a function of the mixed strategy of P_1 for the two cases of P_2 playing the 1st column ($z_1 = 1$) or the 2nd column ($z_2 = 1$). The worst payoffs for P_1 (the minimizer) are given by the point-wise maximum, which is emphasized in boldface and whose expression is

$$\max_z y^T A z = \max_{z_1, z_2} z_1(-3y_1 + 4y_2) + z_2(8y_1 + 4y_2).$$

The point-wise maximum is the line drawn in boldface. The best payoff for P_1 is the minimum among the point-wise maximum solutions, which is obtained for

$$y^* = \arg\min_y \max_z y^T A z, \quad y^{*T} = \begin{bmatrix} 0 & 1 \end{bmatrix}.$$

Thus, we have found the conservative strategy y^* of P_1 (the minimizer) by solving a minimax problem.

Note that the 2nd column is weak dominant for P_2, from which we have that the point-wise maximum corresponds to the case $z_2 = 1$ (see the boldface line in Fig. 3.3) and is monotonic decreasing. Also note that the intersection between the two lines capturing the plot of the average payoff for $z_1 = 1$ and $z_2 = 1$ is at the extreme point $y_2 = 1$ and that this is also the minimum of the point-wise maximum plot.

Let us replicate the same analysis for P_2 (the maximizer) and find his conservative strategy. Depending on whether $y_1 = 1$ or $y_2 = 1$, we have the following values for the mean payoff:

$$\begin{cases} J_m(A) = y^T A z = -3z_1 + 8z_2, & y_1 = 1, \\ J_m(A) = y^T A z = 4z_1 + 4z_2, & y_2 = 1. \end{cases} \tag{3.4}$$

Note that, once we fix y, the above expressions return a function of the only z. This is illustrated in Fig. 3.4, which plots the mean payoff as a function of the mixed strategy of P_2 for the two cases of P_1 playing the first row ($y_1 = 1$) or the second row ($y_2 = 1$).

Actually, the worst payoff for P_2, which is a maximizer, is given by the point-wise minimum curve. This is the line emphasized in boldface and which is given by the expression

$$\min_y y^T A z = \min_{y_1, y_2} y_1(-3z_1 + 8z_2) + y_2(4z_1 + 4z_2).$$

We need to isolate the best among the worst cases, which corresponds to the maximum of the boldface line. Remarkably, the point-wise minimum returns infinite local maxima, i.e., all the points in the segment from $z_2 = 0.65$ to $z_2 = 1$. Therefore we have an infinite number of saddle-points. One saddle-point is, for instance,

$$z^* = \arg\max_z \min_y y^T A z, \quad z^{*T} = \begin{bmatrix} 0.35 & 0.65 \end{bmatrix}.$$

Then, we have computed the conservative strategy for the maximizer via a maximin problem. ∎

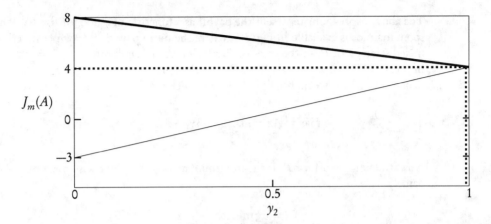

Figure 3.3. *Graphical resolution for P_1: average payoff $J_m(A)$ (vertical axis) as a function of y_2 (horizontal axis).*

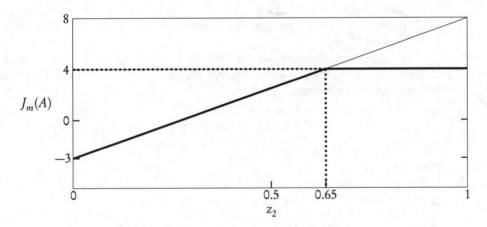

Figure 3.4. *Graphical resolution for P_2: average payoff $J_m(A)$ (vertical axis) as a function of z_2 (horizontal axis).*

Example 3.2. This third example presents a two-person zero-sum game whose conservative strategies are pure strategies. As a consequence the saddle-point is in pure strategies as well. We apply the same procedure illustrated at the beginning of this chapter. Let the following matrix game be given:

		P_2	
		L	R
P_1	T	-3	8
	B	2	4

For this game we wish to find the saddle-point. To do this, let us start by computing the conservative strategy for the minimizer P_1. Depending on whether P_2 plays $z_1 = 1$ or $z_2 = 1$, we get two payoff expressions:

$$\begin{cases} J_m(A) = y^T A z = -3y_1 + 2y_2, & z_1 = 1, \\ J_m(A) = y^T A z = 8y_1 + 4y_2, & z_2 = 1. \end{cases} \tag{3.5}$$

The above expressions represent the payoff as a function of y and for fixed z. The corresponding plot is available in Fig. 3.5, where the mean payoff is drawn as function of the mixed strategy of P_1 in the cases of P_2 playing the left column ($z_1 = 1$) or the right column ($z_2 = 1$). The worst payoff for P_1, which is the minimizer, is given by the point-wise maximum, which we highlight in boldface. Such a point is obtained by solving

$$\max_z y^T A z = \max_{z_1, z_2} z_1(-3y_1 + 2y_2) + z_2(8y_1 + 4y_2).$$

Now, the best payoff for P_1 is the minimum among the point-wise maximum solutions, namely

$$y^* = \arg\min_y \max_z y^T A z, \quad y^{*^T} = \begin{bmatrix} 0 & 1 \end{bmatrix}.$$

This completes the minimax procedure leading to the conservative strategy y^* for P_1.

That the point-wise maximum corresponds to the case $z_2 = 1$ depends on the fact that the 2nd column is dominant for P_2 (see the boldface line in Fig. 3.5), which is monotonic decreasing. Also note that the intersection between the two lines falls outside the interval $[0, 1]$ in the y_2 axis.

By reiterating the same analysis for P_2 (the maximizer) we can find his conservative strategy. In particular, assuming first $y_1 = 1$ and then $y_2 = 1$, we have the following two values for the mean payoff:

$$\begin{cases} J_m(A) = y^T A z = -3z_1 + 8z_2, & y_1 = 1, \\ J_m(A) = y^T A z = 2z_1 + 4z_2, & y_2 = 1. \end{cases} \tag{3.6}$$

Fig. 3.6 plots the mean payoff as a function of the mixed strategy of P_2 for the two cases of P_1 playing the top row ($y_1 = 1$) or the down row ($y_2 = 1$).

Observe that the worst payoff for P_2, which is a maximizer, is given by the point-wise minimum plot. This is the line emphasized in boldface, obtained by solving

$$\min_y y^T A z = \min_{y_1, y_2} y_1(-3z_1 + 8z_2) + y_2(2z_1 + 4z_2).$$

We need to isolate the best among the worst cases, which corresponds to the maximum of the boldface line. The point-wise minimum returns a unique maximum for $z_2 = 1$:

$$z^* = \arg\max_z \min_y y^T A z, \quad z^{*^T} = \begin{bmatrix} 0 & 1 \end{bmatrix}.$$

This concludes the search of the conservative strategy for the maximizer via a maximin problem. The saddle-point is obtained when both players play conservatively. ∎

So far, we have discussed the minimax and maximin procedures that lead to saddle-points and have illustrated such procedures on simple examples that can be solved graphically. In the next section we adapt the procedure to a more general example involving a 3×4 matrix game. After doing this, we present a general formulation of the linear program leading to the computation of saddle-points.

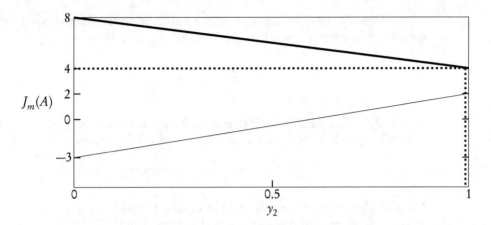

Figure 3.5. *Graphical resolution for P_1: average payoff $J_m(A)$ (vertical axis) as a function of y_2 (horizontal axis).*

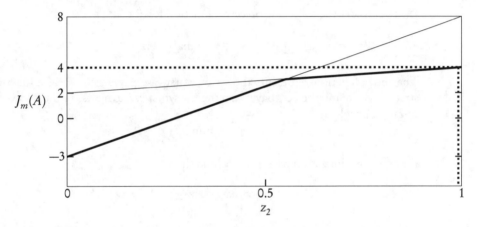

Figure 3.6. *Graphical resolution for P_2: average payoff $J_m(A)$ (vertical axis) as a function of z_2 (horizontal axis).*

3.3 ▪ Saddle-points via linear programming

Let the following two-player zero-sum game be given, where the minimizer P_1 has three strategies and the maximizer P_2 has four strategies:

P_2

P_1	6	0	5	6
	−3	3	−4	3
	8	1	2	2

The mixed strategy for P_1 is a probability distribution over a support set involving three actions, namely $y^T = [y_1 \quad y_2 \quad y_3] \in Y$, where

$$Y = \left\{ y \in \mathbb{R}^3 : \sum_{i=1}^{3} y_i = 1, \, y_i \geq 0, \forall i = 1, \ldots, 3 \right\}.$$

Analogously, the mixed strategy for P_2 is a probability distribution over a support set involving four strategies, i.e., $z^T = [z_1 \quad z_2 \quad z_3 \quad z_4] \in Z$, where

$$Z = \left\{ z \in \mathbb{R}^4 : \sum_{j=1}^{4} z_j = 1, z_j \geq 0, \forall j = 1, \dots, 4 \right\}.$$

Recall that in a saddle-point both players play conservative strategies, which yields the following equivalence between the minimax and maximin payoffs:

$$J_m(A) = \min_{y \in Y} \max_{z \in Z} y^T A z = \max_{z \in Z} \min_{y \in Y} y^T A z.$$

Let us recall that the two payoffs must be equal as the duality gap in a linear program is zero (see the *strong duality property* on p. 236, Chap. 6 of [117]).

3.3.1 ▪ Saddle-point computation via linear programming

Let us consider the following objective function for P_1:

$$v_1(y) = \max_{z \in Z} y^T A z.$$

The above function represents the set of worst-case payoffs for P_1 as a function of his strategy y. To find the best payoff among the worst-case payoffs, P_1 must solve the minimization problem

$$\min_{y \in Y} v_1(y).$$

From the maximization part, it must hold that

$$v_1(y) = \max_{z \in Z} y^T A z \geq y^T A z \quad \forall z \in Z,$$

which will constitute the constraints of the problem. Let us now transpose the above inequality, and we get

$$z^T A^T y \leq v_1(y) \Longrightarrow [z_1 \quad z_2 \quad z_3 \quad z_4] \begin{bmatrix} 6y_1 - 3y_2 + 8y_3 \\ 0y_1 + 3y_2 + 1y_3 \\ 5y_1 - 4y_2 + 2y_3 \\ 6y_1 + 3y_2 + 2y_3 \end{bmatrix} \leq v_1(y),$$

which must hold for every $z \in Z$. For the latter to be true it suffices that the above holds in each vertex of Z. Thus, by setting one component of z equal to one and the rest equal to zero, we obtain the following set of inequalities:

$$\begin{bmatrix} 6y_1 - 3y_2 + 8y_3 \\ 0y_1 + 3y_2 + 1y_3 \\ 5y_1 - 4y_2 + 2y_3 \\ 6y_1 + 3y_2 + 2y_3 \end{bmatrix} \leq \begin{bmatrix} v_1(y) \\ v_1(y) \\ v_1(y) \\ v_1(y) \end{bmatrix}.$$

Let us now normalize the variables with respect to $v_1(y)$ and take

$$\tilde{y} = \frac{1}{v_1(y)} y.$$

We can do this, as we can suppose $v_1(y) > 0$ without loss of generality. We can rewrite the above set of inequalities by using $\tilde{y}$ rather than y, and we obtain

$$
\begin{bmatrix}
6\tilde{y}_1 - 3\tilde{y}_2 + 8\tilde{y}_3 \\
0\tilde{y}_1 + 3\tilde{y}_2 + 1\tilde{y}_3 \\
5\tilde{y}_1 - 4\tilde{y}_2 + 2\tilde{y}_3 \\
6\tilde{y}_1 + 3\tilde{y}_2 + 2\tilde{y}_3
\end{bmatrix}
\leq
\begin{bmatrix}
1 \\
1 \\
1 \\
1
\end{bmatrix}.
$$

From the constraint $\tilde{y}_1 + \tilde{y}_2 + \tilde{y}_3 = \frac{1}{v_1(y)}$, which holds true as y_i are probabilities, we note that minimizing $v_1(y)$ is equivalent to maximizing $\tilde{y}_1 + \tilde{y}_2 + \tilde{y}_3$. Then, the latter problem turns into the following linear programming problem:

$$
\max \tilde{y}_1 + \tilde{y}_2 + \tilde{y}_3
$$
$$
\begin{bmatrix}
6\tilde{y}_1 - 3\tilde{y}_2 + 8\tilde{y}_3 \\
0\tilde{y}_1 + 3\tilde{y}_2 + 1\tilde{y}_3 \\
5\tilde{y}_1 - 4\tilde{y}_2 + 2\tilde{y}_3 \\
6\tilde{y}_1 + 3\tilde{y}_2 + 2\tilde{y}_3
\end{bmatrix}
\leq
\begin{bmatrix}
1 \\
1 \\
1 \\
1
\end{bmatrix},
$$
$$
\tilde{y}_1 \geq 0, \tilde{y}_2 \geq 0, \tilde{y}_3 \geq 0.
$$

Analogously, by replicating the analysis for P_2, we have the dual problem

$$
\min \tilde{z}_1 + \tilde{z}_2 + \tilde{z}_3 + \tilde{z}_4
$$
$$
\begin{bmatrix}
6\tilde{z}_1 + 0\tilde{z}_2 + 5\tilde{z}_3 + 6\tilde{z}_4 \\
-3\tilde{z}_1 + 3\tilde{z}_2 - 4\tilde{z}_3 + 3\tilde{z}_4 \\
8\tilde{z}_1 + 1\tilde{z}_2 + 2\tilde{z}_3 + 2\tilde{z}_4
\end{bmatrix}
\geq
\begin{bmatrix}
1 \\
1 \\
1
\end{bmatrix},
$$
$$
\tilde{z}_1 \geq 0, \tilde{z}_2 \geq 0, \tilde{z}_3 \geq 0, \tilde{z}_4 \geq 0.
$$

In conclusion, the saddle-point computation involves the formulation and solution of two linear programming problems of the form

$$
\begin{array}{ll}
& \max \mathbf{1}^T \tilde{y} \qquad\qquad\qquad \min \mathbf{1}^T \tilde{z} \\
(P_1) & \quad A^T \tilde{y} \leq 1, \quad (P_2) \quad A\tilde{z} \geq 1, \\
& \quad \tilde{y} \geq 0. \qquad\qquad\qquad \tilde{z} \geq 0.
\end{array}
\tag{3.7}
$$

This concludes the treatment on the computation of saddle-points for two-person zero-sum games. In the next section we address the computation of Nash equilibrium solutions via linear complementarity programming for nonzero-sum games.

3.4 ▪ Nash equilibrium via linear complementarity programming

First, we elucidate the computation of Nash equilibrium solutions using three simple examples introduced in Section 1.6. Then, we generalize the discussion and derive the linear complementarity program in abstract terms.

3.4.1 ▪ The *Indifference Principle* illustrated on simple examples

The whole procedure to find Nash equilibrium solutions is based on the *Indifference Principle*, which we have already mentioned below Theorem 1.9 in Chapter 1.

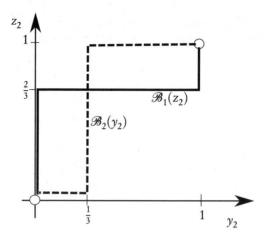

Figure 3.7. *Best-response curves for the* Battle of the Sexes.

Consider the *Battle of the Sexes* game, and let us compute the Nash equilibrium solutions in mixed strategy. Recall that the game is described by the bimatrix

	S	C
S	(2,1)	(0,0)
C	(0,0)	(1,2)

To start with, let us also recall that the mixed Nash equilibrium lies at the intersection of the best-response curves. Such curves are depicted in Fig. 3.7, where in the horizontal axis we plot the action set of player 1 and in the vertical axis the one of player 2. Note that the value on the horizontal axis is y_2 (probability of 2nd row), while the value on the vertical axis is z_2 (probability of 2nd column). The two points $(0,0)$ and $(1,1)$ (circles) are the Nash equilibrium solutions in pure strategies, (S, S) and (C, C), respectively. The best-response curve of player 1 $\mathscr{B}_1(z_2)$ is S (probability $y_2 = 0$) for any probability $z_2 \in [0, \frac{2}{3}]$ and is C (probability $y_2 = 1$) for any probability $z_2 \in (\frac{2}{3}, 1]$ (see the solid curve). In accordance with the *Indifference Principle*, when $z_2 = \frac{2}{3}$, player 1 can indifferently play S or C, as the payoff is $\frac{2}{3}$ in both cases. Conversely, the best-response curve of player 2 $\mathscr{B}_2(y_2)$ is S (probability $z_2 = 0$) for any probability $y_2 \in [0, \frac{1}{3}]$ and is C (probability $z_2 = 1$) for any probability $y_2 \in (\frac{1}{3}, 1]$ (see the dashed curve). In accordance with the *Indifference Principle*, when $y_2 = \frac{1}{3}$, player 2 can indifferently play S or C, as the payoff is $\frac{2}{3}$ in both cases. The action profile at the intersection between the two reaction curves is $y^* = [\frac{2}{3}, \frac{1}{3}]$ and $z^* = [\frac{1}{3}, \frac{2}{3}]$. Finally, note also that for player 1 playing the mixed strategy $y^* = [\frac{2}{3}, \frac{1}{3}]$ or any action in the support, namely S or C, is indifferent if player 2 is playing the mixed strategy $z^* = [\frac{1}{3}, \frac{2}{3}]$ (the payoff is $\frac{2}{3}$ in all three cases).

More formally the *Indifference Principle* says that at a Nash equilibrium, playing mixed or changing to any strategy in the support is equivalent, namely,

$$u_1(S, z^*) = u_1(C, z^*) = u_1(y^*, z^*),$$

where S and C are the two strategies in the support of y^*. We revisit now two examples introduced in Section 1.6 before developing the linear complementarity program.

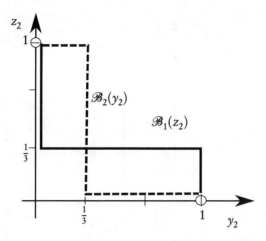

Figure 3.8. *Best-response curves for the* Hawk and Dove *game.*

Example 3.3. In this first example we consider the *Hawk and Dove* or *Chicken game*:

	Hawk	*Dove*
Hawk	$\left(\frac{V-C}{2}, \frac{V-C}{2}\right)$	$(V,0)$
Dove	$(0,V)$	$\left(\frac{V}{2}, \frac{V}{2}\right)$

For the sake of simplicity we set $C = 6 > V = 4 > 0$; then we have

	Hawk	*Dove*
Hawk	$\left(-1,-1\right)$	$(4,0)$
Dove	$(0,4)$	$\left(2,2\right)$

The best-response curves $\mathscr{B}_1(z_2)$ (solid) of P_1 and $\mathscr{B}_2(y_2)$ (dashed) of P_2 are displayed in Fig. 3.8, where in the horizontal axis we plot the action set of player 1 and in the vertical axis the one of player 2. Note that the value on the horizontal axis is y_2 (probability of 2nd row), while the value on the vertical axis is z_2 (probability of 2nd column). The two points $(0,1)$ and $(1,0)$ (circles) are the Nash equilibrium solutions in pure strategies, $(Hawk, Dove)$ and $(Dove, Hawk)$, respectively. The best-response curve of player 1 $\mathscr{B}_1(z_2)$ returns $Hawk$ (probability $y_2 = 0$) if the probability $z_2 \in (\frac{1}{3}, 1]$ and yields $Dove$ (probability $y_2 = 1$) for any probability $z_2 \in [0, \frac{1}{3}]$ (see the solid curve). In accordance with the *Indifference Principle*, when $z_2 = \frac{1}{3}$, player 1 can indifferently play $Hawk$ or $Dove$, as the payoff is $\frac{2}{3}$ in both cases. Likewise, the best-response curve of player 2 $\mathscr{B}_2(y_2)$ is $Hawk$ (probability $z_2 = 0$) for any probability $y_2 \in (\frac{1}{3}, 1]$ and is $Dove$ (probability $z_2 = 1$) for any probability $y_2 \in [0, \frac{1}{3}]$ (see the dashed curve). In accordance with the *Indifference Principle*, when $y_2 = \frac{1}{3}$, then player 2 can indifferently play $Hawk$ or $Dove$, as the payoff is $\frac{2}{3}$ in both cases. The action profile at the intersection between the two reaction curves is $y^* = [\frac{2}{3}, \frac{1}{3}]$ and $z^* = [\frac{2}{3}, \frac{1}{3}]$. ∎

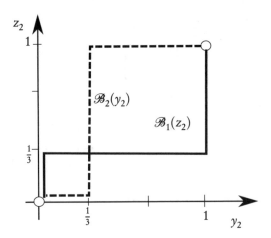

Figure 3.9. *Best-response curves for the* Stag-Hunt game.

Example 3.4. This example is about the *Stag-Hunt game*, which is described by the following bimatrix:

	Stag	*Hare*
Stag	$\left(\frac{3}{2},\frac{3}{2}\right)$	$(0,1)$
Hare	$(1,0)$	$\left(1,1\right)$

The best-response curves $\mathscr{B}_1(z_2)$ (solid) of P_1 and $\mathscr{B}_2(y_2)$ (dashed) of P_2 are displayed in Fig. 3.9, where the horizontal axis is y_2 (probability of 2nd row) and the vertical axis is z_2 (probability of 2nd column). The two points $(0,0)$ and $(1,1)$ (circles) are the Nash equilibrium solutions in pure strategies, $(Stag, Stag)$ and $(Hare, Hare)$, respectively. The best-response curve of player 1 $\mathscr{B}_1(z_2)$ yields $Stag$ (probability $y_2 = 0$) if the probability $z_2 \in [0, \frac{1}{3}]$ and yields $Hare$ (probability $y_2 = 1$) for any probability $z_2 \in (\frac{1}{3}, 1]$ (see the solid curve). In accordance with the *Indifference Principle*, when $z_2 = \frac{1}{3}$, player 1 can indifferently play $Stag$ or $Hare$, as the payoff is 1 in both cases. Likewise, the best-response curve of player 2 $\mathscr{B}_2(y_2)$ is $Stag$ (probability $z_2 = 0$) for any probability $y_2 \in [0, \frac{1}{3}]$ and is $Hare$ (probability $z_2 = 1$) for any probability $y_2 \in (\frac{1}{3}, 1]$ (see the dashed curve). In accordance with the *Indifference Principle*, when $y_2 = \frac{1}{3}$, player 2 can indifferently play $Stag$ or $Hare$, as the payoff is 1 in both cases. The action profile at the intersection between the two reaction curves is $y^* = [\frac{2}{3}, \frac{1}{3}]$ and $z^* = [\frac{2}{3}, \frac{1}{3}]$. ∎

Next we extend our discussion to any bimatrix game and derive the linear complementarity program in its abstract form.

3.4.2 ▪ Linear complementarity programming

In this section we generalize the procedure illustrated in the previous section. To do this, consider a bimatrix game (A, B) where player 1 has m strategies and player 2 has n strategies. Here $A \in \mathbb{R}^{m \times n}$ is the matrix collecting the payoffs of player 1 and $B \in \mathbb{R}^{m \times n}$ is the matrix involving the payoffs of player 2. For any Nash equilibrium (y^*, z^*), we have

$$y^{*^T} A z^* \geq y^T A z^* \quad \forall y \quad \Rightarrow \quad y^{*^T} A z^* \geq \sum_{j=1}^n a_{ij} z_j^* \quad \forall i = 1, \ldots, m,$$
$$y^{*^T} B z^* \geq y^{*^T} B z \quad \forall z \quad \Rightarrow \quad y^{*^T} B z^* \geq \sum_{i=1}^m b_{ij} y_i^* \quad \forall j = 1, \ldots, n.$$

The inequalities on the left derive from both players being maximizers. Therefore, y^* solves $\max_y y^T A z^*$ and z^* solves $\max_z y^{*^T} B z$. The inequalities on the right are obtained by specializing the inequalities on the left in each vertex of the simplex for y (top) and for z (bottom). In other words, the inequality $y^{*^T} A z^* \geq y^T A z^* \forall y$ holds true if the same inequality is satisfied in each vertex of the simplex in $\mathbb{R}^m$, namely for any y such that $y_i = 1$ and $y_j = 0$ for all $j \neq i$ (one component of y is one, and the other components are zeros). Essentially, we are restricting the inequality to the pure strategies of player 1. We can replicate the same reasoning for player 2 and rewrite the inequality $y^{*^T} B z^* \geq y^{*^T} B z \ \forall z$ only considering the pure strategies of player 2.

With the above inequalities in mind, from the *Indifference Principle* we have that for any pure strategy in the support of y^* and in the support of z^*, the corresponding payoff must be equal to the one at the equilibrium. This corresponds to the conditions

$$y_i^* > 0 \Rightarrow \sum_{j=1}^n a_{ij} z_j^* = y^{*^T} A z^*; \quad z_j^* > 0 \Rightarrow \sum_{i=1}^m b_{ij} y_i^* = y^{*^T} B z^*.$$

Now, let us normalize the strategies of both players and take $u_j = \frac{z_j^*}{y^{*^T} A z^*}$ and $v_i = \frac{y_i^*}{y^{*^T} B z^*}$. Then we have

$$\begin{cases} \sum_{j=1}^n a_{ij} u_j \leq 1 & \forall i = 1, \ldots, m, \\ \sum_{i=1}^m b_{ij} v_i \leq 1 & \forall j = 1, \ldots, n, \\ \forall i \ v_i > 0 \ \Rightarrow \ \sum_{j=1}^n a_{ij} u_j = 1, \\ \forall j \ u_j > 0 \ \Rightarrow \ \sum_{i=1}^m b_{ij} v_i = 1. \end{cases} \tag{3.8}$$

To turn the inequalities into equalities, let us introduce the slack variables $r \in \mathbb{R}^m$ and $t \in \mathbb{R}^n$ and rewrite the problem above as

$$\begin{cases} Au + r = 1, \\ B^T v + t = 1, \\ v^T r = 0, \\ u^T t = 0, \\ r \geq 0, t \geq 0, \end{cases} \quad \text{or} \quad \begin{cases} \overbrace{\begin{bmatrix} 0 & A \\ B^T & 0 \end{bmatrix}}^{H} \overbrace{\begin{bmatrix} v \\ u \end{bmatrix}}^{x} + \overbrace{\begin{bmatrix} r \\ t \end{bmatrix}}^{s} = 1, \\ \begin{bmatrix} v^T & u^T \end{bmatrix} \begin{bmatrix} r \\ t \end{bmatrix} = 0, \begin{bmatrix} r \\ t \end{bmatrix} \geq 0. \end{cases} \tag{3.9}$$

The above problem is a *linear complementarity program* because of the complementarity relationships that either $v_i = 0$ or $r_i = 0$ (or both) for each $i = 1, 2, \ldots, m$ and $u_j = 0$ or $t_j = 0$ (or both) for each $j = 1, 2, \ldots, n$ (see Section B.2.2 in the Appendix, Chapter B). Finally, from setting $s := [r^T \quad t^T]^T \geq 0$, the above problem is equivalent to

$$\begin{cases} 1 - Hx \geq 0, \\ x^T(1 - Hx) = 0, \\ x \geq 0. \end{cases}$$

A solution to the above problem can be obtained via the following quadratic bilinear program:

$$\begin{cases} \min_x x^T(1 - Hx) \\ 1 - Hx \geq 0, \\ x \geq 0, \end{cases}$$

and this concludes our discussion on the computation of Nash equilibrium solutions.

3.5 ▪ Notes and references

For a complete treatment of linear programming we refer the reader to [117, Chap. 3]. Duality theory is developed in [117, Chap. 6]. Linear complementarity programming is examined in [117, Sect. 13.3, p. 670]. The computation approach for saddle-points and Nash equilibrium solutions presented in this chapter is based on [239, Chaps. 6 and 7] and on [23, Sects. 2.3 and 3.4]. Computation approaches are studied in algorithmic game theory, for which we refer the reader to [190].

Chapter 4

Refinement on Nash Equilibrium Solutions, Stackelberg Equilibrium, and Pareto Optimality

4.1 ▪ Introduction

This chapter analyzes properties of Nash equilibrium solutions such as *payoff dominance*, *risk dominance*, or *subgame perfectness*. These properties may help a game designer distinguish between socially desirable and not desirable Nash equilibria. Such a distinction is particularly important in the design of incentives that may enforce the agents or players to converge to certain equilibria rather than others. The analysis of Nash equilibrium solutions and their properties is generally referred to as *refinement on Nash equilibrium solutions*. The second part of this chapter deals with another equilibrium concept, known as *Stackelberg equilibrium*, which arises when there is a hierarchy between the players or asynchrony in the learning process. This chapter also examines *Pareto optimality* and *social optimality*.

This chapter is organized as follows. Section 4.2 deals with refinement on Nash equilibrium solutions. Section 4.3 introduces the Stackelberg equilibrium. Section 4.4 examines Pareto optimality. Finally, Section 4.5 provides concluding remarks and references for this chapter.

4.2 ▪ Refinement on Nash equilibrium solutions

Nash equilibrium solutions may enjoy several properties, which can make some equilibria more interesting than others in terms of their impact on the global welfare of the system. Let us look at two main characteristics of Nash equilibrium solutions, *payoff dominance* and *risk dominance*.

4.2.1 ▪ Payoff dominant Nash equilibrium

Payoff dominant Nash equilibrium solutions, also called admissible equilibria, produce higher payoffs for all the players.

Definition 4.1. *A Nash equilibrium strategy pair is* admissible *or* payoff dominant *if there exists no better (for all players) Nash equilibrium strategy pair.*

The above concept is elucidated in the following example.

Example 4.2. Consider the two-player game in Fig. 4.1. Player 1 can play T (*top*) or B (*bottom*), while player 2 can play L (*left*), M (*middle*), or R (*right*). There exist three Nash equilibrium solutions: (T,L), (T,M), and (B,R). The Nash equilibrium (B,R) is payoff dominant, as both players get more than what they would get if they played one of the other two Nash equilibrium solutions, namely (T,L) or (T,M). Actually, (B,R) produces a payoff of 1 to both players which exceeds what both players get in (T,L), that is, 0, and also what both players get in (T,M), namely -1 for player 1 and 0 for player 2.   ∎

	L	M	R
T	$(0,0)^*$	$(-1,0)^*$	$(-3,-1)$
B	$(-2,1)$	$(-2,0)$	$(1,1)^*$

Figure 4.1. *Payoff dominant (or admissible) Nash equilibrium (gray).*

4.2.2 ▪ Risk dominant Nash equilibrium

Another interesting property of a Nash equilibrium is known as *risk dominance*. Let us look at the parametrized *Coordination game* shown below:

	H	G
H	(A,a)	(C,b)
G	(B,c)	(D,d)

Here we assume that $A > B$, $D > C$ and that $a > b$, $d > c$. With the above game in mind, the formal definition of risk dominance is given as follows.

Definition 4.3 (Risk dominance). *A Nash equilibrium strategy pair (G,G) risk dominates the other Nash equilibrium strategy pair (H,H) if the product of the deviation losses is highest for (G,G), namely if the following inequality holds:*

$$(C-D)(c-d) \geq (B-A)(b-a).$$

Evolutionary biologists have noticed that while payoff dominance is in principle more convenient to all the players, in most experimental situations in nature, animals or individuals usually converge to risk dominant equilibria [138, 254]. In other words, risk dominance arises in nature more often than payoff dominance. The difference between risk and payoff dominance is discussed further in the *Stag-Hunt game*.

Example 4.4. Consider the *Stag-Hunt game* described by the bimatrix

	Stag	Hare
Stag	$\left(\frac{3}{2},\frac{3}{2}\right)$	$(0,1)$
Hare	$(1,0)$	$(1,1)$

The action profile $(Hare, Hare)$ is risk dominant, while $(Stag, Stag)$ is payoff dominant.

It is worth noting that if one player assumes that the other player plays the two available actions with equal probability, namely $(\frac{1}{2},\frac{1}{2})$, then the action $Hare$ has a higher expected payoff. This introduces an alternative and equivalent way to identify risk dominant Nash equilibrium solutions. ∎

4.2.3 ▪ Subgame perfect Nash equilibrium

Subgame perfect Nash equilibrium solutions arise in extensive games, where there is an explicit order of events. Subgame perfectness accounts for credible threats or robust solutions. The definition of subgame perfect Nash equilibrium makes use of the concept of *subgame*, which we explain next.

Definition 4.5 (Subgame). *A* subgame *is a subset of nodes that still forms a game.*

With the above definition in mind, a subgame perfect Nash equilibrium is a Nash equilibrium for the subgame.

Definition 4.6 (Subgame perfect Nash equilibrium). *A Nash equilibrium is* subgame perfect *if when played from any point in the game, such a solution is a Nash equilibrium.*

To compute subgame perfect Nash equilibrium solutions one can use backwards dynamic programming. This is illustrated in the following examples.

Example 4.7. This example involves a two-player extensive game. Fig. 4.2 depicts its extensive representation (left) and its normal form representation (right). Both players have two actions: A or B for player 1 and C or D for player 2.
   A typical dynamic programming algorithm proceeds as follows. Let us consider the last stage, namely stage 2. Here the rational choice of player 2 is (D), as it produces a payoff of 1, which is better than the payoff 0 corresponding to the other choice, C. We use a dashed line to store in memory that D is the rational choice of player 2 in stage 2. Going back to stage 1 and using backwards induction, we infer for player 1 that his best choice is B, where he gets 5 rather than 4 if he played A. Obviously, this is valid as long as player 2 is rational and plays D in stage 2. If we look at the normal form representation, we note that (A, C) is also a Nash equilibrium. Indeed, if player 1 deviates and plays B, he is not better off. Analogously, if player 2 deviates and plays D, he is not better off. However, such a Nash equilibrium is not subgame perfect, as player 2 playing C is not a credible choice. Subgame perfectness gives a measure of how *robust* a Nash equilibrium is. Indeed, being a Nash equilibrium for any subgame implies that if the past actions have not been rational, the solution continues to be a Nash equilibrium. Nash equilibrium solutions that are not subgame perfect do not guarantee to still be equilibrium solutions when some of the past actions have not been rational. We can put it differently and say that Nash equilibria that are not subgame perfect involve *noncredible threats*. ∎

Example 4.8. This example deals with a three-stage two-player extensive game. Fig. 4.3 depicts its extensive representation (left) and its normal form representation (right). The game is played in three stages, it has four different states, and the players always have two actions in each state. Using dynamic programming backwards, let us start with the last stage, stage 3. Here we have one state, state 4 (node at level three), in which the rational choice for P_1 is R_4, which produces a payoff of 8 (greater than 7 produced by L_4). Note that the index refers to the label of the state. We draw a dotted line to store in memory the rational choice R_4. At stage 2, we have two states, states 2 and 3. In state 2 (left

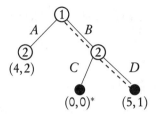

	C	D
A	$(4,2)^*$	$(4,2)$
B	$(0,0)$	$(5,1)^*$

Figure 4.2. (B,D) *is a subgame perfect Nash equilibrium obtained via dynamic programming (see the dashed lines in the tree and the gray solution in the bimatrix).*

node, second level) the rational choice of P_2 is L_2 (dotted line, left, level two). This is obtained by backwards induction comparing the payoff of player 2 given by L_2, which is 7, to the one associated to R_2, which is 6. In state 3 (right node, level two), both actions are equivalently rational, namely L_3 (dotted line, right, level two) or R_3 (dashed line, level two). Actually the payoff for P_2 is 5 in both cases. Going back to stage 1, the best response of P_1 when P_2 plays L_3 is L_1 (dotted line, level one). Conversely, the best response to R_3 is R_1 (dashed line, level one). Both $(L_1 R_4, L_2 L_3)$ and $(R_1 R_4, L_2 R_3)$ are subgame perfect Nash equilibrium solutions. From the normal form representation we have several other Nash equilibrium solutions which are not subgame perfect. ∎

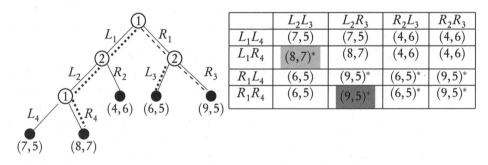

	$L_2 L_3$	$L_2 R_3$	$R_2 L_3$	$R_2 R_3$
$L_1 L_4$	$(7,5)$	$(7,5)$	$(4,6)$	$(4,6)$
$L_1 R_4$	$(8,7)^*$	$(8,7)$	$(4,6)$	$(4,6)$
$R_1 L_4$	$(6,5)$	$(9,5)^*$	$(6,5)^*$	$(9,5)^*$
$R_1 R_4$	$(6,5)$	$(9,5)^*$	$(6,5)^*$	$(9,5)^*$

Figure 4.3. *Two subgame perfect Nash equilibrium solutions:* $(L_1 R_4, L_2 L_3)$ *(dotted edges in the tree and light gray cell in the bimatrix) and* $(R_1 R_4, L_2 R_3)$ *(dashed edges in the tree and dark gray cell in the bimatrix). Both can be computed via dynamic programming.*

4.3 ▪ Stackelberg equilibrium

The Stackelberg equilibrium refers to a game with a hierarchical structure. That the game admits a hierarchical structure means that a leader announces and enforces his best strategy by taking into account the rational reactions of the followers. The game is played in one shot. The formal definition of Stackelberg equilibrium builds upon the concept of best-response set, as in Definition 1.3, which is reiterated as

$$\mathscr{B}_i(a_{-i}) := \{a_i^* \in \mathscr{A}_i \mid u_i(a_i^*, a_{-i}) = \max_{a_i \in \mathscr{A}_i} u_i(a_i, a_{-i})\}.$$

Having introduced the best-response set, the definition of Stackelberg equilibrium is given as follows.

Definition 4.9. *An action profile* (a_1^S, a_2^S) *is a Stackelberg equilibrium (SE) for player 1 if* $a_2^S \in \mathscr{B}_2(a_1^S)$ *and*

$$u_1(a_1^S, a_2^S) \geq u_1(a_1, a_2) \quad \forall a_1 \in \mathscr{A}_1, a_2 \in \mathscr{B}_2(a_1).$$

In the following we analyze a few examples that shed light on the relation between Nash equilibrium solutions and Stackelberg equilibrium solutions. It turns out that occasionally a Stackelberg equilibrium is better than a Nash equilibrium, though this is not a rule. In general, a Stackelberg equilibrium differs from a Nash equilibrium and may yield better payoffs for all the players or for only a subset of players.

Example 4.10. The first example is the *Prisoner's dilemma*, as displayed in Fig. 4.4. First, assume that P_1 is the leader. Then P_1 knows that

- if he plays C, then P_2 (follower) will respond D (light gray cell) and the payoff of P_1 will be 0;

- if he plays D, then P_2 will still respond D and the payoff of P_1 will be 1.

We conclude that D is the rational choice of P_1. In summary, the Stackelberg equilibrium, denoted by SE_1 when P_1 is the leader, is (D, D). The game is symmetric, so if we repeat the same analysis for P_2, we arrive at the same result. To see this, consider P_2 as the leader. Then P_2 knows that

- if he plays C, then P_1 (follower) will respond D (dark gray cell) and the payoff of P_2 will be 0;

- if he plays D, then P_1 will still respond D (dark gray cell) and the payoff of P_2 will be 1.

We conclude that D is the rational choice of P_2. The Stackelberg equilibrium SE_2 when P_2 is the leader is (D, D), as in the previous case. Note that in this example the two Stackelberg equilibria are the same and also coincide with the unique pure Nash equilibrium solution of the game. ∎

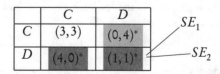

Figure 4.4. *Stackelberg equilibrium for the* Prisoner's dilemma.

Example 4.11. This example deals with the two-player nonzero-sum game whose bimatrix is in Fig. 4.5. Both players have three actions. First, consider P_1 as the leader. P_1 knows that

- if he plays L, then P_2 (follower) will respond L (light gray cell, 1st row) and the payoff of P_1 will be 0;

- if he plays M, then P_2 will respond M (light gray cell, 2nd row) and the payoff of P_1 will be -1;

- if he plays R, then P_2 will respond R (light gray cell, 3rd row) and the payoff of P_1 will be -2.

Then L is the rational choice of P_1. The Stackelberg equilibrium SE_1 when P_1 is the leader is then (L, L). Analogously, if the leader is P_2, then he knows that

- if he plays L, then P_1 (follower) will respond R (dark gray cell, 1st column) and the payoff of P_2 will be 0;

- if he plays L, then P_1 will respond M (dark gray cell, 2nd column) and the payoff of P_2 will be 0;

- if he plays R, then P_1 will respond L (dark gray cell, 3rd column) and the payoff of P_2 will be $\frac{2}{3}$.

Then R is the rational choice of P_2. The Stackelberg equilibrium SE_2 when P_2 is the leader is again (L, R). In this example, SE_1 is better than the Nash equilibrium for both players, and SE_2 is better than the Nash equilibrium only for P_2 (leader). The Nash equilibrium is not a Stackelberg equilibrium. ∎

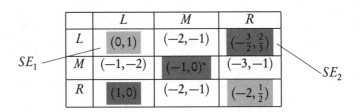

Figure 4.5. *Example of Stackelberg equilibrium.*

Example 4.12. A last example involves the infinite game whose level curves are as in Fig. 4.6. The horizontal coordinate is the action of player 1, while the vertical coordinate is the action of player 2. From the graph we understand that point P is the global maximum for player 1 and point Q the global maximum for player 2. The best-response curve of player 2 to player 1 is the curve $\beta(a)$ (dash-dot). Given this, the Stackelberg equilibrium when player 1 is the leader is (a_s, b_s). Note that this point is crossed by the tangent to player 1's level curves. ∎

4.3.1 ▪ Nonuniqueness

Similarly to what we saw for the Nash equilibrium, the Stackelberg equilibrium may also not be unique. A way to deal with such a case is to introduce additional criteria. A possible idea is to select the equilibrium that minimizes the risk. The following example elaborates this idea in detail.

Example 4.13. The example of Fig. 4.7 shows that when the leader is P_1, he knows that

- if he plays L, then P_2 (follower) will respond indifferently L or M (light gray cell, 1st row) and the payoff of P_1 will be 0 or -1, respectively;

- if he plays R, then P_2 will respond indifferently L or R (light gray cell, 2nd row) and the payoff of P_1 will be -2 or 1, respectively.

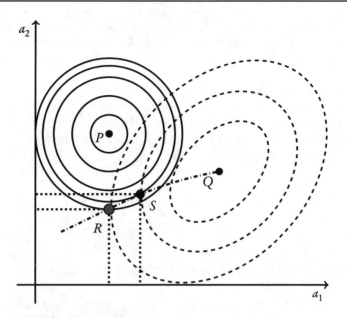

Figure 4.6. *Two-player continuous infinite game. Level curves of player 1 (solid) and player 2 (dashed); action spaces of player 1 (horizontal axis) and player 2 (vertical axis). Global maximum is P for player 1 and Q for player 2, while the Nash equilibrium is point R and the Stackelberg equilibrium is point S. Reprinted with permission from A. Bressan and Springer Science+Business Media [65, 66].*

SE_1

	L	M	R
L	(0,0)	(−1,0)*	(−3,−1)
R	(−2,1)	(−2,0)	(1,1)*

Figure 4.7. *Nonunique Stackelberg equilibrium and risk minimization.*

Note that the expected payoff by playing L or R of P_1 is $-\frac{1}{2}$. However, the worst payoff is better in the first case (L), so minimizing the risk yields (L, M) as SE_1. Also note that SE_1 is worse than the Nash equilibrium for both players. ∎

4.4 ▪ Pareto optimality

The last part of this chapter deals with Pareto optimality. Equilibria that are also Pareto optimal represent stable solutions in that not only is no player better off by changing actions, but also no players can be better off by jointly deviating without causing a loss for at least one player. The definition of Pareto optimality for a two-player game is as follows.

Definition 4.14. *A pair of strategies (a_1^{PO}, a_2^{PO}) is said to be Pareto optimal if there exists no other pair (a_1, a_2) such that for $i = 1, 2$*

$$u_i(a_1, a_2) > u_i(a_1^{PO}, a_2^{PO}) \quad \text{and} \quad u_{-i}(a_1, a_2) \geq u_{-i}(a_1^{PO}, a_2^{PO}).$$

In other words, given a Pareto optimal solution, it is not possible to strictly increase the payoff of one player without strictly decreasing the payoff of the other. We analyze Pareto optimality in a few examples, which are already familiar to the reader.

Example 4.15. This example answers the following question: What solutions are Pareto optimal in the *Prisoner's dilemma*? Recall that this game is described by the bimatrix displayed in Fig. 4.8. There, we have three different Pareto optimal solutions: (C,D), (D,C), and (C,C). Indeed, for each of these action profiles, any deviation causes a loss for at least one of the players. ∎

	C	D
C	$(3,3)$	$(0,4)$
D	$(4,0)$	$(1,1)$

Figure 4.8. *In the* Prisoner's dilemma (C,D), (D,C), *and* (C,C) *are all Pareto optimal solutions.*

In the following we explore Stackelberg equilibria and Pareto optimal solutions for a few classical games introduced earlier, such as the *Battle of the Sexes*, the *Coordination game*, and the *Hawk and Dove game*.

Example 4.16 (*Battle of the Sexes*). If player 1 is leader, then he knows that

- if he plays S, then player 2 will respond S and the payoff of player 1 will be 2;

- if he plays C, then player 2 will respond C and the payoff of player 1 will be 1.

Then, the Stackelberg equilibrium when player 1 is leader is (S,S).
Reiterating for player 2 leader, we have that

- if he plays S, then player 1 will respond S and the payoff of player 2 will be 1;

- if he plays C, then player 1 will respond C and the payoff of player 2 will be 2.

The Stackelberg equilibrium when player 2 is the leader is (C,C). The two Stackelberg equilibria are also the only Pareto optimal solutions of the game.

	S	C
S	$(2,1)$	$(0,0)$
C	$(0,0)$	$(1,2)$

∎

Example 4.17 (*Coordination game*). This example deals with the *Coordination game*. If player 1 is leader, then he knows that

- if he plays $Mozart$, then player 2 will respond $Mozart$ and the payoff of player 1 will be 2;

- if he plays $Mahler$, then player 2 will respond $Mahler$ and the payoff of player 1 will be 1.

Then, the Stackelberg equilibrium when player 1 is leader is $(Mozart, Mozart)$. The game is symmetric, so the latter is the Stackelberg equilibrium also when player 2 is leader. Note that the Stackelberg equilibrium is also the only Pareto optimal solution of the game.

	$Mozart$	$Mahler$
$Mozart$	$(2, 2)$	$(0, 0)$
$Mahler$	$(0, 0)$	$(1, 1)$

Example 4.18 (*Hawk and Dove game*). This last example develops the analysis for the *Hawk and Dove game* described by the following bimatrix:

	$Hawk$	$Dove$
$Hawk$	$\left(\frac{V-C}{2}, \frac{V-C}{2}\right)$	$(V, 0)$
$Dove$	$(0, V)$	$\left(\frac{V}{2}, \frac{V}{2}\right)$

Under the assumption that the cost of fighting exceeds the prize of victory, we have $C > V > 0$. To find the Stackelberg equilibrium when player 1 is leader we apply the following reasoning. Player 1 knows that

- if he plays $Hawk$, then player 2 will respond $Dove$ and the payoff of player 1 will be V;

- if he plays $Dove$, then player 2 will respond $Hawk$ and the payoff of player 1 will be 0.

When player 1 is the leader the Stackelberg equilibrium is $(Hawk, Dove)$, whereas when player 2 is the leader the Stackelberg equilibrium is $(Dove, Hawk)$. All solutions are Pareto optimal except $(Hawk, Hawk)$. ∎

4.5 ▪ Notes and references

The definition of admissible or payoff dominant Nash equilibrium is presented in [23, Def. 3.3, p. 69]. Risk dominance was formulated by Harsanyi and Selten [107, Lemma 5.4.4]. Risk dominant equilibria in biology and nature are analyzed in [138, 254]. For more details on subgame perfect equilibria we refer the reader to [196, Sect. 6.2]. The Stackelberg equilibrium was first introduced in [249]. Example 4.8 is borrowed from [239, Exercise 1.1, p. 7]. Fig. 4.6 is courtesy of Bressan, *Noncooperative Differential Games: A Tutorial* (2010) [65].

Chapter 5

Coalitional Games

5.1 ▪ Introduction

This chapter develops the theory of *coalitional games with transferable utility*, also known as *TU games*. After formulating such games, the chapter presents some examples of operations research games, such as the *minimum spanning tree game*, the *permutation game*, and the *max-flow game*. The discussion proceeds with the definition of *imputation set*. The chapter concludes with the formalization of *cooperative differential games*. In this context, we review the notion of *dynamic stability or time consistency*.

In Section 5.2 we formulate coalitional games with transferable utility. In Section 5.3 we present the operations research examples. In Section 5.4 we introduce the imputation set. In Section 5.5 we examine properties of coalitional games. In Section 5.6 we reframe coalitional games in a dynamic context. Finally, in Section 5.7 we provide notes, conclusions, and references.

5.2 ▪ Coalitional games with transferable utility (TU games)

A *coalitional game with transferable utility*, in short *TU game*, is defined by a tuple $\langle N, v \rangle$, where

- $N = \{1, \ldots, n\}$ is the set of players, and

- $v : 2^N \to \mathbb{R}$ is the *characteristic function*.

For any coalition $S \subseteq 2^N$ of players, the characteristic function $v(S)$ returns the *value* of that coalition. To compute such a value we assume that all the players that do not join the coalition will play joint actions against the coalition. This corresponds to solving a maximin optimization problem. In other words, the (worst-case) value of a coalition is computed as the maximum over the set of joint actions of all the players of the coalition and as the minimum over the set of joint actions of all the players who are not in the coalition. We can think of this value as the amount of money that the players of the coalition can get by themselves with no help from people outside the coalition.

Example 5.1. Let the *Prisoner's dilemma* be given as in Fig. 5.1. We can construct a TU game version of the *Prisoner's dilemma* as follows. The value of a coalition is given by the total years of freedom that the players of a coalition can get against any possible play of the players who are not in the coalition. Having said this, every player alone can get at least 1 year of freedom by playing D. At the same time, the two players playing jointly can get a total of 6 years of freedom, which is the total payoff of (C,C). Thus, the TU game formulation is given by the set $N = \{1,2\}$, and the characteristic function is given by

$$v(\{1\}) = v(\{2\}) = 1, \quad v(\{1,2\}) = 6. \quad \blacksquare$$

	C	D
C	(3,3)	(0,4)
D	(4,0)	(1,1)

Figure 5.1. *The* Prisoner's dilemma *as a TU game. Reprinted with permission from Hindustan Book Agency* [239].

Example 5.2. A second example involves the two-player extensive game in Fig. 5.2. The game involves two stages, and both players can go *left* or *right*. This game can be turned into a TU game as follows. The set of players is $N = \{1,2\}$, and the characteristic function is given by

$$v(\{1\}) = 4, \quad v(\{2\}) = 2, \quad v(\{1,2\}) = 12.$$

To see why we get the above values, consider $v(S)$ as the money that each coalition can get in the worst case. Actually, if player 1 plays R, he will get 4 in the worst case (when P_2 responds l_2). Player 2 can get at least 2 by choosing r_1 in state 1 (light gray node, stage 2) and can get at least 8 by playing l_2 in state 2 (dark gray node, stage 2). The state in stage 2 depends on the action taken by player 1 in stage 1. So in the worst case, player 2 can get 2. If both players form a coalition, they can select the joint actions (R, l_2) and obtain a total payoff of 12. This explains why we get the above values of the coalitions. $\blacksquare$

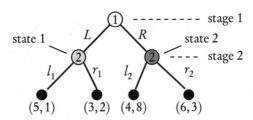

Figure 5.2. *Two-person extensive game as a TU game. Reprinted with permission from Hindustan Book Agency* [239].

Example 5.3. This example develops a TU game formulation for the three-player extensive game displayed in Fig. 5.3. The example admits a TU game formulation as follows: the set of players is $N = \{1,2,3\}$, and the characteristic function is given by

$$v(\{1\}) = 10, \quad v(\{2\}) = 0, \qquad v(\{3\}) = 0,$$
$$v(\{1,2\}) = 14, \quad v(\{1,3\}) = 11, \quad v(\{2,3\}) = 0,$$
$$v(\{1,2,3\}) = 16.$$

To see why we obtain the above values, consider that player 1 can play M_1 and get 10 and the game terminates. In this case, players 2 and 3 do not take any actions. This also justifies why the same players can get by themselves no more than 0. The same result applies to the coalition made by players 2 and 3. Differently, if players 1 and 2 form a coalition, they will agree on player 1 playing R_1, which yields a total payoff equal to 14. This is the sum of payoffs 8 and 6 produced by (R_1, L_3). Note that this is the worst-case payoff, as in the other scenario, that is, (R_1, R_3), the total payoff of coalition $\{1, 2\}$ is 15, which is obtained by summing 11 and 4. For the coalition $\{1, 3\}$, we know that players 1 and 3 can select the joint actions (R_1, R_3), which yield a total payoff of 11. This is obtained as the sum of 11 and 0. Finally, all players forming a grand coalition can agree on (R_1, L_3) and get a total of 16, which results from $6 + 8 + 2$. ∎

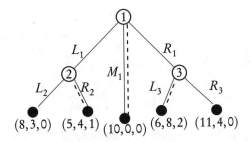

Figure 5.3. *Three-person extensive game as a TU game. Reprinted with permission from Hindustan Book Agency* [239].

5.3 ▪ Game-theoretic examples of operations research problems

We browse here three stylized operations research problems. The idea is to reformulate them as game-theoretic examples involving multiple decision makers. The decision makers now turn into players. When all the players join in a grand coalition, the solution collapses into the optimal solution of the operational research counterpart problem.

5.3.1 ▪ Minimum spanning tree game

This game involves three communities and a power source. The communities are labeled 1, 2, and 3, and the power source is labeled s. They wish to have a direct or indirect connection to the power source, and for this there are a few alternatives. Fig. 5.4 summarizes the available transmission links and the corresponding costs. In particular, direct transmission links to the source costs 100, 90, and 80, respectively. If communities 1 and 2 team up, they can use the transmission links that correspond to the edges of the tree $\{(s, 2), (2, 1)\}$ and both be connected to the source. This yields a minimum total cost of 130 obtained by summing the payoffs 90 of edge $(s, 2)$ and 40 of edge $(2, 1)$. Analogously, communities 1 and 3 can form a coalition and use the links that correspond to the edges of the tree $\{(s, 3), (3, 1)\}$. Both communities are then connected to the source at the minimum total cost of 110 obtained by summing the payoff 80 of edge $(s, 3)$ and the payoff 30 of edge $(3, 1)$. Replicating the same analysis for communities 2 and 3, we get that both can connect to the source using the edges of the tree $\{(s, 3), (3, 2)\}$. This solution has a minimum total cost of 110, this being the sum of 80 for the edge $(s, 3)$ and 30 for the edge $(3, 2)$. If the three communities team up in a grand coalition, the minimum total cost is 140 corresponding to the spanning tree $\{(s, 3), (3, 1), (3, 2)\}$. This cost is obtained as the sum of 80 for edge $(s,3)$, 30 for edge $(3,1)$, and 30 for edge $(3,2)$. The resulting cost game is given by $\langle N,c\rangle$, where $N=\{1,2,3\}$ is the set of players, and c is the characteristic function (here this function represents a cost and not a value) which satisfies

$$c(\{1\})=100, \quad c(\{2\})=90, \qquad c(\{3\})=80,$$
$$c(\{1,2\})=130, \; c(\{1,3\})=110, \; c(\{2,3\})=110,$$
$$c(\{1,2,3\})=140.$$

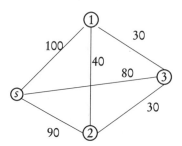

Figure 5.4. *Minimum spanning tree problem as a TU game. Reprinted with permission from Hindustan Book Agency* [239].

To derive the values of the coalitions, we need to consider the money saved by all the players of a coalition when they team up. This is given by the expression

$$v(S)=\sum_{i\in S}c(\{i\})-c(S).$$

Then, the cost game mentioned above can be turned into a *cost-savings game*. The latter is defined by the tuple $\langle N,v\rangle$, where $N=\{1,2,3\}$ is the set of players, and v is the new characteristic function representing the values of the coalitions and satisfying

$$v(\{1\})=0, \quad v(\{2\})=0, \qquad v(\{3\})=0,$$
$$v(\{1,2\})=60, \; v(\{1,3\})=70, \; v(\{2,3\})=60,$$
$$v(\{1,2,3\})=130.$$

It is worth noting that when all the players form the grand coalition, the problem coincides with the classical minimum spanning tree problem [117, Chap. 9.4], which is among the foundations of network flow optimization [117, Chap. 9].

5.3.2 ▪ Permutation game

In this example we have n players $i=1,2,\ldots,n$. Each one owns a machine M_i and has to process a job J_i. Any machine M_j can process any job J_i, but each machine can process at most one job. Forming coalitions implies that the players can agree on processing other players' jobs. Thus, if a player does not form coalitions with other players, he will process his job by himself, namely using his own machine. The cost k_{ij} refers to job J_i processed on machine M_j. It turns out that, for each coalition $S\in 2^N\setminus\emptyset$, the cost of the coalition is the minimum of the costs considering any possible permutation of the elements of that coalition. This is given by the optimization problem

$$c(S)=\min_{\sigma}\sum_{i\in S}k_{i\sigma(i)}, \qquad \sigma \text{ is any permutation of } 1,\ldots,n.$$

Table 5.1. *Coalition values for the permutation game.*

S	$\emptyset$	$\{1\}$	$\{2\}$	$\{3\}$	$\{1,2\}$	$\{1,3\}$	$\{2,3\}$	$\{1,2,3\}$
$c(S)$	0	1	6	12	5	7	17	13
$v(S)$	0	0	0	0	2	6	1	6

In the above expression, $\sigma(i)$ is the ith element of the permutation. As an example, consider the three-person permutation game represented by the cost matrix

$$
\begin{array}{c}
\text{Machine} \\
\begin{bmatrix}
1 & 2 & 3 \\
3 & 6 & 9 \\
4 & 8 & 12
\end{bmatrix} \text{ Job,}
\end{array} \tag{5.1}
$$

where the rows correspond to jobs and the columns correspond to machines. After some calculations, the corresponding costs and coalition values are listed in Table 5.1. It is worth noting that when all the players team up in a grand coalition, the problem takes the form of a classical assignment problem [117, Chap. 8.3].

5.3.3 ▪ Max-flow game

This example explores a situation where we have a flow from a *source* to a *sink*. The flow traverses the edges of a network. The topology of the network is displayed in Fig. 5.5. Different owners (the players) possess different edges. Edges have capacity constraints. In particular, edge 1 has a maximum capacity equal to 4, and the owner is player P_1. Edge 2 has a maximum capacity of 5, and the owner is player P_2. Edge 3 has a maximum capacity equal to 10, and the owner is player P_3. The maximum flow capable of flowing from the source to the sink using the edges of the only members of a coalition gives the value of that coalition.

We can reformulate the problem as a TU game $\langle N, v \rangle$, where $N = \{1, 2, 3\}$ is the set of players, and v is the characteristic function given by

$$
v(\{1\}) = 0, \quad v(\{2\}) = 0, \quad v(\{3\}) = 0,
$$
$$
v(\{1,2\}) = 0, \quad v(\{1,3\}) = 4, \quad v(\{2,3\}) = 5,
$$
$$
v(\{1,2,3\}) = 9.
$$

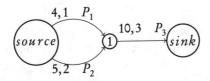

Figure 5.5. *Max-flow problem as a TU game: the labels 4, 1, and P_1 on one of the edges mean that this is edge 1 with maximum capacity equal to 4 and whose owner is player P_1. Reprinted with permission from Hindustan Book Agency [239].*

Remarkably, the game turns into a max-flow optimization problem [117, Chap. 9.5] when all the players join in a grand coalition and act as if there were just one single decision maker.

5.4 ▪ Imputation set

Having formulated TU games in the first part of this chapter, we now turn to the solution of such games. A first solution concept is the one of *imputation set*. The imputation set answers the challenging question about how to divide the costs or rewards produced by a coalition among the players of the same coalition.

The imputation set, denoted by $I(v)$, is a convex polyhedral set. The set contains all the allocations that satisfy the following properties:

- *efficiency or Pareto optimality*, that is, all the components of the allocation vector sum up to the value of the grand coalition, and

- *individual rationality*, namely no player benefits from quitting the grand coalition and playing alone.

A formal definition of the imputation set is as follows.

Definition 5.4 (Imputation set). *Let a TU game $\langle N, v \rangle$ be given. The* imputation set *is the set of allocations given by*

$$I(v) = \left\{ x \in \mathbb{R}^n \mid \overbrace{\sum_{i \in N} x_i = v(N)}^{\text{efficiency}}, \ \underbrace{x_i \geq v(\{i\}), \forall i \in N}_{\text{individual rationality}} \right\}.$$

In general, the above set may be empty. This occurs when, for every efficient allocation, there is at least one player who benefits from quitting the grand coalition. It can be shown that a necessary and sufficient condition for the imputation set to be nonempty states that the sum of the values of the single players must not exceed the value of the grand coalition. More formally,

$$I(v) \neq \emptyset \quad \text{if and only if} \quad v(N) \geq \sum_{i \in N} v(\{i\}).$$

To compute the imputation set, one generally uses the property according to which the imputation set $I(v)$ is the convex hull of the points $f^1, f^2, \ldots, f^n$, where

$$f_k^i = \begin{cases} v(N) - \sum_{k \subset N \setminus \{i\}} v(\{k\}), & k = i, \\ v(\{k\}), & k \neq i. \end{cases} \tag{5.2}$$

Here, the generic vector f^i has a straightforward interpretation. Such a vector is a tentative bid submitted by player i about how to allocate the revenue of the coalition. In particular, the component f_k^i is the revenue that player i suggests to allocate to player k. The idea is the following: player i will allocate to the other players their own values and will keep the rest for himself. In other words, for $k \neq i$ the revenue is $f_k^i = v(\{k\})$. For $k = i$, $f_i^i = v(N) - \sum_{k \in N \setminus \{i\}} v(\{k\})$. The computation of the imputation set based on this procedure is elucidated further in the following example.

Example 5.5. Let a three-person game $\langle N, v \rangle$ be given, where the characteristic function satisfies

$$v(\{1\}) = v(\{3\}) = 0, \quad v(\{2\}) = 3, \quad v(\{1, 2, 3\}) = 5.$$

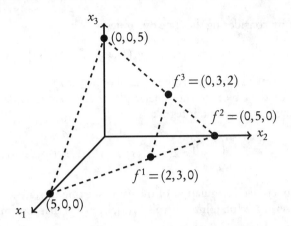

Figure 5.6. *The imputation set $I(v)$ for the game in Example 5.5 is the convex hull of points f^1, f^2, and f^3 computed according to (5.2). Reprinted with permission from Hindustan Book Agency* [239].

Based on what we have said in the previous section, the imputation set $I(v)$ is the triangle with vertices $f^1 = (2,3,0)$, $f^2 = (0,5,0)$, and $f^3 = (0,3,2)$, as displayed in Fig. 5.6.

The vertices have the following interpretation. Vertex $f^1 = (2,3,0)$ is the bid of player 1, who says, "I will allocate to player 2 and 3 their own values, which are 3 and 0, respectively, and will keep the rest for me, namely 2." The same approach can be reiterated for players 2 and 3 to obtain $f^2 = (0,5,0)$ and $f^3 = (0,3,2)$. ∎

TU games can be classified in such a way that it is immediate to realize whether stable allocation rules exist or not.

5.5 ▪ Properties

TU games can be assigned to each one of the following classes. A first class is the one of *superadditive games*. These are games which satisfy the property

$$v(S \cup T) > v(S) + v(T) \quad \forall S, T \in 2^N : S \cap T = \emptyset.$$

A second class involves so-called *subadditive games*. These games are characterized by the condition

$$v(S \cup T) < v(S) + v(T) \quad \forall S, T \in 2^N : S \cap T = \emptyset.$$

A third class of games is made by the *additive games*, for which it holds that

$$v(S \cup T) = v(S) + v(T) \quad \forall S, T \in 2^N : S \cap T = \emptyset.$$

Additive games are also referred to as *inessential games*. The reason is that there exists a unique policy to divide earnings in such games, which is straightforward. Such a policy requires that every player gets exactly his own value.

5.6 ▪ Cooperative differential games

So far, we have mentioned TU games in a static framework. The discussion now turns to dynamic scenarios where the values of the coalitions evolve as in differential games. Let us start by considering the state dynamics

$$\dot{x}(s) = f[s, x(s), u_1(s), u_2(s), \ldots, u_n(s)], \quad x(t_0) = x_0.$$

In the above dynamics, u_i is the control of player $i \in N$.

As for the payoff of the generic player $i \in N$, we assume that this is given by the following finite horizon integral:

$$\int_{t_0}^{T} g^i[s, x(s), u_1(s), u_2(s), \ldots, u_n(s)]ds + q^i(x(T)).$$

In a cooperative formulation of the game starting in state x_0 at time t_0, we assume that the players agree on joint controls $u_1^*(s), \ldots, u_n^*(s)$ that maximize the total payoff

$$v(N; x_0, T - t_0) = \sum_{i \in N} \int_{t_0}^{T} g^i[s, x(s), u_1^*(s), \ldots, u_n^*(s)]ds + q^i(x(T))$$

and on a mechanism to distribute the total payoff among the players. We compactly refer to such a game as *cooperative differential game* in characteristic function form and denote it $\Gamma_v(x_0, T - t_0)$. In this context we denote by $\xi_i(x_0, T - t_0)$ the share of the players $i \in N$ obtained from the total payoff $v(N; x_0, T - t_0)$.

Similarly to the static case, the value of a single player $i \in N$, denoted by $v(\{i\}; x_0, T - t_0)$, is the amount of money that the player can get for himself if all the others play against him. This is formally given by the maximization problem

$$\max_{u_i} \min_{u_j, j \neq i} \int_{t_0}^{T} g^i[s, x(s), u_1(s), u_2(s), \ldots, u_n(s)]ds + q^i(x(T)).$$

The notion of *imputation* can be adapted to such a dynamic scenario, as shown next.

Definition 5.6 (Imputation in dynamic setting). *A vector of shares*

$$\xi(x_0, T - t_0) = [\xi_1(x_0, T - t_0) \ldots \xi_n(x_0, T - t_0)]$$

is called imputation *if*

(i) $\xi_i(x_0, T - t_0) \geq v(\{i\}; x_0, T - t_0) \quad \forall i \in N \quad$ *(rational)*,
(ii) $\sum_{j \in N} \xi_j(x_0, T - t_0) = v(N; x_0, T - t_0) \qquad$ *(efficient)*.   (5.3)

Now, given the optimal trajectory $\{x^*(s)\}_{s=t_0}^{T}$, let the family of games along such a trajectory be

$$\{\Gamma_v(x^*(t), T - t), \quad t_0 \leq t \leq T\}.$$

It is possible to separate the total share ξ_i in two parts: the current share ω_i (the amount allocated so far) and the future share η_i:

$$\eta_i[\xi_i(x_0, T - t_0); x^*(t), T - t]$$
$$= \xi_i(x_0, T - t_0) - \omega_i[\xi_i(x_0, T - t_0); x^*(\cdot), t - t_0].$$ (5.4)

This leads to the following property, which represents a landmark in the context of dynamic TU games.

Definition 5.7 (Dynamic stability and time consistency). *An imputation*

$$\xi(x_0, T - t_0) = [\xi_1(x_0, T - t_0) \dots \xi_n(x_0, T - t_0)]$$

is said to be dynamically stable or time consistent *if*

(i) $\eta_i[\xi_i(x_0, T - t_0); x^*(t), T - t] \geq v(\{i\}; x^*(t), T - t) \quad \forall i \in N,$

(ii) $\sum_{j \in N} \eta_j[\xi_j(x_0, T - t_0); x^*(t), T - t] = v(N; x_0, T - t).$

In a nutshell, we are dealing with a stream of infinite TU games. Each game is played starting from time t, for any t in the interval $[t_0, T)$. For each such game, we search allocations that lie in the corresponding imputation set. This requires the computation of the full Pareto curve.

In the next chapter we continue the exploration of other seminal solution concepts, such as the *core*, the *Shapley value*, and the *nucleolus*.

5.7 ▪ Notes and references

The formulation of TU games in Section 5.2 and the operations research examples of Section 5.3 are based on [239, Chap. 10]. Example 5.3 is discussed on p. 3 in [239]. For the operations research counterpart problems we refer the reader to a classical textbook in operations research [117]. The introduction of the imputation set in Section 5.4 is heavily based on [239, Chap. 11]. The presentation of *cooperative differential games* in Section 5.6 is subsumed by [253].

For the computation of Pareto curves we refer the reader to [92] and [204]. The literature offers other formulations of dynamic cooperative games that involve also a detailed analysis and design of robust allocation policies (see [44] and [45]). In addition, if the allocation process is distributed, connections with consensus problems are explored in [185]. Further investigations on social optimal equilibria and their use in multi-inventory applications are available in [31, 32].

Connections between cooperative game theory and bargaining can be found in the Kalai–Smorodinski solution in [137]. The egalitarian solution in bargaining is axiomatized in [133] and extended to games with nontransferable utility in [136].

The material of this chapter is integrated with some lecture notes taken by the author during a Summer School on Game Theory and Operations Research held in Lavagna, Italy, in September 2003, and organized by Fioravante Patrone and Vito Fragnelli.

Chapter 6

Core, Shapley Value, Nucleolus

6.1 ▪ Introduction

Following the formulation of TU games presented in the previous chapter, and with in mind the definition of imputation set from Section 5.4, we now review other solution concepts, such as the *core*, the *Shapley value*, and the *nucleolus*. As for the imputation set, these solution concepts yield stable allocation rules. The chapter concludes with a discussion on computational issues for the nucleolus.

This chapter is organized as follows. In Section 6.2 we introduce the core. In Section 6.3 we present the Shapley value. In Section 6.4 we deal with convex coalitional games. In Section 6.5 we discuss the nucleolus. Finally, in Section 6.6 we provide notes, conclusions, and references.

6.2 ▪ Core

Given a TU game, a common solution concept is the one of *core*. The core provides allocations that are stable with respect to any sub-coalitions. In this sense, it represents a refinement of the imputation set. To define the core, the conditions valid for the imputation set are strengthened, as not only do the players have no incentives to split from the grand coalition and play individually, but they also do not benefit from forming any sub-coalition. As a consequence, the core is still a polyhedral set and is a subset of the imputation set.

In particular, the core of a game is the set of allocations that satisfy

- efficiency,
- individual rationality, and
- stability with respect to sub-coalitions.

The three properties mentioned above lead to the following definition of core.

Definition 6.1 (Core). *Let a TU game $\langle N, v \rangle$ be given, and let $I(v)$ be the imputation set. The* core *is the set of allocations given by*

$$C(v) = \left\{ x \in I(v) \Big| \underbrace{\sum_{i \in S} x_i \geq v(S)}_{\text{stability w.r.t. sub-coalitions}} \quad \forall S \in 2^N \setminus \emptyset \right\}.$$

As an example of how the core can be described in matrix form, consider a three-player game. Then, the following set of inequalities applies:

$$C(v) = \left\{ x \in \mathbb{R}^3 \Big| \right.$$
$$x^T \begin{bmatrix} 1 & 0 & 0 & 1 & 1 & 0 & 1 & -1 \\ 0 & 1 & 0 & 1 & 0 & 1 & 1 & -1 \\ 0 & 0 & 1 & 0 & 1 & 1 & 1 & -1 \end{bmatrix} \qquad (6.1)$$
$$\left. \geq \left[v(\{1\}) \, v(\{2\}) \ldots v(N) - v(N) \right] \right\}.$$

We can interpret the above conditions by saying that if $x \in C(v)$, then no coalition has an incentive to split off.

As for the imputation set, there exist conditions that guarantee that such stable allocations exist. These are necessary and sufficient conditions for the core to be nonempty and have been proved by Bondareva (1963) and Shapley (1967). We restate such conditions in the following theorem.

Theorem 6.2 (Bondareva and Shapley theorem). *Given a game $\langle N, v \rangle$, the following are equivalent:*

1. *the core $C(v) \neq \emptyset$;*

2. *$\langle N, v \rangle$ is a balanced game.*

We omit the proof for the sake of conciseness and as a few versions of the proof are available in other textbooks. We simply highlight that such a proof is based on the duality theorem from linear programming theory.

In order to understand what is meant by *balanced game*, consider the characteristic vector $\mathbf{1}_S$ of coalition S,

$$[\mathbf{1}_S]_i = \begin{cases} 1, & i \in S, \\ 0, & i \notin S, \end{cases} \qquad (6.2)$$

where $[\mathbf{1}_S]_i$ denotes the ith component of the characteristic vector $\mathbf{1}_S$. In essence, the characteristic vector has as many components as the number of players. Given coalition S, the ith component of the characteristic vector $\mathbf{1}_S$ is 1 if the corresponding player is in the coalition and 0 otherwise. It is worth noting that the characteristic vectors are the columns of the constraint matrix except for the last column in (6.1). Balanced games build upon the notion of a balanced map, which we review below.

Definition 6.3 (Balanced map). *A map $\lambda : 2^N \setminus \emptyset \to \mathbb{R}$ is* balanced *if*

$$\sum_{S \subseteq N, S \neq \emptyset} \lambda(S) \mathbf{1}_S = \mathbf{1}_N.$$

The interpretation of the concept of a balanced map is that it is a rule expressing the portion of unitary time that a player dedicates to each coalition S to which he belongs. We illustrate this idea in the following example.

Example 6.4. For a three-player game the map $\lambda(S) = 1/2$ if $|S| = 2$ and 0 otherwise is a balanced map. Analogously, also the map $\lambda(S) = 1$ if $|S| = 1$ and 0 otherwise is a balanced map. In other words, every single player can devote half of his time to any of the two two-player coalitions he forms with the other players, or he could spend all his time working in the grand coalition. ∎

With in mind the notion of a balanced map as provided above, we are ready to define balanced games.

Definition 6.5 (Balanced game). *A game is* balanced *if for every balanced map* λ

$$\sum_{S \subseteq N, S \neq \emptyset} \lambda(S) v(S) \leq v(N).$$

The following example shows that the core may coincide with the imputation set.

Example 6.6. Given the game $\langle N, v \rangle$, where $v(S) = 0$ for all $S \neq N$, and $V(N) = 1$, the core coincides with the simplex,

$$C(v) = \left\{ x \in \mathbb{R}^n \,|\, x_i \geq 0 \quad \forall i \in N, \sum_{i \in N} x_i = 1 \right\}. ∎$$

The next example is known as the *Gloves game*. It simulates the scenario where one player owns a left glove and two other players own a right glove. Obviously, this attributes a higher contractual power to the player with the left glove (player 3 in the example). This example shows that the core may include only one point.

Example 6.7 (*Gloves game*). Given the *Gloves game* $\langle N, v \rangle$, where $N = \{1, 2, 3\}$ is the set of players and v is the characteristic function given by

$$v(\{1\}) = v(\{2\}) = v(\{3\}) = v(\{1, 2\}) = 0,$$
$$v(\{1, 3\}) = v(\{2, 3\}) = v(\{1, 2, 3\}) = 1,$$

the core is a singleton and is obtained by solving

$$\begin{aligned} C(v) &= \{ x \in \mathbb{R}^3_+ \,|\, x_1 + x_3 \geq 1, x_2 + x_3 \geq 1, \textstyle\sum_{i \in N} x_i = 1 \} \\ &= (0, 0, 1). \end{aligned} \quad ∎ \tag{6.3}$$

6.3 ▪ Shapley value

The *Shapley value* was first formalized by Lloyd Shapley in 1953. It is based on an extremely simple idea which has contributed to its popularity even among nonexperts of cooperative game theory. The Shapley value is also very intuitive and enjoys useful properties. As a drawback it does not always provide stable allocations. We use a simple story to introduce the Shapley value.

Imagine a group of players that enter a room according to a predefined sequence. Indicate by $\sigma : N \to N$ the ordering of the entries. By this we mean that $\sigma(k)$ is the player who enters as kth, and $\sigma^{-1}(i)$ is the entry number of player i. As an example, if we have $\sigma = (3, 2, 1)$, then this means that

$$\sigma(1) = 3, \quad \sigma(2) = 2, \quad \sigma(3) = 1,$$
$$\sigma^{-1}(1) = 3, \quad \sigma^{-1}(2) = 2, \quad \sigma^{-1}(3) = 1.$$

Given a predefined ordering, we can define the set of *predecessors* of i. Let us denote it by

$$P_\sigma(i) = \{k \in N \,|\, \sigma^{-1}(k) < \sigma^{-1}(i)\}.$$

Once we know the predecessors, we also know the marginal value of player i. This is obtained as

$$m_i^\sigma(v) = v(P_\sigma(i) \cup \{i\}) - v(P_\sigma(i)).$$

In other words, the marginal value captures the value added to a coalition by a player. After storing the marginal values in a vector, we obtain the *marginal vector*

$$m^\sigma(v) = \{m_i^\sigma(v), i \in N\}.$$

It is worth noting that there exist $n!$ marginal values for player i. This is the exact number of permutations of n objects.

The above preamble takes us to the following definition of Shapley value.

Definition 6.8 (Shapley value). *The* Shapley value *is the average of the marginal vector over all possible permutations, namely*

$$\Phi(v) = \frac{1}{n!} \sum_\sigma m^\sigma(v).$$

In the following example we mention the steps necessary to compute the Shapley vector.

Example 6.9. Let the game $\langle N, v \rangle$ be given where the characteristic function satisfies

$$v(\{1\}) = v(\{2\}) = v(\{3\}) = 0,$$
$$v(\{1,2\}) = 4, \quad v(\{1,3\}) = 7, \quad v(\{2,3\}) = 15, \quad v(\{1,2,3\}) = 20.$$

Fig. 6.1 lists the marginal values and the resulting Shapley vector. This is given by $\Phi(v) = \frac{1}{6}(21, 45, 54) \in C(v)$. To understand the computation of the Shapley vector, let us look at the column of $m_1^\sigma(v)$ (1st column). Consider the ordering $(1, 2, 3)$ (1st row). According to this, player 1 enters first. The marginal value of player 1 is zero, as the value of the coalition $\{1\}$ is zero. A similar consideration applies if we consider the ordering $(1, 3, 2)$ (2nd row). Consider now the ordering $(2, 1, 3)$ (3rd row). Here, player 1 enters after player 2. The coalition $\{2\}$ turns into $\{1, 2\}$, and the value increases from 0 to 4. This scenario yields a marginal value of player 1 equal to 4. Consider the ordering $(2, 3, 1)$ (4th row). Here, player 1 enters the room last. Actually, coalition $\{2, 3\}$ turns into the grand coalition, and the value increases from 15 to 20. This scenario yields a marginal value of player 1 equal to 5. Consider the ordering $(3, 1, 2)$ (5th row). Here, player 1 enters the room after player 3. Coalition $\{3\}$ turns into $\{1, 3\}$, and the value increases from 0 to 7. This scenario yields a marginal value of player 1 equal to 7. Finally, the last row considers the ordering $(3,2,1)$. In this case, player 1 enters the room last. The coalition $\{2,3\}$ turns into the grand coalition, and the value goes from 15 to 20, which yields a marginal value of 5. The interpretation of the values in columns 2 and 3 is analogous. The columns collect the marginal values $m_2^\sigma(v)$ and $m_3^\sigma(v)$ of players 2 and 3, respectively. These are obtained by repeating the same procedure for these players. ∎

σ	$m_1^\sigma(v)$	$m_2^\sigma(v)$	$m_3^\sigma(v)$
$(1,2,3)$	0	4	16
$(1,3,2)$	0	13	7
$(2,1,3)$	4	0	16
$(2,3,1)$	5	0	15
$(3,1,2)$	7	13	0
$(3,2,1)$	5	15	0

Figure 6.1. *Marginal values. Reprinted with permission from Hindustan Book Agency* [239].

There are a few reasons why the Shapley value has become popular; it is intuitive, and it verifies properties like (i) the *efficiency* property and (ii) the *dummy player* property. A player is *dummy* if the marginal contribution he adds to any coalition is equal to $v(\{i\})$. The dummy player property deals with the fact that each dummy player i is rewarded with a value $\Phi(v)_i = v(\{i\})$. In other words he receives a reward which is equal to the marginal contribution he adds to any coalition. The drawback of using the Shapley value is that such a value is not always in the core of the game. A solution concept which is always in the core of the game is the *nucleolus*. In the rest of this chapter we introduce this concept and show that, despite its computation being a bit troublesome, it provides allocation in the core, provided that the latter is nonempty. However, before introducing the nucleolus, we wish to discuss an interesting category of games known as convex games.

6.4 ▪ Convex games

There exists a category of games for which computing the core is simple. This is the category of *convex games*.

Definition 6.10 (Convex games). *We say that a TU game is convex if it satisfies the property*

$$v(S \cup T) + v(S \cap T) > v(S) + v(T) \quad \forall S, T \in 2^N.$$

Recall the category of superadditive games introduced in Section 5.5. Superadditive games are characterized by the following property:

$$v(S \cup T) > v(S) + v(T) \quad \forall S, T \in 2^N : S \cap T = \emptyset.$$

Then convex games constitute a subset of superadditive games. This is clear by noting that in convex games the marginal contribution of any player i or coalition of players T to coalition S increases with the number of players in S. It can be shown that, given a convex game, any allocation which gives each player a reward equal to his marginal value is in the core. The resulting allocation vector x is constituted by the components $x_i = m_i^\sigma$ for any σ. Such a vector yields a point in the core. We are ready to illustrate the nucleolus in the remaining part of this chapter.

6.5 ▪ Nucleolus

Let the following lexicographic order $\leq_L$ be given. For instance, let us think of the order used to store words in a dictionary. We say that $x \in \mathbb{R}^p$ is *lexicographically smaller* than $y \in \mathbb{R}^p$ if $x = y$ or there exists an $s = 1, 2, \ldots, p$ such that $x_i = y_i$ for all $i < s$ and $x_s < y_s$. As an example, we have that $(0, 100, 100) \leq_L (1, -10, -10)$, and $(10, 4, 100) \leq_L (10, 5, 6)$.

Definition 6.11 (Excess vector). *Given an allocation $x \in I(v)$, we define* excess vector *as the vector*

$$\theta(x) = \left\{ e(S, x) := v(S) - \sum_{i \in S} x_i \quad \forall S \in 2^N \setminus \emptyset \right\}.$$

Furthermore, the quantity $e(S, x) := v(S) - \sum_{i \in S} x_i$ is said to be the excess of coalition S.

Based on the above definition, the excess vector has a number of components equal to the number of coalitions. For each coalition the excess vector captures the discrepancy between the value of the coalition and the total amount given to the members of the coalition. Then, the coalition is stable if such a difference is negative. In other words stability arises when the total amount exceeds the value of the coalition. With in mind the lexicographic ordering defined above, the nucleolus is defined as the solution that minimizes such an excess vector. This is formalized in the next definition.

Definition 6.12 (Nucleolus). *The nucleolus is the* lexicographic minimizer *of any excess vector:*

$$\theta(Nu(v)) \leq_L \theta(x) \quad \forall x \in I(v).$$

It can be proved that the nucleolus is always in the core of the game whenever the latter is nonempty, as established in the next theorem.

Theorem 6.13. *If $C(v) \neq \emptyset$, then $Nu(v) \in C(v)$.*

From a different perspective, we can say that the nucleolus minimizes the maximal excess. We can also say that if the core is nonempty, the nucleolus always belongs to it.

As an example, for the *Gloves game* introduced in Example 6.7, we saw that the core is a singleton, and therefore the nucleolus corresponds to the core. We repropose the example in the following.

Example 6.14. Given the *Gloves game* $\langle N, v \rangle$, where

$$v(\{1\}) = v(\{2\}) = v(\{3\}) = v(\{1, 2\}) = 0,$$
$$v(\{1, 3\}) = v(\{2, 3\}) = v(\{1, 2, 3\}) = 1,$$

the nucleolus is $Nu(v) = (0, 0, 1) = C(v)$. ∎

If the game admits a symmetric structure, the computation of the nucleolus can be enormously simplified, as shown next.

Example 6.15. Given the game $\langle N, v \rangle$, $N = \{1, 2, 3, 4\}$ and $v(S) = |S|^2$ for all $S \in 2^N$:

$$C(v) = \{x \in \mathbb{R}_+^4 \mid x_i + x_j \geq 4 \quad \forall i, j, \ x_i + x_j + x_k \geq 9, \forall i, j, k, \\ x_1 + x_2 + x_3 + x_4 = 16\}. \tag{6.4}$$

Note that the constraints are symmetric; thus $Nu(v) = (a,a,a,a)$. The only solution verifying efficiency is then $Nu(v) = (4,4,4,4)$. ∎

6.5.1 ▪ Computation through sequence of linear programs

This chapter concludes with a description of a computational technique to compute the nucleolus. This technique is based on a sequence of linear programming problems. The technique is recursive, and we comment next only on the first two steps. The first step involves minimizing the maximal excess. This corresponds to solving the following linear programming problem:

$$\begin{aligned}
&\theta_1 := \min t \\
&e(S,x) \le t \quad \forall S \subseteq N, \\
&\mathbf{1}^T x(N) = v(N), \\
&x_i \ge v(\{i\}) \quad \forall i \in N.
\end{aligned} \tag{6.5}$$

Let us denote $X_1 = \{x | e(S,x) \le \theta_1 \; \forall S \subseteq N\}$. In other words, X_1 is the set of allocations that attain the minimum for the maximal excess. In addition to this, let us denote $\Sigma_1 = \{S \subseteq N | e(S,x) = \theta_1 \; \forall x \in X_1\}$. Set Σ_1 is the set of all coalitions at which the maximal excess is attained at all $x \in X_1$.

In the second step we wish to minimize the second-largest excess. This boils down to the following linear programming problem:

$$\begin{aligned}
&\theta_2 := \min t \\
&e(S,x) = \theta_1 \quad \forall S \in \Sigma_1, \\
&e(S,x) \le t \quad \forall S \notin \Sigma_1, \\
&\mathbf{1}^T x(N) = v(N), \\
&x_i \ge v(\{i\}) \quad \forall i \in N, \quad \text{and so forth.}
\end{aligned} \tag{6.6}$$

The algorithm for the computation of the nucleolus continues recursively until a solution is obtained. Note that the computation procedure that leads to the nucleolus is more cumbersome than the one which leads to the Shapley value. This probably justifies the large popularity of the Shapley value among the nonexperts of game theory.

6.6 ▪ Notes and references

The material of this chapter is inspired by [239, Chap. 11], [196, Chaps. 13–14] and [173, Chaps. 16–18].

For a formal proof of the Bondareva and Shapley theorem we refer the reader to the original works [63, 221] as well as to [196, Sect. 13.3]. The introduction of the core in Section 6.2 is heavily based on [239, Chap. 11] and [196, Chap. 13]. The Shapley value were first formulated in [219], and it is also discussed in [196, Chap. 14] and [108]. Example 6.9 is borrowed from [239, Chap. 14]. More details on the nucleolus can be found in [196, Chap. 14].

The material of this chapter is also integrated with some lecture notes taken by the author during a Summer School on Game Theory and Operations Research held in Lavagna, Italy, in September 2003, and organized by Fioravante Patrone and Vito Fragnelli.

For a tutorial on the use of coalitional game theory in communication we refer the reader to [209].

Chapter 7
Evolutionary Game Theory

7.1 ▪ Introduction

This chapter covers basic concepts in evolutionary game theory. In a nutshell, the theory of evolutionary games deals with the study of cooperation and competition in evolutionary biology. The theory goes further and studies the impact that evolutionary biology has on social science. This is illustrated in the classical model delineated throughout this chapter.

A fundamental concept is the one of *evolutionarily stable strategy*, which we develop at the beginning of the chapter. Different perspectives will lead to two equivalent definitions of evolutionarily stable strategies. The first definition is developed in the context of evolutionary biology. The second definition arises in economics and presents similarities with the notion of Nash equilibrium. Evolutionarily stable strategies are explored in the context of prototypical games like the *Prisoner's dilemma*, the *Hawk and Dove game*, and the *Coordination game*, just to name a few.

This chapter is organized as follows. Section 7.2 introduces the model involving a population of incumbents and mutants. Section 7.3 examines relations between evolutionarily stable strategies, dominance, and equilibrium solutions. Section 7.4 introduces the two equivalent definitions of evolutionarily stable strategies. Section 7.5 discusses implications of stability and illustrates the latter on a number of examples. Finally, Section 7.6 provides notes, conclusions, and references.

7.2 ▪ Population of incumbents and mutants

Imagine a population of individuals, henceforth called *incumbents*. These individuals are "designed" so that they can play a given strategy. Individuals are subjected to random matchings at every time. Such a scenario can be described using a two-player symmetric game. Here a single individual, the row player, seeks to maximize his average payoff based on the distribution of the population behavior. The population is fictitiously represented by the column player, and the distribution of the population is modeled by the mixed strategy of the column player. Furthermore, the strategies represent the *genes*; and the payoffs indicate the *fitness* of the individuals, i.e., the expected number of offsprings. Together with the incumbents, in the population there may also be *mutants*. These are offsprings that play randomly any feasible strategy in the set of strategies. We call *successful* those strategies that tend to grow and *unsuccessful* those strategies that tend to extinguish.

This preamble takes us to the following main question: What conditions guarantee that the given strategy is robust against mutations? Under what circumstances will the mutants die or thrive? The theory of evolutionary games essentially gives an answer to the above questions.

Example 7.1 (*Prisoner's dilemma*). This example reframes the *Prisoner's dilemma* in an evolutionary game-theoretic context. Recall that the *Prisoner's dilemma* is described by the bimatrix displayed in Fig. 7.1. The scenario we are considering is as follows. There is a group of lions on a hunt. Every lion can *cooperate* (C) and go on a hunt together with the group or can *defect* (D) and go on a hunt alone. Likewise, we can think of a group of ants defending a nest. Every ant can cooperate in defending the nest from a spider or can defect and flee away from the danger.

	C	D
C	$(3,3)$	$(0,4)$
D	$(4,0)$	$(1,1)$

Figure 7.1. Prisoner's dilemma *as an evolutionary game. Reprinted with permission from Benjamin Polak and Yale University* [202].

Fig. 7.1 displays the bimatrix containing the payoff of each individual—think of it as the row player—when such an individual plays against a random opponent extracted from the population. Here the opponent is the column player, which plays a mixed strategy that simulates the distribution of the population. Given the setup as illustrated above, let us consider a mixed strategy $(1-\epsilon,\epsilon)$ for the column player. This strategy admits the following interpretation: a small portion ϵ of mutants play D, while the rest of the population, constituted by the incumbents, plays C. The question we wish to answer is whether the strategy *cooperation* is an evolutionarily stable strategy or not. We will see that *cooperation* is not evolutionarily stable. To obtain an answer, note that the mutant performs better than the incumbent on a random matching. To see this, we need to consider two different scenarios: one contemplates an incumbent playing against the population, and the other involves a mutant playing against the population.

- **Case 1: Incumbent against the population.** When an incumbent (row player) meets an opponent (column player) randomly extracted from the population, the opponent plays C with probability $1-\epsilon$ and D with probability ϵ. This corresponds to saying that the column player adopts a mixed strategy $(1-\epsilon,\epsilon)$. For the expected payoff we then obtain $(1-\epsilon)3+\epsilon0 = 3(1-\epsilon)$. To indicate the random matching between an incumbent and an individual from the population we shortly write C vs. $[(1-\epsilon)C+\epsilon D]$. The expected payoff computation yields $(1-\epsilon)[3]+\epsilon0 = 3(1-\epsilon)$. Such a scenario is shortly described in one line as shown below:

$$C \text{ vs. } [(1-\epsilon)C+\epsilon D] \to (1-\epsilon)[3]+\epsilon0 = 3(1-\epsilon),$$

where vs. means *versus*.

- **Case 2: Mutant against the population.** Differently, imagine that a mutant (row player) fights against an opponent (column player), which is randomly extracted from the population. The opponent plays C with probability $1-\epsilon$ and plays D with probability ϵ. As a result, the expected payoff is $(1-\epsilon)4+\epsilon = 4(1-\epsilon)+\epsilon$.

Briefly, we can indicate such a scenario in one line as displayed below:

$$D \text{ vs. } [(1-\epsilon)C + \epsilon D] \rightarrow (1-\epsilon)[4] + \epsilon 1 = 4(1-\epsilon) + \epsilon.$$

Given both scenarios, we need to compare the expected payoffs of incumbents and mutants when involved in random matchings with other individuals. It is clear that the mutant performs better than the incumbent, from which we conclude that *cooperation* is not an evolutionarily stable strategy. In addition to this, it is worth noting that *cooperation* is a strictly dominated strategy. The challenge now is about generalizing the result provided by the example under study. We do this in the next section. ∎

7.3 ▪ Evolutionarily stable strategy, dominance, and equilibrium

In this section we highlight connections between evolutionarily stable strategies, dominance, and equilibrium solutions.

7.3.1 ▪ A strictly dominated strategy is not an evolutionarily stable strategy

From the analysis of the *Prisoner's dilemma* in the preceding section, we understand that *cooperation*, which is a strictly dominated strategy, is not evolutionarily stable. The idea now is to generalize this result. Actually, the main message of this section is essentially that *strictly dominated strategies are not evolutionarily stable*. To understand this, let us go back to the *Prisoner's dilemma* example. Imagine that the population is now "designed" so that most individuals *defect*, while there is a minority who plays C. This corresponds to assuming a mixed strategy $(\epsilon, 1-\epsilon)$ for the column player. We can interpret such a mixed strategy as if a small percentage ϵ of mutants will play C within a population of incumbents that play D. We see here that the strategy *defection* is evolutionarily stable, as a mutant performs worse than an incumbent on a random matching. This is clear once we analyze the two scenarios involving first an incumbent playing against the population and then a mutant playing against the population. Both scenarios are discussed next.

- **Case 1: Incumbent against the population.** Let us suppose that an incumbent (row player) plays against an opponent (column player) randomly extracted from the population. We shortly indicate such a scenario as D vs. $[(1-\epsilon)D + \epsilon C]$. The incumbent's expected payoff is given by $(1-\epsilon)[1] + \epsilon[4] = (1-\epsilon) + 4\epsilon$. In short, we have

$$D \text{ vs. } [(1-\epsilon)D + \epsilon C] \rightarrow (1-\epsilon)[1] + \epsilon[4] = (1-\epsilon) + 4\epsilon.$$

- **Case 2: Mutant against the population.** Now, suppose that a mutant (row player) faces an opponent (column player) randomly extracted from the population, namely C vs. $[(1-\epsilon)D + \epsilon C]$. The mutant's expected payoff is then $(1-\epsilon)[0] + \epsilon[3] = 3\epsilon$. More compactly, we have

$$C \text{ vs. } [(1-\epsilon)D + \epsilon C] \rightarrow (1-\epsilon)[0] + \epsilon[3] = 3\epsilon.$$

We infer that the incumbent is more successful than the mutant on random matchings. This implies that mutations from D tend to extinguish. In general, we can state that a strictly dominated strategy is not evolutionarily stable. This is easily proven by showing that the strictly dominant strategy turns out to be a successful mutation.

7.3.2 ▪ From evolutionarily stable strategy to symmetric Nash equilibrium

This section develops another general consideration, namely that evolutionarily stable strategies yield symmetric Nash equilibrium solutions. Recall that strategies represent genes, and therefore a symmetric Nash equilibrium corresponds to a so-called *monomorphic* population. This is a population with a unique gene. We develop further this concept in the bimatrix example provided in Fig. 7.2. The example shows a two-player nonzero-sum game where both players can play b or c. The game represents a variant of the *Coordination game* where one player is better off by playing b if the other player is playing c and vice versa.

	b	c
b	$(0,0)$	$(1,1)$
c	$(1,1)$	$(0,0)$

Figure 7.2. *Example showing that an evolutionarily stable strategy yields a Nash equilibrium. Reprinted with permission from Benjamin Polak and Yale University* [202].

Given the above games, we wish to analyze strategy c and answer the question of whether such a strategy is evolutionarily stable or not. We anticipate that the answer is negative, as we will see that the mutants who select b perform better than the incumbents who are programmed to play c. The two scenarios and the corresponding average payoffs are summarized in the next two lines:

$$c \text{ vs. } [(1-\epsilon)c + \epsilon b] \to (1-\epsilon)[0] + \epsilon[1] = \epsilon,$$
$$b \text{ vs. } [(1-\epsilon)c + \epsilon b] \to (1-\epsilon)[1] + \epsilon[0] = 1 - \epsilon.$$

The first line above contemplates the case where an incumbent picking strategy c faces an opponent who plays the mixed strategy $[(1-\epsilon)c + \epsilon b]$. As a result the expected payoff is $(1-\epsilon)[0] + \epsilon[1] = \epsilon$. Analogously, the second line simulates the case where a mutant selecting strategy b fights against an opponent from the population who plays the mixed strategy $[(1-\epsilon)c + \epsilon b]$. The expected payoff is then $(1-\epsilon)[1] + \epsilon[0] = 1 - \epsilon$. We can conclude that the mutant's payoff is better than the incumbent's payoff, and therefore c is not an evolutionarily stable strategy.

From a deeper exploration, we also observe that the mutants who play b, as a consequence of their being more successful than the incumbents playing c, grow from a smaller percentage of ϵ to $\frac{1}{2}$. In addition to this, observe that strategy b, which is the mutant gene or the *invader*, is itself not evolutionarily stable. Despite this, strategy b still avoids dying out.

Another question we can pose at this point is whether strategy c is a Nash equilibrium or not. To put it differently, we wish to know whether the symmetric profile (c, c) is a symmetric Nash equilibrium. Again, we see that this is not the case, since strategy b turns out to be a strict profitable deviation. The conclusion is that, given a strategy s, if such a strategy does not yield a symmetric Nash equilibrium, then the same strategy is not evolutionarily stable. From a different angle, this is equivalent to saying that a necessary condition for a strategy to be evolutionarily stable is that such a strategy is a Nash equilibrium strategy. This is stated shortly in the following implication:

If s is evolutionarily stable $\Rightarrow$ (s, s) is a Nash equilibrium.

7.3.3 ▪ A Nash equilibrium strategy is not necessarily an evolutionarily stable strategy

The fundamental message of the previous section is that an evolutionarily stable strategy yields a symmetric Nash equilibrium. In this section, we turn to analyze the opposite case and show that the converse is not true. Actually, we show that a Nash equilibrium strategy is not necessarily evolutionarily stable. To see this, let us look at the example displayed in Fig. 7.3. The example is a two-player nonzero-sum game where both players can play a or b. The example shows two Nash equilibrium solutions, namely (a,a) and (b,b). However, it is worth noting that strategy b, which is a Nash equilibrium strategy as it appears in the Nash equilibrium (b,b), is not evolutionarily stable. To understand this, let us observe that the mutants who play the strategy a perform better than the incumbents who are programmed to play b. This is evident by considering the two scenarios summarized below:

$$b \text{ vs. } [(1-\epsilon)b + \epsilon a] \rightarrow (1-\epsilon)[0] + \epsilon[0] = 0,$$
$$a \text{ vs. } [(1-\epsilon)b + \epsilon a] \rightarrow (1-\epsilon)[0] + \epsilon[1] = \epsilon.$$

The first line represents the case where an incumbent playing b fights against an opponent randomly extracted from the population and therefore is playing the mixed strategy $[(1-\epsilon)b + \epsilon a]$. The resulting expected payoff is $(1-\epsilon)[0] + \epsilon[0] = 0$. The second line describes another scenario, where a mutant who plays strategy a faces an opponent randomly extracted from the population and therefore is characterized by a mixed strategy $[(1-\epsilon)b + \epsilon a]$. As a consequence, the expected payoff is given by $(1-\epsilon)[0] + \epsilon[1] = \epsilon$. We conclude that the mutant performs better than the incumbent, and therefore the strategy b is not evolutionarily stable. This is true despite the symmetric profile (b,b) being a Nash equilibrium. The main justification for this is that the symmetric profile (b,b) is not a strict Nash equilibrium. In conclusion, we can observe the following fact: If (s,s) is a strict Nash equilibrium, then s is an evolutionarily stable strategy.

	a	b
a	$(1,1)$	$(0,0)$
b	$(0,0)$	$(0,0)$

Figure 7.3. *Example showing that a Nash equilibrium strategy is not necessarily evolutionarily stable. Reprinted with permission from Benjamin Polak and Yale University* [202].

7.4 ▪ Formal definition of evolutionarily stable strategy

There exist two equivalent definitions of evolutionarily stable strategies: the first one has been developed in Evolutionary Biology, while the second one arises in the Economics literature. The first definition uses the notion of "small" perturbation ϵ, and this makes the conditions stated in the first definition difficult to be checked. The verification of the second definition, the one developed in the Economics literature, is more straightforward and simpler.

7.4.1 ▪ A first definition in Biology

The setup is as follows. Let a two-player symmetric game be given, where Δ is the set of mixed strategies for the row player, and $u(a, b)$ is his payoff resulting from the row player playing the strategy $a \in \Delta$ against a population playing the strategy $b \in \Delta$.

Definition 7.2. *A mixed strategy $s^* \in \Delta$ is evolutionarily stable if there exists $\bar{\epsilon} > 0$ such that for any $s \in \Delta$ and $\epsilon \leq \bar{\epsilon}$, we have*

$$\underbrace{u(s^*, \epsilon s + (1-\epsilon)s^*)}_{\text{payoff to incumbent } s^*} > \underbrace{u(s, \epsilon s + (1-\epsilon)s^*)}_{\text{payoff to mutant } s}.$$

Two interpretations are available for the above definition. According to a first interpretation, we can state that the incumbents perform better than the mutants on random matchings. Another interpretation is that the strategy s^* cannot be invaded by s.

7.4.2 ▪ A second definition in Economics

Evolutionarily stable strategies have also been studied by economists, who provide for them an equivalent definition. Such a definition reminds us of the first and second optimality conditions.

Definition 7.3. *A mixed strategy $s^* \in \Delta$ is evolutionarily stable if for any $s \in \Delta$ the following two conditions hold:*

(a) $u(s^*, s^*) \geq u(s, s^*)$;

(b) *if $u(s^*, s^*) = u(s, s^*)$, then $u(s^*, s) > u(s, s)$.*

Condition (a) in the above definition states essentially that the symmetric profile (s^*, s^*) is a symmetric Nash equilibrium. This condition contemplates two possibilities, namely that (s^*, s^*) be a strict Nash equilibrium or not. Condition (b) says that if the symmetric profile (s^*, s^*) is not a strict Nash equilibrium, then the mutant must perform poorly when playing against another mutant.

In plain words, condition (a) requires that the mutant perform poorly against the masses. In addition, condition (b) states that the mutant performs reasonably well against the masses but poorly against itself. The two definitions are proven to be equivalent.

7.5 ▪ Implications and examples

After introducing the formal definition of evolutionarily stable strategy, we are in a position to comment on a few direct consequences of such a definition. We also illustrate the aforementioned stability concept on a number of examples.

7.5.1 ▪ A nonstrict Nash equilibrium can be an evolutionarily stable strategy

This section examines condition (b) in Definition 7.3. In particular we show that such a condition implies that a nonstrict Nash equilibrium can be evolutionarily stable. We illustrate this in the example provided in Fig. 7.4. The example deals with a two-player

	a	b
a	$(1,1)$	$(1,1)$
b	$(1,1)$	$(0,0)$

Figure 7.4. *Example showing that a nonstrict Nash equilibrium can be evolutionarily stable. Reprinted with permission from Benjamin Polak and Yale University* [202].

nonzero-sum game where both players can play a or b. We immediately note that the symmetric profile (a,a) is a symmetric Nash equilibrium. However, such an equilibrium is not a strict Nash equilibrium, as player 1, by deviating, is not worse off, namely, $u(a,a) = u(b,a) = 1$. The point we wish to develop here is about strategy a being evolutionarily stable or not. A possible way to proceed to provide an answer is to check condition (b) in Definition 7.3. To do this, let us compare $u(a,b)$ and $u(b,b)$, and so we have

$$u(a,b) > u(b,b).$$

The conclusion is that strategy a is evolutionarily stable.

Example 7.4 (Evolution of social convention). This example deals with a game that presents multiple evolutionarily stable strategies. The example builds on the well-known *Coordination game*. This game, whose bimatrix is displayed in Fig. 7.5, is useful to describe the evolution of social conventions. The setup of the game involves two players that can drive *left* or *right*. In an evolutionary context, the row player is an individual, and the column player is a fictitious opponent whose mixed strategy simulates the distribution of the population.

	L	R
L	(2,2)	(0,0)
R	(0,0)	(1,1)

Figure 7.5. Coordination game *describing the evolution of social convention. Reprinted with permission from Benjamin Polak and Yale University* [202].

The profiles (L,L) (both players driving *left*) and (R,R) (both players driving *right*) are two strict Nash equilibrium solutions. From what we have said so far, both L and R are evolutionarily stable strategies. Then we can infer that there may exist multiple evolutionarily stable strategies. It turns out that these strategies need not be equally good. As an example, the profile (L,L) returns better payoffs than the profile (R,R). ∎

Example 7.5 (Battle of the Sexes). In this example we show that there may exist evolutionarily stable mixed strategies. Recall that strategies correspond to genes. Then mixed strategies correspond to a *polymorphic* population. In the parlance of evolutionary biology, this is a population with multiple genes. To illustrate this idea let us introduce the bimatrix in Fig. 7.6. The game is obtained by swapping the columns of the *Battle of the Sexes'* bimatrix. In the above game, both players have two strategies. A first strategy is a, which corresponds to playing aggressively, and a second strategy is b, which corresponds to playing nonaggressively. There exist no symmetric pure Nash equilibrium solutions for this game. This means that we have no monomorphic population.

	a	b
a	(0,0)	(2,1)
b	(1,2)	(0,0)

Figure 7.6. Battle of the Sexes *showing evolutionarily stable mixed strategies. Reprinted with permission from Benjamin Polak and Yale University* [202].

Despite the above considerations, we note that the solution $[(\frac{2}{3}, \frac{1}{3}), (\frac{2}{3}, \frac{1}{3})]$ is a symmetric mixed-strategy Nash equilibrium, which yields a polymorphic population. Such an equilibrium cannot be strict, as it is a mixed Nash equilibrium, and therefore from the *Indifference Principle* we know that any strategy in the support returns a same payoff.

From condition (b) in Definition 7.3, we have that $u(s^*, s) > u(s, s)$ for all possible mutations $s \in \Delta$. Actually, let us take $s = a$ and $s = b$ and obtain for the row player

$$u(s^*, a) = 1/3 > u(a, a) = 0, \quad u(s^*, b) = 4/3 > u(b, b) = 0. \quad \blacksquare$$

Example 7.6 (Monomorphic evolutionarily stable strategy (case 1)). In this example we develop a game characterized by *monomorphic evolutionarily stable strategies*. Consider a two-player nonzero-sum game whose bimatrix is shown in Fig. 7.7. The game is the classical *Hawk and Dove game*.

	Hawk	*Dove*
Hawk	$\left(\frac{V-C}{2}, \frac{V-C}{2}\right)$	$(V, 0)$
Dove	$(0, V)$	$\left(\frac{V}{2}, \frac{V}{2}\right)$

Figure 7.7. *The* Hawk and Dove game *where* $V > C$ *shows monomorphic evolutionarily stable strategies. Reprinted with permission from Benjamin Polak and Yale University* [202].

Recall that V is essentially the prize of victory and that C is the cost of fight. If we take $V > C$, then the game can be assimilated to the *Prisoner's dilemma* for which there exists a unique strict Nash equilibrium, which is $(Hawk, Hawk)$. Consequently, $Hawk$ is also an evolutionarily stable strategy. Under the assumption that the prize of victory is higher than the cost of fighting, then the evolutionary interpretation of the game suggests that all individuals will end up selecting an aggressive behavior, namely strategy $Hawk$. $\blacksquare$

Example 7.7 (Monomorphic evolutionarily stable strategy (case 2)). The same example as in the previous section but with different parameters shows that behaving aggressively, namely the strategy $Hawk$, leads to a monomorphic evolutionarily stable strategy. The bimatrix is displayed in Fig. 7.8. Differently from the previous section, let us assume now that the prize of victory is equal to the cost of fighting. This is equivalent to setting $V = C$. In this case, we have that $(Hawk, Hawk)$ is still a Nash equilibrium solution but not strict, as deviations lead to a same payoff. To see this, note that $u(D, H) = u(H, H) = 0$. In this context, we are interested in answering the question of whether $Hawk$ is an evolutionarily stable strategy or not. After noting that

	Hawk	*Dove*
Hawk	$\left(\frac{V-C}{2}, \frac{V-C}{2}\right)$	$(V, 0)$
Dove	$(0, V)$	$\left(\frac{V}{2}, \frac{V}{2}\right)$

Figure 7.8. *The* Hawk and Dove game *where* $V = C$ *shows monomorphic evolutionarily stable strategies. Reprinted with permission from Benjamin Polak and Yale University* [202].

$u(H,D) = V > u(D,D) = \frac{V}{2}$ for any mutation D, we can conclude with a positive answer, namely *Hawk* is evolutionarily stable. In conclusion, if the prize of victory is equal to the cost of fighting, then the whole population will end up fighting. ∎

Example 7.8 (Polymorphic evolutionarily stable strategy (case 3)). As in the previous two cases, this example develops the *Hawk and Dove game* but under a further different assumption on the parameters. The example, whose bimatrix is as in Fig. 7.9, shows that we can have polymorphic evolutionarily stable strategies. To understand this, note that if the prize of victory is less than the cost of fight, namely $V < C$, then the profiles $(Hawk, Dove)$ and $(Dove, Hawk)$ are two nonsymmetric Nash equilibrium solutions. In addition to these pure Nash equilibrium solutions, we also have a mixed Nash equilibrium, which is given by $[(\frac{V}{C}, 1 - \frac{V}{C}), (\frac{V}{C}, 1 - \frac{V}{C})]$. We also realize that the strategy *Hawk* is not evolutionarily stable, and that the strategy *Dove* is not evolutionarily stable as well. The question is then whether the mixed strategy $s^* = (\frac{V}{C}, 1 - \frac{V}{C})$ is evolutionarily stable. As $[(\frac{V}{C}, 1 - \frac{V}{C}), (\frac{V}{C}, 1 - \frac{V}{C})]$ is a mixed Nash equilibrium, then such an equilibrium cannot be strict. Indeed from the *Indifference Principle* we have that the mixed strategy or any strategy in the support must return the same payoff. This corresponds to saying that $u(D, s^*) = u(H, s^*) = u(s^*, s^*)$. Given this, we then need to check condition (b) in Definition 7.3. By doing this we have that $u(s^*, H) > u(H, H)$ for a mutant H, and also that $u(s^*, D) > u(D, D)$ for a mutant D. We conclude that $s^* = (\frac{V}{C}, 1 - \frac{V}{C})$ is evolutionarily stable. Furthermore, as V increases, more players playing *Hawk* are in the evolutionarily stable strategy. This is in accordance with the intuition that says that the higher the prize of victory the higher the percentage of individuals who play aggressively at the equilibrium. ∎

	Hawk	*Dove*
Hawk	$\left(\frac{V-C}{2}, \frac{V-C}{2}\right)$	(V,0)
Dove	(0,V)	$\left(\frac{V}{2}, \frac{V}{2}\right)$

Figure 7.9. *The* Hawk and Dove game *where $V < C$ shows monomorphic evolutionarily stable strategies. Reprinted with permission from Benjamin Polak and Yale University* [202].

Example 7.9 (*Rock-Paper-Scissors*: No evolutionarily stable strategy). In this last example, we present a game for which there exist no evolutionarily stable strategies. The game is the well-known *Rock-Paper-Scissors game*, which is modeled by the bimatrix in Fig. 7.10. In accordance with what we saw earlier, the game is a further example that

	R	P	S
R	(γ, γ)	$(-1, 1)$	$(1, -1)$
P	$(1, -1)$	(γ, γ)	$(-1, 1)$
S	$(-1, 1)$	$(1, -1)$	(γ, γ)

Figure 7.10. *The* Rock-Paper-Scissors game *shows no evolutionarily stable strategies. Reprinted with permission from Benjamin Polak and Yale University* [202].

Nash equilibrium strategies are not necessarily evolutionarily stable strategies. In particular, for this game we have one mixed Nash equilibrium, which is $s^* = (\frac{1}{3}, \frac{1}{3}, \frac{1}{3})$. From the *Indifference Principle* such an equilibrium cannot be a strict Nash equilibrium. Indeed, it must hold that any strategy in the support is equivalent in terms of payoffs to the mixed strategy. This is captured by the equalities below:

$$u(R, s^*) = u(P, s^*) = u(S, s^*) = u(s^*, s^*).$$

With the above consideration in mind, we proceed with the verification that the mutant R is successful. In other words we check the condition

$$u(s^*, R) = \frac{\gamma}{3} < u(R, R) = \gamma.$$

As the above inequality holds, we conclude that there exists no evolutionarily stable strategy. What happens in an evolutionarily dynamic scenario is that the strategies keep cycling around. The theory of learning in games makes use of such an example to introduce the concept of *uncoupled dynamics* and to prove that convergence through learning is not always possible. This is the subject of the next chapter. ∎

7.6 • Notes and references

The layout of this chapter follows the open course by Polak at Yale [202]. The roots of evolutionary games can be traced back to the work of Smith, *Game Theory and the Evolution of Fighting* (1972) [223]. This paper was followed one year later by the article "The Logic of Animal Conflict" (1973) in *Nature*, coauthored by Price [225]. The foundations of the theory are also in the book *Evolution and the Theory of Games* (1982) [224]. Further results on mixed evolutionarily stable strategy are provided in the Bishop–Cannings theorem in [55]. For a comprehensive treatment of the topic we refer the reader to the classical books by Weibull [250] and Sandholm [211].

Chapter 8

Replicator Dynamics and Learning in Games

8.1 • Introduction

All the different types of equilibria we have encountered so far represent stationary solutions. This chapter turns the attention to the dynamics that can lead to some of such equilibria. It is in this chapter that we introduce for the first time the *replicator dynamics*. Such a dynamics describes the evolution of the strategies under the assumption that the players are reactive to the environment. By that we mean that the players observe the population behavior and adopt their best-response strategies. The replicator dynamics provides also an opportunity to look into asymptotic stability and to analyze the link with the evolutionarily stable strategies defined in the previous chapter. In the second part of this chapter, we focus on *learning in games*. A comprehension of the different techniques in learning in games involves a few dynamic aspects that we borrow from systems theory. A classical situation we will consider is the one in which the players construct an *empirical frequency distribution* from observations of the past opponents' plays. By empirical frequency we mean a probability distribution over the opponents' action spaces that says how many times in the past a given choice has been played by a player. Empirical frequency introduces to the reader the broad area of *fictitious games*.

Section 8.2 introduces the replicator dynamics. Section 8.3 deals with stationarity, Nash equilibrium solutions, and stability. Section 8.4 introduces learning in games and fictitious play. Section 8.5 provides notes and references.

8.2 • Replicator dynamics

The evolutionary models encountered in the previous chapter do not consider any explicit dynamics. In contrast to this, we here focus on how the players change dynamically their strategies. The first evolution dynamics we consider is known as *replicator dynamics*. Let us enumerate the strategies using an index $s = 1, 2, \ldots, K$. Let us indicate with x_s the percentage of the population playing a strategy s. Obviously, it must hold that the sum of the percentages over the whole action space must sum up to one, i.e.,

$$\sum_{s=1}^{K} x_s = 1.$$

Also, let us use the symbol $x = (x_s)_{s \in 1,2,\ldots,K}$ to denote the population distribution. From the previous chapter, we know that this corresponds to a polymorphic strategy profile in the language of evolutionary biology. At the same time, from a game-theoretic perspective, x can also be viewed as a mixed strategy over $1, 2, \ldots, K$. Now, let us suppose that individuals are subjected to random matchings. Consider the *expected fitness* of playing a generic strategy s against a population playing x. In the previous chapter the fitness resulting from such strategies has been denoted by $u(s, x)$.

We are ready to introduce the replicator dynamics in its generic form.

Definition 8.1 (Replicator dynamics). *For each* $s = 1, 2, \ldots, K$ *and for all* t *and* τ *the* replicator dynamics *is given by*

$$x_s(t + \tau) - x_s(t) = x_s(t) \frac{\tau[u(s, x(t)) - \bar{u}(x(t))]}{\bar{u}(x(t))},$$

where $\bar{u}(x(t))$ *is the average fitness at time* t *resulting from a population distribution given by* $x(t)$*. Such an average fitness is given by*

$$\bar{u}(x(t)) := \sum_{s=1}^{K} x_s u(s, x(t)).$$

The replicator dynamics concedes the following interpretation: *The greater the fitness of a strategy relative to the average fitness, the greater its relative increase in the population.* It is worth noting that from the conservation of the mass we have $\sum_{s=1}^{K} x_s(t + \tau) = 1$. In addition, we also note that by setting $\tau = 1$, we find the well-known *discrete-time replicator equation.*

To derive the *replicator dynamics in continuous time* we need to divide both sides of the equation by τ. By taking the limit as $\tau \to 0$ we obtain

$$\lim_{\tau \to 0} \frac{x_s(t + \tau) - x_s(t)}{\tau} = x_s(t) \frac{[u(s, x(t)) - \bar{u}(x(t))]}{\bar{u}(x(t))}.$$

Definition 8.2 (Continuous-time replicator dynamics). *The continuous-time version of the dynamics takes the form*

$$\dot{x}_s(t) = x_s(t) \frac{[u(s, x(t)) - \bar{u}(x(t))]}{\bar{u}(x(t))}.$$

The above dynamics is also referred to as the *continuous replicator.*

Introducing the replicator dynamics paves the way to a few questions, such as the following:

- Given a vector of distribution x^*, is such a vector a *stationary state*? Recall that from systems theory a stationary state is a state for which the first-order derivative is null, namely $\dot{x}^*(t) = 0$.

- Is a given vector of distribution x^* *asymptotically stable*? This is equivalent to saying that there exists a neighborhood of x^* such that any trajectory starting from any x_0 in this neighborhood is such that the continuous replicator dynamics provides a trajectory that converges to x^*.

In the next section, we shall deal with stationarity for the continuous replicator dynamics.

8.3 ▪ Stationarity, equilibria, and asymptotic stability

This section deals with stationarity, Nash equilibrium solutions, and stability. In particular we show that a Nash equilibrium is a stationary state. Then, we highlight that asymptotic stable solutions are Nash equilibria. Finally, we point out that evolutionarily stable strategies imply asymptotic stable solutions.

8.3.1 ▪ A Nash equilibrium is a stationary state

The core message of this section is that a Nash equilibrium is a stationary state for the replicator dynamics. This is stated in the following theorem.

Theorem 8.3. *If the vector x^* is a Nash equilibrium, then it is a stationary state.*

Proof (Sketch). A possible proof of the above theorem builds upon the following reasoning. Assume that x^* is a Nash equilibrium. Consequently, x^* must also be a best-response to itself. This is equivalent to saying that

$$u(s, x(t)) - \bar{u}(x(t)) \leq 0 \qquad \forall s$$
$$u(s, x(t)) - \bar{u}(x(t)) = 0 \ \forall \ s \text{ in the support of } x^*.$$

From the above inequalities we derive that for any s, either $u(s, x(t)) - \bar{u}(x(t)) = 0$ or $x_s(t) = 0$, and hence $\dot{x}_s(t) = 0$ for all s. □

Remarkably, the converse is not true. To see this, let x^* be a non-Nash pure strategy. Consequently, $x_s(t) = 0$ for all s other than the pure strategy in question. This implies that x^* is stationary.

8.3.2 ▪ Asymptotic stable solutions are Nash equilibria

This section shows another important fact; that is, if a solution is asymptotic stable, then such a solution is also a Nash equilibrium. We state this formally in the next theorem.

Theorem 8.4. *If x^* is asymptotically stable, then it is a Nash equilibrium.*

Proof (Sketch). For a monomorphic population, namely when x^* yields a pure strategy, the proof is straightforward. In the other case, in which x^* corresponds to a mixed-strategy Nash equilibrium, the proof is also straightforward but a bit longer. We mention the underlying idea briefly. The continuous replicator equation implies that the population distribution is evolving along the direction of the better responses, these being computed looking at the average fitness. If the resulting dynamics converges, then there cannot exist any other strict better responses at the point of convergence. This means that we must be at a Nash equilibrium. □

Again, the converse is not true. To see that the converse is not true, consider the following example.

Example 8.5. (A Nash equilibrium is not necessarily an asymptotically stable solution). Let the bimatrix game be given as in Fig. 8.1. The pair (b, b) is a Nash equilibrium, but such an equilibrium is not asymptotically stable. Indeed, any perturbation away from (b, b) will start a process in which the fraction of agents playing a steadily increases. To

	a	b
a	$(1,1)$	$(0,0)$
b	$(0,0)$	$(0,0)$

Figure 8.1. *Example showing that a Nash equilibrium is not necessarily asymptotically stable.*

see this, take $x(t) = (\epsilon, 1 - \epsilon)$; then

$$x_a = \epsilon, \quad u(a, x(t)) - \bar{u}(x(t)) = \epsilon - \epsilon^2 > 0.$$

From the above we understand that playing a returns a payoff higher than the average payoff computed over the population, and therefore the percentage of people playing a increases. ∎

8.3.3 ▪ Evolutionarily stable strategies imply asymptotic stable solutions

Next we turn our attention to the following fact. Strategies that are evolutionarily stable lead to asymptotical stable solutions. This is formally stated as follows.

Theorem 8.6. *If the strategy x^* is evolutionarily stable, then it is asymptotically stable.*

Proof **(Sketch).** The proof builds upon the definition itself of evolutionarily stable strategy which we recall next.
    A mixed strategy $s^* \in \Delta$ is evolutionarily stable if there exists $\bar{\epsilon} > 0$ such that for any $s \in \Delta$ and $\epsilon \leq \bar{\epsilon}$, we have

$$\underbrace{u(s^*, \epsilon s + (1 - \epsilon)s^*)}_{\text{payoff to incumbent } s^*} > \underbrace{u(s, \epsilon s + (1 - \epsilon)s^*)}_{\text{payoff to mutant } s}.$$

Now, let us denote by A the first matrix of the game. By this we mean the matrix collecting the payoffs of the row player. Recall that in this chapter we consider only symmetric games if not differently specified. The basic idea is then as follows. We can rewrite the above condition as

$$x^{*^T} A x > x^T A x \quad \forall x \text{ in neighborhood of } x^*.$$

The above condition implies that $\underbrace{(x^* - x)}_{z} A(x - x^*) > 0$ or $V(z) = zAz < 0$, where

$\sum_i z_i = 0$. In this case we call the game a *negative definite game*. The thesis derives from observing that $\dot{V}(z) = zA\dot{z} > 0$ for all $z \neq 0$ when $\dot{z}$ is as in the left-hand side of the replicator dynamics. □

8.4 ▪ Learning in games

The topic of learning in games exploits several aspects of systems theory and stability theory. Actually, learning presumes that the game is repeated in time and that the players respond to what they have observed in the past. A main issue is about the convergence properties of the game to a Nash equilibrium or to any other type of solution, such as

	$Stag$	$Hare$
$Stag$	$(10, \times)$	$(0, \times)$
$Hare$	$(2, \times)$	$(3, \times)$

Figure 8.2. Stag-Hunt game *simulating a learning process.*

Pareto optimal solutions, social optimal solutions, and Stackelberg equilibrium solutions, just to name a few. The underlying assumption is that the players have no knowledge of the opponents' payoffs. This takes us to a second type of dynamics (other than the replicator dynamics) which is known in the literature as *uncoupled dynamics*. Roughly speaking, in the uncoupled dynamics, the updates of a player do not depend on others' payoffs. We cannot stress enough at this point that in the replicator dynamics, the players need to know the average payoff. The example below replicates a typical learning process in the context of the *Stag-Hunt game* introduced in the first chapter.

Example 8.7 (How would you play?). Consider the *Stag-Hunt game* whose bimatrix is copied in part in Fig. 8.2. Imagine that the row player has to select an action knowing his payoffs but ignoring the payoffs and the next action of his opponent. Suppose that the game has a past history. In particular, in the previous 10 iterations the column player has played $Stag$ for 7 times and the same player has played $Hare$ for 3 times. Past observations lead the row player to assume that the column player is playing the mixed strategy $z = (0.7, 0.3)$. The values 0.7 and 0.3 are essentially the empirical frequency distribution of the column player. Based on such an empirical frequency distribution, the best response of the row player is given by

$$\begin{aligned} u_1(Stag, z) &= 0.7 \cdot 10 + 0.3 \cdot 0 &= 7, \\ u_1(Hare, z) &= 0.7 \cdot 2 + 0.3 \cdot 3 &= 2.3. \end{aligned} \tag{8.1}$$

Under the hypothesis that the column player is playing the stationary mixed strategy $z = (0.7, 0.3)$, then for the row player the action $Stag$ is better than the action $Hare$. ∎

8.4.1 ▪ Fictitious play

Fictitious play occupies a relevant part in the theory of learning in games. In fictitious play, the players are assumed to be *myopic*. This means that they use their best responses to their best guesses of the opponent's mixed strategy. This assumption raises the question of what the *best guess* must be. A partial answer exploits the concept of *stationarity*. By stationarity we mean that every player takes his opponent's action as the result of a randomization starting from a time-invariant mixed strategy. Roughly speaking, the players assume that their opponents' strategies do not change with time.

Let us now derive a mathematical model for the learning process that fits well with the setup. To this purpose, let the tuple $\langle N, (\mathscr{A}_i)_{i \in N}, (u_i)_{i \in N} \rangle$ be given, where N is the set of players, and $\mathscr{A}_i$ is the set of pure actions of player i. The set of action profiles is $A := \{a = (a_i)_{i \in N}, a_i \in \mathscr{A}_i(s)\}$. The players play repeatedly in time, and the time is indexed by $t = 1, 2, \ldots$. For each time, player i has a payoff function given by $u_i : A \to \mathbb{R}$. For each $a_{-i} \in A_{-i}$, let $x_i^t(a_{-i})$ be the number of iterations in which player i has observed his opponent playing a_{-i}. Furthermore, let $x_i^0(a_{-i})$ be the starting point, also called the *fictitious past*.

	L	R
U	$(3,3)$	$(0,0)$
D	$(4,0)$	$(1,1)$

Figure 8.3. *Example of a learning process in fictitious play.*

Example 8.8. As an example, let us consider a two-player game, where $\mathscr{A}_2 = \{L, R\}$. Assuming that $x_1^0(L) = 3$ and $x_1^0(R) = 5$, and that player 2 plays L, L, R in the first three periods, then we have $x_1^3(L) = 5$ and $x_1^3(R) = 6$. ∎

The core concept in fictitious play is about each player modeling his opponent's strategy as a stationary mixed strategy. Each player then updates his beliefs at each step when new information becomes available. Players use their prediction on their opponent's strategies in order to choose their best responses. Formally, in each period (or stage) they maximize the period's expected payoff, which they form according to

$$\mu_i^t(a_{-i}) = \frac{x_i^t(a_{-i})}{\sum_{a_{-i} \in \mathscr{A}_{-i}} x_i^t(a_{-i})}.$$

For instance, given a two-player game, player i forecasts player $-i$'s strategy at time t using the empirical frequency distribution of past play. Then, player i selects his best response to $\mu_i^t(a_{-i})$, which is given by

$$a_i^t = \arg\max_{a_i \in \mathscr{A}_i} u_i(a_i, \mu_i^t(a_{-i})) \in \mathscr{B}_i(\mu_i^t(a_{-i})).$$

For this example, the *beliefs' iteration* are as follows:

$$
\begin{aligned}
\mu_i^t(a_{-i}) & = \frac{x_i^t(a_{-i})}{t} = \frac{x_i^{t-1}(a_{-i}) + 1_{\{a_{-i}^t = a_{-i}\}}}{t} \\
& = \left(\frac{t-1}{t}\right)\frac{x_i^{t-1}(a_{-i})}{t-1} + \frac{1_{\{a_{-i}^t = a_{-i}\}}}{t} \quad\quad (8.2) \\
& = \left(\frac{t-1}{t}\right)\mu_i^{t-1}(a_{-i}) + \frac{1_{\{a_{-i}^t = a_{-i}\}}}{t}.
\end{aligned}
$$

Remarkably, it is not necessary to keep in memory all past plays. Furthermore, it is worth noting that the entire learning process can be run online, as the current belief depends on the previous belief and the current observation. In the following example we illustrate the few steps of a learning process, arising in a fictitious play context.

Example 8.9. Let a two-person nonzero-sum game be given, and let this game have the payoffs as in the bimatrix illustrated in Fig. 8.3. Imagine that the game is played repeatedly and that the players adopt a classical fictitious play model. A first thing to observe is that this game is dominant solvable. In other words, D is a strictly dominant strategy for the row player. Furthermore, there exists a unique Nash equilibrium which is (D, R). After the above preamble, let us set $x_1^0 = (3, 0)$ and $x_2^0 = (1, 2.5)$. As a result, the fictitious play proceeds as follows:

- Period 1: Then, $\mu_1^0 = (1, 0)$ and $\mu_2^0 = (1/3.5, 2.5/3.5)$, so play follows $a_1^0 = D$ and $a_2^0 = L$.

- Period 2: We have $x_1^1 = (4, 0)$ and $x_2^1 = (1, 3.5)$, so play follows $a_1^1 = D$ and $a_2^1 = R$.

- Period 3: We have $x_1^2 = (4, 1)$ and $x_2^2 = (1, 4.5)$, so play follows $a_1^2 = D$ and $a_2^2 = R$, and so forth. ∎

The example above shows that convergence may occur. In the above example, the row player always plays D, as D is a dominant strategy for him. Consequently, μ_2^t converges to $(0, 1)$ with probability 1. This also implies that player 2 will eventually converge to the action R. Remarkably, fictitious play does not require that the players know anything about their opponents' payoffs. The players only form beliefs about how their opponents will play.

In general, we can say that convergence occurs if $\mu^t = (\mu_i^t)_{i \in N} \to \mu^*$. It is worth noting that if convergence occurs, then μ^* must be a Nash equilibrium. Games in which convergence must occur are said to have the *fictitious play property*. However, in general, convergence is not guaranteed.

8.5 ▪ Notes and references

The replicator dynamics is analyzed in detail in the classical books by Weibull [250] and Sandholm [211]. Convergence in negative definite games is discussed in [119].

We present the topic of learning in games following the open MIT course *Game Theory with Engineering Applications* by Ozdaglar [197]. A comprehensive treatment of learning in games is provided in the classical book by Fudenberg and Levine [99]. Uncoupled dynamics is studied in [109, 110, 111]. For further details on convergence in fictitious play we refer the reader to [220] and [130]. Fictitious play in multi-agent systems is studied in [217] and [218]. Fictitious play in potential games is discussed in [168]. Convergence of *aspiration learning* in coordination games is studied in [77]. *Log-linear learning*, originally introduced in [62], is studied further in [169]. Learning in potential games and cooperative control is investigated in [167]. Learning in *weakly acyclic games* is investigated in [171]. For a brief discussion on multi-agent learning for engineers we highlight the paper [166]. Convergence for several classes of reinforcement learning schemes is explored in [78].

The reader is referred to the Kalai–Lehrer model of rational learning in repeated games [135] to better understand how rational players with truth-compatible beliefs eventually learn to play Nash equilibria.

Chapter 9

Differential Games

9.1 ▪ Introduction

This chapter deals with *noncooperative differential games*. In short, differential games are characterized by a state variable, whose evolution follows a differential equation. Such a differential equation is subject to controlled inputs representing the players' actions. The players' payoffs are then influenced not only by the players' actions but also by the state. Differential games can be reviewed as a generalization of optimal control problems. Having said this, we shall first survey the basics of optimal control theory. More specifically, we shall introduce the *Pontryagin Maximum Principle* and the *Hamilton–Jacobi–Bellman equation*. After introducing these concepts, we present differential games under open-loop or closed-loop strategies. We then mention the *Hamilton–Jacobi–Isaacs equation*. The chapter concludes with a survey on linear-quadratic differential games and H^∞-optimal control.

The structure of this chapter is as follows. Section 9.2 introduces the optimal control problem, the *Pontryagin Maximum Principle*, and the *Hamilton–Jacobi–Bellman equation*. Section 9.3 deals with differential games, as well as open-loop and closed-loop Nash equilibrium solutions. Section 9.4 discusses linear-quadratic differential games. Section 9.5 casts H^∞-optimal control within the framework of differential games. Section 9.6 provides notes and references for this chapter.

9.2 ▪ Optimal control problem

Let vector $x(t) \in \mathbb{R}^m$ be the state and vector $u(t) \in U$ be the control, where $U \subseteq \mathbb{R}^m$ is a compact set. Compactness is necessary to guarantee the existence of an optimal solution; see [159, Chap. 4.5]. Also, let function f be continuous with respect to (x, u, t) and differentiable with respect to x. Consider the controlled dynamics

$$\dot{x}(t) = f(t, x(t), u(t)), \quad u(t) \in U.$$

Let us make the assumption that the solutions in finite time are not unbounded. Such an assumption is described by the condition

$$|f(t, x, u)| \le C(1 + |x|) \quad \forall (t, x, u) \in [0, T] \times \mathbb{R}^m \times U.$$

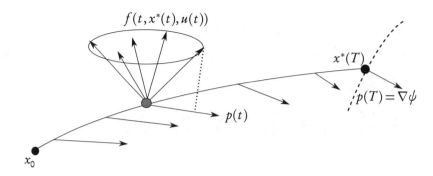

Figure 9.1. *Graphical illustration of the PMP when the running cost is null. Reprinted with permission from AIMS* [67].

For any initial state $x(t_0) = x_0$, let us indicate by

$$t \mapsto x(t) = (t; t_0, x_0, u)$$

the trajectory starting from state x_0 at time t_0.

Given the above preamble, let us formulate the optimization problem:

$$\max J(u; t_0, x_0) := \psi(x(T)) - \int_{t_0}^{T} L(t, x(t), u(t)) dt,$$

where the "max" operation is over measurable functions $u : [t_0, T] \mapsto U$. The integrand function $L(\cdot)$ is called *running cost*, and function $-\psi(\cdot)$ is called *terminal penalty*.

9.2.1 ▪ Pontryagin Maximum Principle

A first breakthrough in the theory of differential games is the result available in the Russian literature of the 1950s known as the *Pontryagin Maximum Principle (PMP)*. This principle, which provides a necessary condition for a control to be optimal, is illustrated next.

Let a problem with a free terminal point be given. This is a problem where there are no constraints on the final state. Also, let us denote by $u^*(t)$ the optimal control; let us use the symbol $x^*(t)$ to mean the corresponding trajectory, and let us define $p(t)$ as the *adjoint variable*, sometimes also referred to as *co-state*. The adjoint variable satisfies the differential equation

$$\dot{p}(t) = -p(t)\frac{\partial f}{\partial x}(t, x^*(t), u^*(t)) + \frac{\partial L}{\partial x}(t, x^*(t), u^*(t)), \; p(T) = \nabla \psi(x^*(T)).$$

Let us indicate with ∇ the gradient operator. The maximality condition for the problem under study is given by

$$u^*(t) = \arg\underbrace{\max_{u \in U}\left\{p(t) \cdot f(t, x^*(t), u(t)) - L(t, x^*(t), u(t))\right\}}_{H(t, x, p) \text{ is the maximized Hamiltonian}}.$$

To gain insights on the meaning of the PMP, Fig. 9.1 provides a graphical illustration under the assumption of null running cost. The maximality condition yields essentially the control corresponding to the maximal inner product between the right-hand side of the dynamics and the adjoint variable at every time.

9.2.2 ▪ Two-point boundary value problem

A stronger result can be obtained if the *maximized Hamiltonian* is concave in x, where the maximized Hamiltonian is given by

$$H(t,x,p) := \max_{u \in U} \left\{ p(t) \cdot f(t,x(t),u(t)) - L(t,x(t),u(t)) \right\}.$$

In this case, the PMP provides also a sufficient condition for a control to be optimal. With this in mind, one can formulate the problem in two steps. We shall first solve

$$\tilde{u}(t,x,p) = \arg\max_{u \in U} \left\{ p(t) \cdot f(t,x(t),u(t)) - L(t,x(t),u(t)) \right\}.$$

Then we turn to the *two-point boundary value problem*

$$\begin{cases} \dot{x} = f(t,x,\tilde{u}(t,x,p)), \quad x(t_0) = x_0, \\ \dot{p} = -p\frac{\partial f}{\partial x}(t,x,\tilde{u}(t,x,p)) + \frac{\partial L}{\partial x}(t,x,\tilde{u}(t,x,p)), \quad p(T) = \nabla \psi(x(T)). \end{cases}$$

A common technique to solve two-point boundary value problems makes use of *shooting methods*. These methods involve two steps as follows:

(i) guess an initial $\overline{p}$ and solve the Cauchy problem involving the above set of differential equations where the boundary condition at final time $p(T) = \nabla\psi(x(T))$ is replaced by the boundary condition at the initial time $p(t_0) = \overline{p}$;

(ii) readjust $\overline{p}$ so to minimize $\Lambda(\overline{p}) := p(T) - \nabla\psi(x(T))$.

The next section develops a parallel approach to the one provided by Pontryagin.

9.2.3 ▪ Hamilton–Jacobi–Bellman equation

The *Hamilton–Jacobi–Bellman (HJB) equation* constitutes a milestone in the theory of differential games. This equation was published in the American literature on optimal control theory back in the 1950s. The equation builds on the *dynamic programming (DP)* principle. A graphical illustration of the DP principle, also known as *Principle of Optimality*, is shown in Fig. 9.2. According to the DP principle, *an optimal policy has the property that whatever the initial state and initial decision are, the remaining decisions must constitute an optimal policy with regard to the state resulting from the first decision* (see [47, Chap. III.3]). That is to say that given an optimal trajectory, any subtrajectory must be optimal as well. The importance of the DP principle lies in the fact that it allows the decomposition of the original problem into infinite subproblems, one for each time t.

To decompose the problem let us define the function

$$V(t_0,x_0) := \sup_{u(\cdot)} J(u;t_0,x_0),$$

which is commonly referred to as the *value function*. Roughly speaking, the DP principle states that the *value function today is the optimal running cost until tomorrow plus the value function tomorrow*. By translating this into a mathematical expression we get

$$V(t_0,x_0) = \sup_{u(\cdot)} \left\{ V(t_1,x(t_1;t_0,x_0,u)) - \int_{t_0}^{t_1} L(t,x(t;t_0,x_0,u),u(t))dt \right\}. \tag{9.1}$$

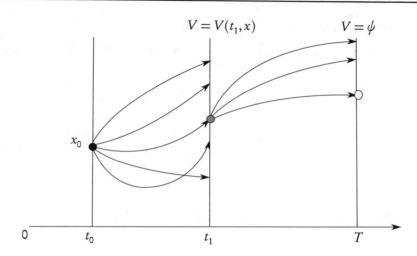

$$V = V(t_1, x) \qquad\qquad V = \psi$$

Figure 9.2. *Graphical illustration of the HJB equation and the DP principle. Reprinted with permission from AIMS* [67].

Let us now expand according to Taylor function $V(\cdot)$ around point (t_0, x_0). Then, we get the following expression for $V(t_1, x(t_1))$:

$$V(t_1, x(t_1)) = V(t_0, x_0) + \int_{t_0}^{t_1} \partial_t V(t, x(t)) + \nabla V \cdot f(t, x(t), u(t)).$$

In the above expression, $V(t_1, x(t_1))$ is essentially expressed in terms of $V(t_0, x_0)$ and of the first-order derivatives of $V(\cdot)$. By substituting $V(t_1, x(t_1))$ in (9.1), we arrive at the following equation in $[0, T] \times \mathbb{R}^m$:

$$\partial_t V(t, x) + \sup_{u \in U} \left\{ \nabla V \cdot f(t, x, u) - L(t, x, u) \right\} = 0.$$

The above equation is referred to as the HJB equation.

9.3 • Differential game

In this section, we show that a differential game is a generalization of an optimal control problem involving multiple decision makers, the players. Let $x \in \mathbb{R}^m$ be the state, u_i be the control of player $i = 1, 2$, and $U_i \subseteq \mathbb{R}^m$ be compact. Also, let the following controlled dynamics be given:

$$\dot{x}(t) = f(t, x(t), u_1(t), u_2(t)), \quad u_i(t) \in U_i.$$

With the above in mind, the optimization problem for player i takes the form

$$\max_{u_i} J_i(u_1, u_2) := \psi_i(x(T)) - \int_0^T L_i(t, x(t), u_1(t), u_2(t)) dt.$$

In a differential game, one usually makes the following distinction between *open-loop* and *closed-loop* strategies. In the case of open-loop strategies, the players know only the initial state x_0. Conversely, in the case of closed-loop strategies, the players have knowledge of the current state $x(t)$. Closed-loop strategies are also referred to as *Markovian strategies*.

9.3.1 ▪ Open-loop Nash equilibrium

This section deals with the case where the players use open-loop strategies. These strategies are functions of only two variables: the time and the initial state. Once the initial state is given, these strategies are only functions of time. The concept of nonprofitable unilateral deviations which has characterized the Nash equilibrium in sequential and simultaneous games can be extended to open-loop strategies in differential games. The following definition formalizes open-loop Nash equilibrium strategies.

Definition 9.1 (Open-loop Nash equilibrium). *The pair* $(u_1^*(t), u_2^*(t))$ *is a Nash equilibrium if* $u_i^*(\cdot)$, $i = 1, 2$, *is a maximizer of the following cost functional problem:*

$$\begin{cases} J_i(u_i, u_{-i}^*) := \psi_i(x(T)) - \int_0^T L_i(t, x(t), u_i(t), u_{-i}^*(t)) dt, \\ x(0) = x_0, \quad \dot{x}(t) = f(t, x(t), u_i(t), u_{-i}^*(t)), \quad t \in [0, T]. \end{cases} \tag{9.2}$$

That is to say, $u_1^*(t)$ is the best response to $u_2^*(t)$ and vice versa. Recall the PMP, and consider the following one-shot game: for every $(t, x) \in [0, T] \times \mathbb{R}^m$ and vectors $q_1, q_2 \in \mathbb{R}^m$,

$$\tilde{u}_1 = \arg\max_{\omega \in U_1} \{ q_1 \cdot f(t, x, \omega, \tilde{u}_2) - L_1(t, x, \omega, \tilde{u}_2) \},$$

$$\tilde{u}_2 = \arg\max_{\omega \in U_2} \{ q_2 \cdot f(t, x, \tilde{u}_1, \omega) - L_2(t, x, \tilde{u}_1, \omega) \}. \tag{9.3}$$

Under the assumption that the above problem admits a unique solution (cf. Assumption A2 on p. 26 in [65]), the map below is continuous:

$$(t, x, q_1, q_2) \mapsto \left(\tilde{u}_1(t, x, q_1, q_2), \tilde{u}_2(t, x, q_1, q_2) \right).$$

Now, assuming that the pair $(\tilde{u}_1(t), \tilde{u}_2(t))$ is a Nash equilibrium, such a pair must solve the two-point boundary value problem

$$\begin{cases} \dot{x} = f(t, x, \tilde{u}_1, \tilde{u}_2), \quad x(t_0) = x_0, \\ \dot{q}_1 = -q_1 \frac{\partial f}{\partial x}(t, x, \tilde{u}_1, \tilde{u}_2) + \frac{\partial L_1}{\partial x}(t, x, \tilde{u}_1, \tilde{u}_2), \; q_1(T) = \nabla \psi_1(x(T)), \\ \dot{q}_2 = -q_2 \frac{\partial f}{\partial x}(t, x, \tilde{u}_1, \tilde{u}_2) + \frac{\partial L_2}{\partial x}(t, x, \tilde{u}_1, \tilde{u}_2), \; q_2(T) = \nabla \psi_2(x(T)). \end{cases} \tag{9.4}$$

In addition to this, if $x \mapsto H(t, x, q, \tilde{u}_{-i})$ and $x \mapsto \psi_i(x)$ are concave, then the condition illustrated above is also sufficient. In the following example, we analyze a marketing competition scenario using a differential game-theoretic approach.

Example 9.2 (Duopolistic competition). This example presents the renowned *Lanchester* model. This model is commonly used to describe a duopolistic competition scenario. The scenario involves two manufacturers that operate in a same market. The manufacturers produce and sell the same product. Let us use a first variable $x_1(t) = x(t) \in [0, 1]$ to represent the market share of manufacturer 1 at time t. Similarly, let us denote by $x_2(t) = 1 - x(t)$ the market share of manufacturer 2 at time t. The manufacturers have different advertising efforts, which enter the problem as controlled inputs $u_i(t)$, $i = 1, 2$ at time t. The *Lanchester* model simulates the evolution of the market share of manufacturer 1 using the differential equation

$$\dot{x}(t) = (1-x)u_1 - xu_2, \quad x(0) = x_0 \in [0,1].$$

After introducing the above dynamics, let us look at the strategies of the manufacturers. In particular, manufacturer i plays $t \mapsto u_i(t)$ with the aim of maximizing

$$J_i = \int_0^T \left[a_i x_i(t) - c_i \frac{u_i^2(t)}{2} \right] dt + S_i x_i(T)$$

for given parameters $a_i, c_i, S_i > 0$. The optimization method is organized in two steps. First, we compute the optimal controls as functions of the adjoint variables. This leads to the following problem:

$$\begin{cases} \tilde{u}_1(x, q_1, q_2) = \arg\max_{\omega \geq 0} \left\{ q_1 \cdot (1-x)\omega - c_1 \frac{\omega^2}{2} \right\} = (1-x)\frac{q_1}{c_1}, \\[2mm] \tilde{u}_2(x, q_1, q_2) = \arg\max_{\omega \geq 0} \left\{ q_2 \cdot x\omega - c_2 \frac{\omega^2}{2} \right\} = x\frac{q_2}{c_2}. \end{cases}$$

Second, we need to solve the two-point boundary value problem displayed below:

$$\begin{cases} \dot{x} = (1-x)\tilde{u}_1 + x\tilde{u}_2 = (1-x)^2\frac{q_1}{c_1} + x^2\frac{q_2}{c_2}, \quad x(0) = x_0, \\[2mm] \dot{q}_1 = -q_1(\tilde{u}_1 + \tilde{u}_2) - a_1 = -q_1\left[(1-x)\frac{q_1}{c_1} + x\frac{q_2}{c_2}\right] - a_1, \quad q_1(T) = S_1, \\[2mm] \dot{q}_2 = -q_2(\tilde{u}_1 + \tilde{u}_2) - a_2 = -q_2\left[(1-x)\frac{q_1}{c_1} + x\frac{q_2}{c_2}\right] - a_2, \quad q_2(T) = S_2. \end{cases}$$

From the above problem we obtain optimal trajectories and optimal strategies (advertising efforts) for both manufacturers. ∎

9.3.2 ▪ Closed-loop Nash equilibrium

Let us now turn to closed-loop strategies. Recall that the strategies are now functions of time and state. As for the open-loop case, we need first to define closed-loop strategies at a Nash equilibrium.

Definition 9.3 (Closed-loop Nash equilibrium). *The pair $(u_1^*(t,x), u_2^*(t,x))$ is a Nash equilibrium if $(t,x) \mapsto u_i^*(t,x)$ maximizes the following cost functional problem:*

$$\begin{cases} J_i(u_i, u_{-i}^*(t,x)) := \psi_i(x(T)) - \int_0^T L_i(t, x(t), u_i(t), u_{-i}^*(t,x))dt, \\[2mm] x(0) = x_0, \quad \dot{x}(t) = f(t, x(t), u_i(t), u_{-i}^*(t,x)), \quad t \in [0,T]. \end{cases} \tag{9.5}$$

It turns out that to compute Nash equilibrium closed-loop strategies, we need to solve the corresponding HJB equations. These yield a system of partial differential equations in $[0,T] \times \mathbb{R}^m$ of the form

$$\begin{cases} \partial_t V_1 + \nabla V_1 \cdot f(t, x, \tilde{u}_1, \tilde{u}_2) = L_1(t, x, \tilde{u}_1, \tilde{u}_2), \\[2mm] \partial_t V_2 + \nabla V_2 \cdot f(t, x, \tilde{u}_1, \tilde{u}_2) = L_2(t, x, \tilde{u}_1, \tilde{u}_2). \end{cases} \tag{9.6}$$

In the above problem, from the PMP, we have that $(\tilde{u}_1, \tilde{u}_2)$ solve the one-shot game: for every $(t,x) \in [0,T] \times \mathbb{R}^m$ and value functions $V_1, V_2 \in \mathbb{R}^m$,

$$\tilde{u}_1 = \arg\max_{\omega \in U_1}\{\nabla V_1 \cdot f(t,x,\omega,\tilde{u}_2) - L_1(t,x,\omega,\tilde{u}_2)\},$$

$$\tilde{u}_2 = \arg\max_{\omega \in U_2}\{\nabla V_2 \cdot f(t,x,\tilde{u}_1,\omega) - L_2(t,x,\tilde{u}_1,\omega)\}.$$

(9.7)

Remarkably, if the game is a zero-sum differential game, then the above system of partial differential equations collapses into a single partial differential equation:

$$\partial_t V_1 + \max_{\omega_1} \min_{\omega_2}\left\{\nabla V_1 \cdot f(t,x,\omega_1,\omega_2) - L_1(t,x,\omega_1,\omega_2)\right\} = 0.$$

This equation is referred to as the *Hamilton–Jacobi–Isaacs* equation.

9.4 ▪ Linear-quadratic differential games

Linear-quadratic differential games are very popular, as they admit explicit solutions in terms of optimal control strategies. To see this, let $x \in \mathbb{R}^m$ be the state, u_i be the control of player $i = 1,2$, and $U_i \equiv \mathbb{R}^{m_i}$ be compact sets. Let us consider a linear dynamics of the form

$$\dot{x}(t) = A(t)x(t) + B_1(t)u_1(t) + B_2(t)u_2(t), \quad u_i(t) \in \mathbb{R}^{m_i}.$$

For each player i, the optimization problem is given by

$$\max_{u_i} J_i(u_1,u_2) := \psi_i(x(T)) - \int_0^T L_i(t,x(t),u_1(t),u_2(t))dt.$$

Linear-quadratic differential games are such that the terminal penalty is quadratic,

$$\psi_i(x(T)) = \frac{1}{2}x^T \overline{M}_i x,$$

and the running cost is quadratic as well, namely

$$L_i(t,x(t),u_1(t),u_2(t)) = \frac{|u_i|^2}{2} + \frac{1}{2}x^T P_i x + \sum_{1,2} x^T Q_{ij} u_j.$$

To obtain optimal control strategies $\tilde{u}_i$ in closed form, let us proceed as follows. Let us introduce the following expression for $\tilde{u}_i(t,x,q_i)$:

$$\tilde{u}_i(t,x,q_i) = \arg\max_{\omega \in \mathbb{R}^{m_i}} \left\{q_i B_i(t)\omega - \frac{|\omega|^2}{2} - x^T Q_{ii}(t)\omega\right\}$$

$$= (q_i B_i(t) - x^T Q_{ii}(t))^T.$$

(9.8)

Now, let us take for the value function the expression $V_i(t,x) = \frac{1}{2}x^T M_i(t)x$ so that

$$\nabla V_i(t,x) = x^T M_i(t), \quad \partial_t V_i(t,x) = \frac{1}{2}x^T \dot{M}_i(t)x.$$

Using the condition $\partial_t V_i(t,x) = L_i - \nabla V_i \cdot f$, we can write the HJB equation in compact form as

$$\frac{1}{2}x^T \dot{M}_i(t)x = \frac{1}{2}(x^T M_i B_i - x^T Q_{ii})(x^T M_i B_i - x^T Q_{ii})^T + \frac{1}{2}x^T P_i x$$

$$+ \sum_{j=1,2} x^T Q_{ij}(x^T M_j B_j - x^T Q_{jj})^T$$

(9.9)

$$- x^T M_i(Ax + \sum_{j=1,2} B_j(x^T M_j B_j - x^T Q_{jj})^T).$$

From the HJB equation, we can derive the well-known *Riccati differential equation* as follows:

$$\frac{1}{2}x^T \dot{M}_i(t)x = \frac{1}{2}(x^T M_i B_i - x^T Q_{ii})(x^T M_i B_i - x^T Q_{ii})^T + \frac{1}{2}x^T P_i x$$

$$+ \sum_{j=1,2} x^T Q_{ij}(x^T M_j B_j - x^T Q_{jj})^T \qquad (9.10)$$

$$- x^T M_i (Ax + \sum_{j=1,2} B_j (x^T M_j B_j - x^T Q_{jj})^T).$$

Observe that as the HJB equation above must hold for every x, we can drop dependence on x, which yields the following Riccati equation:

$$\frac{1}{2}\dot{M}_i(t) = \frac{1}{2}(M_i B_i - Q_{ii})(M_i B_i - Q_{ii})^T + \frac{1}{2}P_i$$

$$+ \frac{1}{2}\sum_{j=1,2}[Q_{ij}(M_j B_j - Q_{jj})^T + (M_j B_j - Q_{jj})Q_{ij}^T]$$

$$- \frac{1}{2}(M_i A + A^T M_i) - \frac{1}{2}\sum_{j=1,2}[M_i B_j (M_j B_j - Q_{jj})^T \qquad (9.11)$$

$$+ (M_j B_j - Q_{jj})B_j^T M_i].$$

Linear-quadratic differential games play a crucial role in robust control and in particular in H^∞-optimal control. We elaborate more on this in the next section.

9.5 ▪ H^∞-optimal control as linear-quadratic differential game

The relation between two-person zero-sum games and H^∞-optimal control, first introduced in [22], has already been discussed in Section 2.4. Recall that this problem deals with the design of a controller that guarantees a good performance even under the worst-case disturbance. A schematic representation of the system is depicted in Fig. 9.3.

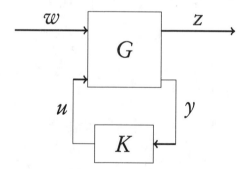

Figure 9.3. *Schematic representation of an H^∞-optimal control problem. Reprinted with permission from Springer Science+Business Media [22].*

In the block system, u is the control input, w is the uncontrolled input referred to as disturbance, and z and y are the controlled and measured outputs, respectively. We assume that all variables are measurable in Hilbert spaces $\mathcal{H}_u, \mathcal{H}_w, \mathcal{H}_z, \mathcal{H}_y$. The variables are mutually dependent, and such dependencies are captured by the system of equations

$$\begin{cases} z = G_{11}(w) + G_{12}(u), \\ y = G_{21}(w) + G_{22}(u), \\ u = K(y). \end{cases} \qquad (9.12)$$

We assume that the operators G_{ij} and the controller $K \in \mathcal{K}$ are bounded causal linear operators. Here $\mathcal{K}$ denotes the controller space.

The main objective, known as *disturbance attenuation*, consists in maintaining the controlled output small despite the presence of the disturbance. In preparation to the formulation of the disturbance attenuation problem, let us introduce bounded causal linear operators $T_K : \mathcal{H}_w \to \mathcal{H}_z$ for every fixed $K \in \mathcal{K}$, namely

$$T_K(w) = G_{11}(w) + G_{12}(I - KG_{22})^{-1}(KG_{21})(w).$$

Recall that *causal* means that the system is *nonanticipative*. That is to say that the output depends on past and current inputs but not on future inputs.

The problem consists in finding the worst-case infimum of the operator norm

$$\begin{cases} \inf\limits_{K \in \mathcal{K}} \langle\langle T_K \rangle\rangle =: \gamma^*, \\ \langle\langle T_K \rangle\rangle = \sup\limits_{w \in \mathcal{H}_w} \dfrac{\|T_K(w)\|_z}{\|w\|_w}. \end{cases} \tag{9.13}$$

It turns out that the above problem turns into a two-person zero-sum game between the controller and the disturbance:

$$\overbrace{\inf_{K \in \mathcal{K}} \sup_{w \in \mathcal{H}_w} \frac{\|T_K(w)\|_z}{\|w\|_w}}^{\text{upper bound}} \geq \overbrace{\sup_{w \in \mathcal{H}_w} \inf_{K \in \mathcal{K}} \frac{\|T_K(w)\|_z}{\|w\|_w}}^{\text{lower bound}}.$$

Given the above problem, it is possible to derive a so-called *soft-constrained game*. To this purpose, consider the attenuation level γ^*, which satisfies

$$\inf_{K \in \mathcal{K}} \sup_{w \in \mathcal{H}_w} \|T_K(w)\|_z^2 - {\gamma^*}^2 \|w\|_w^2 \leq 0.$$

Let us define the parametrized cost (in $\gamma \geq 0$)

$$J_\gamma(K, w) := \|T_K(w)\|_z^2 - \gamma^2 \|w\|_w^2.$$

Consequently, the problem is the one of finding the smallest value of $\gamma \geq 0$ under which the upper value is bounded (by zero).

Furthermore, the above problem can be turned into a linear-quadratic zero-sum differential game. To see this, consider the state space representation

$$\begin{cases} \dot{x}(t) = A(t)x(t) + B(t)u(t) + D(t)w(t), \quad x(0) = x_0, \\ z(t) = H(t)x(t) + G(t)u(t), \\ y(t) = C(t)x(t) + E(t)w(t), \end{cases} \tag{9.14}$$

and for $\gamma \geq 0$ and $Q_T > 0$, let the following cost be given:

$$L_\gamma(u, w) := x(T)^T Q_T x(T) + \int_0^T z(t)^T z(t) dt - \gamma^2 \int_0^T w(t)^T w(t) dt.$$

The zero-sum linear-quadratic differential game is given by

$$\min_{u(\cdot)} \max_{w(\cdot)} L_\gamma(u, w).$$

Given the above game, we only need to solve it using the methods surveyed in the previous sections.

9.6 ▪ Notes and references

This chapter is inspired by the tutorial written by Bressan, *Noncooperative Differential Games*, in 2010 [65]. A comprehensive treatment of dynamic programming is available in the classical book [47]. The Pontryagin Maximum Principle is in [203]. The *Hamilton–Jacobi–Isaacs equation* is due to Isaacs back in 1965 [126]. Section 9.5 is based on the introductory chapter of [22]. Minimax optimal control is developed in [245]. A comprehensive treatment of nonsmooth optimal control is in [246].

Chapter 10

Stochastic Games

10.1 ▪ Introduction

This chapter deals with *stochastic games*. As for differential games, even in stochastic games we have a state variable. However, the state dynamics now is given by a controlled Markov chain rather than by a differential equation. Stochastic games represent a generalization of *Markov decision problems* involving multiple decision makers. This chapter covers the basics of the theory on stochastic games. In particular, after streamlining the game model in general terms, we survey the fundamental results of two-player stochastic zero-sum games. We conclude the chapter with open questions and future directions.

In Section 10.2 we introduce the main ingredients of a stochastic game, define pure and mixed stationary strategies, and discuss alternative formulations involving finite and infinite horizon payoffs. In Section 10.3 we discuss applications. In Section 10.4 we focus on two-player zero-sum stochastic games. In Sections 10.5 and 10.6 we formulate and solve two classical examples. Seminal results and further developments are discussed in Section 10.7. Finally, in Section 10.8 we provide notes and references for this chapter.

10.2 ▪ The model

In stochastic games the environment configuration, captured by the state variable, changes in response to the players' behaviors. These games model the repeated interactions among the players over a time horizon window. We introduce a *stage payoff*, which is the payoff produced at a given stage. The stage payoffs of the players depend on the current behaviors of the players, described through their actions or decisions, and on the environment, the latter described through a state variable.

Let t denote the time, and let S be the state space. The latter can be *countable* or *uncountable*, in which case it is supplemented with a σ-algebra of measurable sets. The mathematical formulation of a stochastic game makes use of the following ingredients:

- the set of players N;

- the set of actions $\mathscr{A}_i$ of player i;

- the valued function $A_i : S \to \mathscr{A}_i$ representing the available actions to player i in a given state;

- the set of action profiles $SA := \{(s,a) : s \in S, a = (a_i)_{i \in N}, a_i \in A_i(s), \forall i \in N\}$;

97

- for every player i, the stage payoff function defined as $u_i : SA \to \mathbb{R}$;

- the transition function by $q : SA \to \Delta(S)$, where $\Delta(S)$ is the space of probability distributions over S.

Keeping in mind the above model, we observe that the class of stochastic games represents a generalization of a few other games or optimization problems as listed below:

(i) games with *finite interactions*; this occurs if the state of the game reaches at time t an absorbing state with null payoff;

(ii) static *matrix games* if we set $t = 1$;

(iii) *repeated games* if the game admits only one state;

(iv) *stopping games* if the stage payoff is null until a player decides to *quit* the game; in consequence of this, the state of the game reaches an absorbing state with nonnull payoff;

(v) *Markov decision problems* if the game involves only one single player.

As for all the types of games encountered so far, the payoff u_i is a profit (to maximize) but can also be a cost (to minimize). Furthermore, note that the actions decide the current payoffs and the future states, and consequently they influence the future payoffs as well. It is worth noting that actions, payoffs, and transitions derive only from the current state.

10.2.1 ▪ Pure and mixed strategies and stationarity

We saw that in extensive games and differential games, the history of the game is captured by a state variable. We also saw that different actions perform differently depending on the state of the game. To capture the connection between state and actions we have introduced a mapping from states to actions which we have defined as *strategy*. We find strategies also in stochastic games. To see this, let us denote by

$$(s^1, a^1, s^2, a^2, \ldots, s^t)$$

the past play at stage t, where (s^k, a^k) is the action profile at time k. Then, we call (pure) *stationary* strategy a strategy that depends only on the current state, that is,

$$\sigma_i(s^1, a^1, s^2, a^2, \ldots, s^t) \in A_i(s^t).$$

That is to say that the past play $\sigma_i(s^1, a^1, s^2, a^2, \ldots, a^{t-1})$ plays no role in the selection of the current action.

The above definition can be extended to include also *mixed* strategies. These strategies consist of probability distributions over the action spaces, which we now indicate by

$$\sigma_i(s^1, a^1, s^2, a^2, \ldots, s^t) \in \Delta(A_i(s^t)),$$

where $\Delta(A_i(s^t))$ is the probability distribution on set $A_i(s^t)$.

With this in mind, we introduce the space of stationary mixed strategies for player i given by

$$X_i = \times_{s \in S} \Delta(A_i(s)).$$

Collecting the strategies of all the players, we obtain the following profile of mixed strategies:

$$\sigma = (\sigma_i)_{i \in N}, \quad \sigma_i \in X_i.$$

Given that the players play repeatedly in time, the game is characterized by the space of infinite plays $H^\infty = SA^{\mathbb{N}}$. Such a space involves the set of all possible infinite sequences and is given by

$$(s^1, a^1, s^2, a^2, \ldots, s^t, a^t, \ldots).$$

Now, it is clear that every profile of mixed strategies $\sigma = (\sigma_i)_{i \in N}$ and every initial state s_1 induce a probability distribution $\mathbf{P}_{s_1, \sigma}$ on $H^\infty = SA^{\mathbb{N}}$. With this in mind, the game results in a finite or infinite ($T \to \infty$) stream of payoffs:

$$u_i(s^t, a^t), \quad t = 1, 2, \ldots, T.$$

Depending on how we combine the above payoffs, we obtain different types of formulations. Three possible formulations are mentioned in the next section.

10.2.2 ▪ Finite and infinite horizon formulations

An element in common between stochastic games and optimal control is that there exist different formulations involving finite or infinite horizons on the one hand, and *myopic* (also called *shortsighted*) players or *patient* (also called *farsighted*) players on the other hand. More specifically, the formulation involving a *finite horizon* implies that the interaction lasts exactly a finite number, say $T > 0$, of stages. Conversely, the formulation involving an infinite horizon calls for a *discounted evaluation*. That is to say that the interaction lasts many stages, and the players discount their stage payoffs. As a result we have that for the players it is better to receive "1 dollar" today than the same dollar tomorrow. This formulation wishes to capture the greedy or shortsighted behavior of the players.

The formulation involving an infinite horizon contemplates another way to combine the payoffs, referred to as *limsup evaluation*. This is the case where the interaction between the players lasts many stages, and the players do not discount their stage payoffs. As a consequence of this, the stage payoff at time t is not relevant when compared with the total payoffs over all the other stages. The scenario at hand sees the players as patient or farsighted decision makers. The instantaneous fluctuations of the payoffs are not relevant. We derive next the formal definition of payoff in each of the above formulations.

The payoff in the finite horizon case is here called *T-stage payoff*. For it we have the expression

$$\gamma_i^T(s_1, \sigma) := \mathbb{E}_{s_1, \sigma}\left[\frac{1}{T} \sum_{t=1}^{T} u_i(s^t, a^t) \right].$$

The formulation involving the infinite horizon with discounted evaluation yields a so-called *λ-discounted payoff*, which is given by

$$\gamma_i^\lambda(s_1, \sigma) := \mathbb{E}_{s_1, \sigma}\left[\lambda \sum_{t=1}^{\infty} (1 - \lambda)^{t-1} u_i(s^t, a^t) \right].$$

Finally, for the infinite horizon case with limsup evaluation, the payoff, which we here refer to as *limsup payoff*, is defined by

$$\gamma_i^\infty(s_1, \sigma) := \mathbb{E}_{s_1, \sigma}\left[\limsup_{T \to \infty} \frac{1}{T} \sum_{t=1}^{T} u_i(s^t, a^t) \right].$$

After introducing the above formations, let us turn to the analysis of *equilibria*. Depending on the formulation, we have different definitions of equilibria. We shall henceforth consider the following types of equilibrium solutions.

Definition 10.1. *We say that σ is a T-stage ε-equilibrium if*

$$\gamma_i^T(s_1,\sigma) \ge \gamma_i^T(s_1,\sigma_i',\sigma_{-i}) - \varepsilon \quad \forall s_1 \in S, \forall i \in N, \forall \sigma_i' \in X_i.$$

Differently, σ is a λ-discounted ε-equilibrium if

$$\gamma_i^\lambda(s_1,\sigma) \ge \gamma_i^\lambda(s_1,\sigma_i',\sigma_{-i}) - \varepsilon \quad \forall s_1 \in S, \forall i \in N, \forall \sigma_i' \in X_i.$$

Finally, σ is a limsup ε-equilibrium *if*

$$\gamma_i^\infty(s_1,\sigma) \ge \gamma_i^\infty(s_1,\sigma_i',\sigma_{-i}) - \varepsilon \quad \forall s_1 \in S, \forall i \in N, \forall \sigma_i' \in X_i.$$

Note that, from the above definitions, the players cannot benefit more than ε by deviating unilaterally from the equilibrium.

10.3 • A brief overview on applications

This section provides a brief overview on application domains. The following list, far from complete, mentions examples of applications that can be described by and studied using stochastic games.

Example 10.2 (Capital accumulation or fishery). The first example deals with capital accumulation. The example admits an interpretation also in the context of fishery. The game involves two players, who jointly own a natural resource or a productive asset. The players have to choose the amount of resource to consume at every period. The residual amount, namely the quantity that is not consumed, grows by a fraction. This fraction can be known or unknown. The state of the game is the current amount of resource. The set of actions involves the quantity of resource that can be exploited in the current period. The decisions of all the players determine the transition between consecutive states. The transition is also a function of the random growth of the resource. ∎

Example 10.3 (Taxation). In this example, we develop a taxation application. The application involves a government that has to set a tax rate at every period. The other players are the citizens who can decide how much to work and how much money to consume at every period. The money that is not consumed grows by a known interest rate at the next period. The game has a state which represents the amount of savings of the citizens. The citizens incur in stage payoffs that depend on (i) the amount of money that they consume, (ii) the free time available to them (free time is a measure of the quality of life), and (iii) the total amount collected by the government from taxes.

Evidently, also the government has a stage payoff which accounts for (i) the average stage payoff of the citizens and (ii) the total amount of tax collected. ∎

Example 10.4 (Communication network). Game theory has a long tradition in communication networks. Here we consider a single-cell system involving one receiver and multiple uplink transmitters who share a single, slotted, synchronous classical collision channel. The players of the game are the transmitters, who decide if and which packet to transmit at each time. The dropped packets are backlogged. The game has a state which is the level of congestion in the channel. The stage payoff includes the probability of successful transmission and the cost of transmission. ∎

Example 10.5 (Queues). Queues is another application where stochastic games can be used. In this example, we have individuals that can opt for a private slow service provider or for a powerful public service provider. The game has a state which describes the current load of public and private service providers. The payoff of the individuals accounts for the time to be served, which the players wish to minimize. ∎

10.4 ▪ Two-player zero-sum stochastic games

This section illustrates two-person zero-sum stochastic games. In particular, it presents solution techniques based on dynamic programming on two classical examples.

In a two-player zero-sum stochastic game the sum of payoffs is zero. That is to say that the following condition holds:

$$u_1(s,a) + u_2(s,a) = 0 \quad \forall (s,a) \in SA.$$

There exists at most one equilibrium payoff at every initial state s_1. This equilibrium payoff is termed the *value* of the game.

It turns out that if a player's strategy σ_1 is at an ε-equilibrium, then such a strategy is ε-optimal. Namely, such a strategy guarantees the value up to a given tolerance ε, as described by the following condition:

$$\gamma_1^T(s_1, \sigma_1, \sigma_2) \geq \underbrace{v^T(s_1)}_{\text{value at } s_1} - \varepsilon \quad \forall \sigma_2 \in X_2.$$

As for zero-sum static games where we have existence results of equilibria provided by von Neumann (cf. Theorem 2.4), even for zero-sum stochastic games there exist existence results provided by Shapley in 1953 [219]. These results have then been extended to multiplayer nonzero-sum games by Fink in 1964 [97]. We state below the theorem by Shapley.

Theorem 10.6 (Shapley 1953 [219]). *If all sets are finite, then for every λ there exists an equilibrium in stationary strategies.*

Proof. We provide here only a sketch of the proof and refer the reader to the original work for a detailed derivation. First, define $\mathcal{V}$ the space of all functions $v : S \to \mathbb{R}$. Consider the zero-sum matrix game $G_s^\lambda(v)$ for all v. Denote by $A_1(s)$ and $A_2(s)$ the space of actions at state s. Here it is useful to think of the payoff as money that player 2 pays to player 1. This payoff is given by

$$\lambda u_1(s,a) + (1-\lambda) \sum_{s' \in S} q(s'|s,a) v(s').$$

Let the value operator $\phi_s(v) = val(G_s^\lambda(v))$ be given. Nonexpansiveness given by $\|\phi(v) - \phi(w)\|_\infty \leq (1-\lambda)\|v - w\|_\infty$ yields a unique fixed point $\hat{v}^\lambda$. ☐

The optimal mixed action σ_i in the matrix game $G_{s^t}^\lambda(\hat{v}^\lambda)$ is also a λ-discounted 0-optimal strategy. It is worth noting that the aforementioned proof is constructive. We will make use of it later to solve the two classical games presented next.

10.5 ▪ The Big Match: "Work hard" or "enjoy life"

The Big Match describes a nice, simple fairy tale involving a king and his trusted minister. One day, the king has to leave for an undefined time and therefore decides to put his trusted minister in charge of the kingdom. The day before leaving, the king informs the minister that he will not hear from the king until his return. On the day the king will return, if the minister will be found *working hard*, the king will award the minister by abducting in favor of him. On the other hand, if on that day the king will find the minister *enjoying life*, the king will put the minister in prison, where he will be tormented for ever and ever. The king is powerful and has informers. Therefore he knows every day whether the minister was at work or not in the past days.

The challenge here is about what the best strategies for the king and ministers are. Obviously, both the king and the minister pursue their own interests. In particular, the minister knows that if he worked hard every day, the king, being informed of this, would not come back. This would mean an everlasting miserable life. The minister also knows that if he did not work at all, the king would come very soon and the minister would be imprisoned.

Evidently, such a fairy tale is a stylized example of competition between two individuals with contrasting goals. We can analyze the optimal strategies of the king and the minister using a stochastic game. To formulate the game, let us introduce the following matrices:

$$
\begin{array}{c}
\begin{array}{cc} L & R \end{array} \\
\begin{array}{c} T \\ B \end{array}
\begin{array}{|cc|cc|}
\hline 0 & s_2 & 1 & s_2 \\ 1 & s_1 & 0 & s_0 \\ \hline
\end{array} \\
\text{State } s_2
\end{array}
\qquad
\begin{array}{c}
L \\
T \;
\begin{array}{|cc|}
\hline 1 & s_1 \\ \hline
\end{array} \\
\text{State } s_1
\end{array}
\qquad
\begin{array}{c}
L \\
T \;
\begin{array}{|cc|}
\hline 0 & s_0 \\ \hline
\end{array} \\
\text{State } s_0
\end{array}
$$

In the above matrices, the row player is the king, and the column player is the minister. The decision of the king not to come back corresponds to action T. Thus, for every day that the king plays T, the state of the game transitions to the same state s_2. This occurs independently of the choice of the minister to be at work, denoted by L, or to rest, denoted by R. The choice of the king of coming back is denoted by the action B. If the king plays B and the minister plays L (the minister is hardworking), the game jumps to state s_1, which implies an everlasting reward for the minister. Conversely, if the king plays B and the minister plays R (the minister is found enjoying life), then the state of the game jumps to state s_0, which implies an everlasting punishment for the minister. With the above game in mind, let us analyze the equilibria. To this purpose, let us apply the dynamic programming principle used in the proof of the theorem exposed earlier. By doing this, we have that for every $v = (v_1, v_2, v_3) \in \mathcal{V} = \mathbb{R}^3$ the games $G_{s_2}^\lambda(v)$, $G_{s_1}^\lambda(v)$, and $G_{s_0}^\lambda(v)$ are given by

$$
\begin{array}{c}
\begin{array}{cc} \qquad L & \qquad\qquad R \end{array} \\
\begin{array}{c} T \\ B \end{array}
\begin{array}{|c|c|}
\hline (1-\lambda)v_2 & \lambda + (1-\lambda)v_2 \\ \hline \lambda + (1-\lambda)v_1 & (1-\lambda)v_0 \\ \hline
\end{array} \\
\text{Game } G_{s_2}^\lambda
\end{array}
\qquad
\begin{array}{c}
\qquad L \\
T \;
\begin{array}{|c|}
\hline \lambda + (1-\lambda)v_1 \\ \hline
\end{array} \\
\text{Game } G_{s_1}^\lambda
\end{array}
$$

$$
\begin{array}{c}
\qquad L \\
T \;
\begin{array}{|c|}
\hline (1-\lambda)v_0 \\ \hline
\end{array} \\
\text{Game } G_{s_0}^\lambda
\end{array}
$$

By imposing the fixed point condition on both states 0 and 1, we obtain

- $\hat{v}_0^\lambda = val(G_{s_0}^\lambda(\hat{v}))$, which yields $\hat{v}_{s_0}^\lambda = 0$;

- $\hat{v}_1^\lambda = val(G_{s_1}^\lambda(\hat{v}))$, which yields $\hat{v}_{s_1}^\lambda = 1$.

After replacing the aforementioned values for state 2 we obtain

	L	R
T	$(1-\lambda)v_2$	$\lambda+(1-\lambda)v_2$
B	1	0

State s_2

From the *Indifference Principle*, the saddle-point of this game is obtained by solving

$$v_2 = y(1-\lambda)v_2 + (1-y)[\lambda+(1-\lambda)v_2] = y,$$
$$v_2 = x(1-\lambda)v_2 + (1-x) = x[\lambda+(1-\lambda)v_2],$$

where y is the probability that player 2 plays L and x the probability that player 1 plays T.

Consequently, we obtain $\hat{v}_2^\lambda = val(G_{s_2}^\lambda(\hat{v}))$, which yields $\hat{v}_{s_2}^\lambda = \frac{1}{2}$. For the best-response strategies we finally obtain

$$\sigma_2 = [\tfrac{1}{2}(L), \tfrac{1}{2}(R)], \quad \sigma_1 = [\tfrac{1}{1+\lambda}(T), \tfrac{\lambda}{1+\lambda}(B)].$$

The interpretation of the above result is as follows. The best strategy for the minister is to work every two days on average. This is equivalent to saying that every day the minister will toss a coin and depending on the result he will work hard or not. The interpretation of the best strategy for the king is as follows. First note that his optimal strategy will depend on the discount factor, that is, on how farsighted he is. The king will return with a probability that increases with the discount factor. That is to say that the more myopic the king is, the sooner he will come back. Conversely, if the king is farsighted, the discount factor is small and tends to zero, and consequently the probability of coming back approaches zero. Note that the discount factor influences only the strategy of the king. The strategy of the minister does not depend on the discount factor. This derives from the fact that only the king can force the state of the game to jump to an absorbing state.

10.6 ▪ The Absorbing game: A variant of the Big Match

This example represents a variant of the Big Match where not only the king but also the minister can force the game to jump to an absorbing state. As a consequence, we will see that the strategy of the minister will also be influenced by the discount factor. The formulation of the game is as follows:

	L		R	
T	0	s_2	1	s_1
B	1	s_1	0	s_0

State s_2

	L	
T	1	s_1

State s_1

	L	
T	0	s_0

State s_0

Differently from the Big Match, the action profile (T,R) makes the state of the game transition to the absorbing state s_1. Using the same approach as for the Big Match, for every $v = (v_1, v_2, v_3) \in \mathcal{V} = \mathbb{R}^3$ the games $G_{s_2}^\lambda(v)$, $G_{s_1}^\lambda(v)$, and $G_{s_0}^\lambda(v)$ take the form

$$
\begin{array}{c|c|c|}
 & L & R \\
\hline
T & (1-\lambda)v_2 & \lambda+(1-\lambda)v_1 \\
\hline
B & \lambda+(1-\lambda)v_1 & (1-\lambda)v_0 \\
\hline
\end{array}
$$
Game $G_{s_2}^{\lambda}$

$$
\begin{array}{c|c|}
 & L \\
\hline
T & \lambda+(1-\lambda)v_1 \\
\hline
\end{array}
$$
Game $G_{s_1}^{\lambda}$

$$
\begin{array}{c|c|}
 & L \\
\hline
T & (1-\lambda)v_0 \\
\hline
\end{array}
$$
Game $G_{s_0}^{\lambda}$

From the fixed point condition on the states, we obtain

- $\hat{v}_0^{\lambda} = val(G_{s_0}^{\lambda}(\hat{v}))$ yields $\hat{v}_{s_0}^{\lambda} = 0$,

- $\hat{v}_1^{\lambda} = val(G_{s_1}^{\lambda}(\hat{v}))$ yields $\hat{v}_{s_1}^{\lambda} = 1$,

- $\hat{v}_2^{\lambda} = val(G_{s_2}^{\lambda}(\hat{v}))$ yields $\hat{v}_{s_2}^{\lambda} = \frac{1-\sqrt{\lambda}}{1-\lambda}$.

The third equation above derives from

$$
v_2 = y(1-\lambda)v_2 + (1-y) = y,
$$
$$
v_2 = x(1-\lambda)v_2 + (1-x) = x,
$$

where we denote by y the probability that player 2 plays L and by x the probability that player 1 plays T.

For the best-response strategies, we obtain

$$
\sigma_2 = [\tfrac{1-\sqrt{\lambda}}{1-\lambda}(L), \tfrac{\sqrt{\lambda}-\lambda}{1-\lambda}(R)], \quad \sigma_1 = [\tfrac{1-\sqrt{\lambda}}{1-\lambda}(T), \tfrac{\sqrt{\lambda}-\lambda}{1-\lambda}(B)].
$$

In accordance with our intuition, now the discount factor affects the strategies of the minister and of the king.

10.7 ▪ Other seminal results and further developments

This section mentions other seminal results and further developments for stochastic games. A first result is due to Mertens and Neyman back in 1981 [179]. This result is about the existence of a *uniform equilibrium* for two-player zero-sum games. We state the result in the following.

Theorem 10.7 (Mertens and Neyman 1981 [179]). *For two-player zero-sum games, each player has a strategy that is ε-optimal for every discount factor sufficiently small.*

Vieille extended the above result to nonzero-sum games in 2000.

Theorem 10.8 (Vieille 2000 [244]). *For every two-player nonzero-sum stochastic game there is a strategy profile that is an ε-equilibrium for every discount factor sufficiently small.*

There are several open questions in the study of stochastic games. A first one is about the search of a strategy profile that is an ε-equilibrium for every discount factor sufficiently small and for every stochastic game.

Other open problems involve the selection of categories of games characterized by a simple strategy profile that is an ε-equilibrium for every discount factor sufficiently small. For instance, let us think of stationary strategy, or periodic strategy.

The study of stochastic games may also involve the development of numerical schemes and algorithms for the computation of equilibria. This is particularly relevant for those games for which explicit solutions are not known. In particular, in the case of two-player zero-sum games, a variety of algorithms are available that use linear programming. Such algorithms have also been extended to nonzero-sum games. Let us think, for instance, of the well-known *Lemke–Howson algorithm*. In addition to this, there are other algorithms that are based on *fictitious play*, *value iterates*, and *policy improvement*.

There are several future directions in the study of stochastic games. These may involve, among other things,

- approximation schemes based on finite games of games with infinite state and action spaces;

- the development of the theory of stochastic games in continuous time;

- the study of existence conditions for uniform equilibrium solutions and limsup equilibrium solutions in multiplayer stochastic games with finite state spaces and action spaces;

- the design of algorithms that efficiently compute the value of two-player zero-sum games;

- the investigation of approachable and excludable sets in stochastic games with vector payoffs.

The last topic in the above list is developed further in the following chapter.

10.8 ▪ Notes and references

This chapter is heavily based on the survey written by Solan in *Stochastic Games* in 2009 [226]. The interested reader is referred to [96, 179, 189, 244] for further reading on the topic. Capital accumulation or fishery applications are discussed in [8, 90, 157, 191]. An example of stochastic games applied to taxation is developed in [76, 200]. Communication networks are the main focus in [210]. Stochastic games to model queues are examined in [7]. Existence results for zero-sum stochastic games were provided by Shapley in 1953 [219]. These results have been extended to multiplayer nonzero-sum games by Fink in 1964 [97]. More details on the Big Match model are in [244]. We refer the reader to [179] for a comprehensive treatment of existence conditions of a uniform equilibrium for two-player zero-sum games by Mertens and Neyman. Extensions of the above results to nonzero-sum games by Vieille are in [244].

Chapter 11

Games with Vector Payoffs: Approachability and Attainability

11.1 ▪ Introduction

This chapter covers the fundamentals of the theory of *games with vector payoffs*. The theory is known as *Approachability Theory*.

Imagine a game where the outcomes produce multiple noninterchangeable items. As an example, in a job interview the employer and the candidate usually bargain over salary, career perspectives, benefits, days off, and several other items. Intuitively, the repetition of such a bargaining process over time produces a flow of instantaneous payoffs. Thus, both the employer and the candidate may wish to regulate the flow of payoffs to their advantage. In case of a lengthy interaction it is reasonable to focus on the average payoff so that it can *approach* a predetermined set. Think of this set as the set of conditions on which the employer and the candidate may find agreement and the contract can be signed. Conditions for this to happen are established in *Blackwell's Approachability Principle*.

The principle is used in several areas of game theory, such as allocation processes in coalitional games, regret minimization, adaptive learning, excludability and bounded recall, and weak approachability, to name just a few. For instance, in coalitional games, one asks whether the core is an approachable set and which allocation processes can drive the excess vector to that set. In regret minimization, we consider the nonpositive orthant in the space of regrets as approachable; then a single player tries to adjust his strategy based on the current regret so as to make that set approachable by the regret vector. It is proven that once the set is reached and the players' regret vectors are nonpositive, the resulting outcome is an equilibrium for the repeated game. This idea of adapting the new action to the current state of the game is common to adaptive learning as well. Still, the approachability principle is behind the notion of *excludability*; along this line some authors investigate which sets are approachable and which ones are excludable under imperfect information (bounded recall, delayed and/or stochastic monitoring).

When dealing with vector payoffs, given a preassigned set in the space of vector payoffs, the challenge is to understand *the conditions that guarantee that there exists a strategy for player 1 such that the average payoff "converges" to that set independently of the strategy used by player 2*. More recently, a new notion has been coined, called *attainability*, which focuses on convergence of the cumulative rather than the average payoff. Convergence has to be intended here in the *limsup* sense.

After providing an illustrative example, Section 11.2 introduces a formal definition of approachable set, discusses *Blackwell's Approachability Principle*, and points out further results. Section 11.3 highlights connections with control theory. Section 11.4 introduces the new notion of attainability and recent results. Section 11.5 points out key research directions. Finally, in Section 11.6 we provide notes and references for this chapter.

11.2 ▪ Approachability theory

This section deals with approachability theory. We shall start with an illustrative example, before providing a formal statement of approachable set and discussing seminal results.

11.2.1 ▪ Illustrative example

Consider a two-player discrete-time repeated game with vector payoffs. Both players have four actions available. Specifically, player 1 can play A, B, C, or D, and player 2 can play a, b, c, or d. Payoffs are three-dimensional vectors, as depicted in Fig. 11.1.

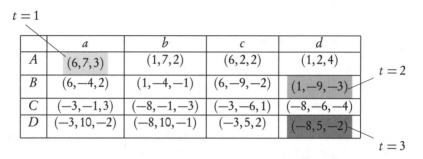

Figure 11.1. *Two-player repeated game with vector payoffs.*

Let $x(t)$ be the cumulative payoff up to time t. Furthermore, let $\bar{x}(t)$ be the average payoff up to time t. Assume that at $t = 1$, player 1 plays A and player 2 plays a. The action profile is then (A,a), and the cumulative payoff, which for $t = 1$ corresponds to the average payoff, is given by $x(1) = \bar{x}(1) = (6,7,3)$. Such a payoff is highlighted using a light gray cell in the bimatrix.

The game proceeds, and at time $t = 2$, player 1 plays B and player 2 plays d. The action profile (B,d) yields an instantaneous payoff of $(1,-9,3)$, a cumulative payoff $x(2) = (7,-2,0)$, and an average payoff $\bar{x}(2) = (\frac{7}{2},-1,0)$. The instantaneous payoff at $t = 2$ is emphasized using a gray cell in the bimatrix.

At the successive iteration, for $t = 3$, we suppose that player 1 plays D and player 2 plays d. Then we have that the instantaneous payoff is $(-8,5,-2)$, while for the cumulative payoff and the average payoff we have $(-1,3,-2)$ and $(-\frac{1}{3},1,-\frac{2}{3})$, respectively. The instantaneous payoff at $t = 3$ is emphasized using a dark gray cell in the bimatrix.

Clearly, in a continuous-time setting, the payoffs in the bimatrix would indicate the integrand and the cumulative payoff would be the integral of the instantaneous payoffs.

11.2.2 ▪ Definition of approachable set

In the following, we introduce Blackwell's definition of *approachable set*.

Definition 11.1 (Approachable set). *A set of payoff vectors A is approachable by player 1 if he has a strategy such that the average payoff up to stage t, $\bar{x}(t) := \frac{x(t)}{t}$, converges to A, regardless of the strategy of player 2.*

	L	R
T	(0,0)	(0,0)
B	(1,1)	(1,0)

Figure 11.2. *Approachability example.*

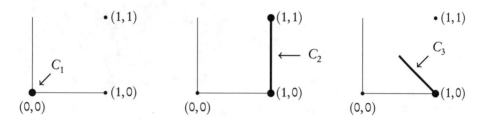

Figure 11.3. *Example of approachable sets (C_1, C_2, and C_3). Reprinted with permission from CUP* [173].

Example 11.2. This example shows different approachable sets for a two-player game with bidimensional payoff. Let the bimatrix in Fig. 11.2 be given. For this game, sets C_1, C_2, and C_3 shown in Fig. 11.3 are approachable sets. Specifically, set C_1 contains one single point, which is the origin $(0,0)$. Set C_1 is approachable, as player 1 can select action T at any time, and this produces the vector payoff $(0,0)$. Also, set C_2 is approachable. To see this, imagine player 1 playing action B at every time. This produces the vector payoff $(1,\$)$, where the second component of the payoff is any value in the interval $[0,1]$. This value will depend on how player 2 will play. Finally, a third approachable set is C_3. This is evident if we think of player 1 playing the strategy illustrated below:

$$\begin{cases} B & \text{if } \bar{x}_1(t-1) + \bar{x}_2(t-1) < 1, \\ T & \text{otherwise.} \end{cases}$$

In other words, if the current average lies in the half-space $\bar{x}_1(t-1) + \bar{x}_2(t-1) < 1$, he will play B so that the new payoff will be a point in C_2. Geometrically, this corresponds to the current average being a point on the left of the segment from $(1,0)$ to $(\frac{1}{2}, \frac{1}{2})$ and the new payoff lying in the segment from $(1,0)$ to $(1,1)$. Conversely, if the current average lies in the half-space $\bar{x}_1(t-1) + \bar{x}_2(t-1) \geq 1$, then player 1 will play T so that the new payoff will be a point in C_1. In the graphical illustration, this corresponds to the current average lying on the right of the segment from $(1,0)$ to $(\frac{1}{2}, \frac{1}{2})$ and the new payoff being $(0,0)$. ∎

11.2.3 ▪ Blackwell's Approachability Principle

After introducing the definition of approachable set, let us discuss *Blackwell's Approachability Principle*. This states conditions for a set to be approachable.

A graphical illustration of such conditions is as in Fig. 11.4. There, we have set A, which is the set player 1 wishes to approach. Assume that the payoffs are d-dimensional vectors. Define $y(t)$ as the projection of $\bar{x}(t)$ on set A. Recall that $\bar{x}(t)$ is the current average payoff. Let us draw the *supporting hyperplane* (dashed line) to A at point $y(t)$. The supporting hyperplane is the set of points satisfying

$$H = \{z \in \mathbb{R}^d \,|\, \langle z - y(t), \bar{x}(t) - y(t) \rangle = 0\}.$$

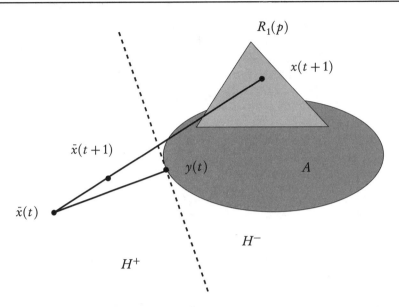

Figure 11.4. *Geometric illustration of* Blackwell's Approachability Principle.

Given the supporting hyperplane H, let us denote by H^+ and H^- the positive and negative half-spaces. That is to say that for H^+ and H^- it holds that

$$H^+ = \{z \in \mathbb{R}^d \mid \langle z - y(t), \bar{x}(t) - y(t)\rangle \geq 0\},$$
$$H^- = \{z \in \mathbb{R}^d \mid \langle z - y(t), \bar{x}(t) - y(t)\rangle \leq 0\}.$$

We are ready to state *Blackwell's Approachability Principle*.

Theorem 11.3 (Blackwell's Approachability Principle). *Set A is* approachable *if for any point $\bar{x}(t) \in H^+$, there exists a p such that $R_1(p) \subset H^-$, where $R_1(p)$ is the set of payoffs when player 1 plays the mixed action p and for all possible actions of player 2.*

Proof (**Sketch**). The proof plays around the idea that if the instantaneous payoff lies in the opposite half-space than the one containing the current average payoff, then the distance of the average payoff from the approachable set decreases monotonically. Recall that the instantaneous payoff is a point in $R_1(p)$. Therefore the existence of a strategy p for player 1 such that $R_1(p)$ is in H^- guarantees that the instantaneous payoff lies in the opposite half-space. The reader familiar with Lyapunov stability may find it useful to interpret the distance as a Lyapunov function.   □

11.2.4 ▪ Further results on approachability

Approachability theory has produced further results. We comment on three of them in order.

- **Approachability in infinite-dimensional spaces.** A first finding deals with the adaptation of *Blackwell's Approachability Principle* to infinite-dimensional vector payoffs. While Blackwell's convergence results make use of the *Euclidean distance*, the new setup plays with a *measure* defined in an infinite-dimensional space [151].

- **Approachability and differential games.** Another result investigates the nature of the approachability problem as a differential game [229] (see also [156]). Specifically, [229] explains how to turn the approachability problem into a zero-sum differential game. To do this, one has to introduce a differential dynamics describing the time evolution of the average payoff. This dynamics is subjected to (i) a controlled input, which is the strategy of player 1, and (ii) an uncontrolled input, which is the strategy of player 2. The problem can be manipulated and turned into an uncertain dynamic system with multiplicative uncertainty. The problem has now the same nature of a *reachability control problem* [53].

- **Approachability in regret learning.** A further result deals with the application of approachability in *regret minimization* [109, 110, 111]. Regret minimization is essentially the topic of learning from previous errors. The learning process is driven by a posteriori observations. Specifically, the players construct their strategy on the basis of a *regret vector*. This vector describes the advantage for a player derived from the same player constantly playing each of his pure strategies rather than the current strategy (this is in general a mixed strategy). Evidently, the regret vector of a player has as many components as the pure strategies of that player. It turns out that if the player is playing his best response, then his regret vector is nonpositive, component-wise. That is to say that at a Nash equilibrium, as all players are playing their best responses, all regret vectors are nonpositive. Clearly, convergence to a Nash equilibrium is equivalent to steering the regret vectors to the negative orthant.

11.3 ▪ A dual perspective: Connection with robust control

This section presents a dual perspective on games with vector payoffs. In particular it throws light on connections with robust control. This dual perspective occupies an important place in the history of attainability. Indeed, the definition of attainability finds its roots in the robust control problems formulated in [26, 27]. The robust control problems we have in mind deal with a flow over a network. For instance, the topology of the network may be as the one depicted in Fig.11.5.

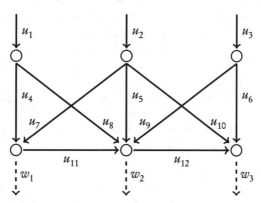

Figure 11.5. *Network robust control problem. Reprinted with permission from IEEE [27].*

The problem involves the following main elements:

- *uncontrolled* flows $w(t) \in \mathcal{W}$ for all t; for instance, the uncertain demand in a market;

- *controlled* flows $u(t) \in \mathcal{U}$ for all t; for instance, the supply in a market.

The excess supply at the nodes accumulates in a buffer. The time evolution of the supply follows a first-order differential equation of type

$$\begin{cases} \dot{x}(t) = Bu(t) - w(t), \\ x(0) = \zeta. \end{cases}$$

In the dynamics mentioned above, B denotes the incidence matrix of the network and ζ is the initial state. The initial state is essentially the initial configuration of the excesses at the nodes. The interpretation of the above dynamics is that at each time the discrepancy between the incoming flow and the outgoing flow accumulates in the buffer.

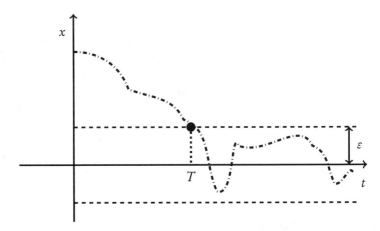

Figure 11.6. *Tube reachability and robustness.*

The robust control problem is about synthesizing a state-feedback control strategy to steer the vector of excesses to zero. This is illustrated in Fig. 11.6. Here, the trajectory of the state is driven to a neighborhood of zero in finite time T and is kept within that neighborhood for the rest of the time. We use the term "robust" to mean the capability to accomplish this task even if the current and future values of the demand w are not known. The only assumption is that such values belong to a predefined set. The knowledge of the only bounding set reframes the problem within the literature on robust control with *unknown but bounded* disturbances [53]. The problem is also known as *tube reachability*.

We can look at this problem from an alternative angle, which provides a dual perspective in terms of repeated games with vector payoffs. To see this, let us suppose that the control input $u(t)$ is selected by player 1. At the same time the demand $w(t)$ is selected by player 2. Review the excess derivative $\dot{x}(t)$ as the instantaneous payoff of the game. Evidently, it turns out that the cumulative payoff at time t is exactly the state variable $x(t)$.

Example 11.4. In this example we shed light on a method to turn a network flow control problem as the one introduced earlier into an attainability problem for a repeated game with vector payoffs.

Consider the topology depicted in Fig. 11.7. Assume that the controlled input $u(t)$ can take a value in a given discrete set, namely

$$u(t) \in \left\{ \begin{bmatrix} 1 \\ -2 \\ 6 \end{bmatrix}, \begin{bmatrix} 1 \\ -2 \\ -5 \end{bmatrix}, \begin{bmatrix} -5 \\ 1 \\ -5 \end{bmatrix}, \begin{bmatrix} -5 \\ 1 \\ 6 \end{bmatrix} \right\}.$$

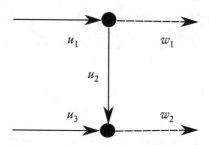

Figure 11.7. *An example of a network flow control problem turned into an attainability problem* [155].

That is to say that the material flows in batches. At the same time, let us assume that for the uncontrolled input $w(t)$ it holds that

$$w(t) \in \left\{ \begin{bmatrix} -3 \\ -3 \end{bmatrix}, \begin{bmatrix} 2 \\ -3 \end{bmatrix}, \begin{bmatrix} -3 \\ 2 \end{bmatrix}, \begin{bmatrix} 2 \\ 2 \end{bmatrix} \right\}.$$

In other words, both flows, controlled and uncontrolled, take values in predetermined discrete sets, which we now reinterpret as discrete action sets. Having said this, the evolution of the excesses is given by

$$\begin{bmatrix} \dot{x}_1(t) \\ \dot{x}_2(t) \end{bmatrix} = \begin{bmatrix} 1 & -1 & 0 \\ 0 & 1 & 1 \end{bmatrix} \begin{bmatrix} u_1(t) \\ u_2(t) \\ u_3(t) \end{bmatrix} - \begin{bmatrix} w_1(t) \\ w_2(t) \end{bmatrix}.$$

We immediately note that if we substitute for $u(t)$ and $w(t)$ the values available in the discrete sets, then we essentially obtain the bimatrix illustrated in Fig. 11.8.

	a	b	c	d
A	(6,7)	(1,7)	(6,2)	(1,2)
B	(6,−4)	(1,−4)	(6,−9)	(1,−9)
C	(−3,−1)	(−8,−1)	(−3,−6)	(−8,−6)
D	(−3,10)	(−8,10)	(−3,5)	(−8,5)

Figure 11.8. *Bimatrix derived from a network flow control problem.*

It is worth noting that the above bimatrix involves the first two payoffs of the three-dimensional payoff matrix introduced at the beginning of this chapter.

In the bimatrix of Fig. 11.8, each entry describes the bidimensional payoff resulting from any feasible action profile. As an example, the entry $(6,7)$ which we find in the 1st row and 1st column can be obtained by substituting

$$u(t) = \begin{bmatrix} 1 \\ -2 \\ 6 \end{bmatrix} \quad \text{and} \quad w(t) = \begin{bmatrix} -3 \\ -3 \end{bmatrix}. \quad \blacksquare$$

After introducing the dual perspective, we are in a position to present the notion of *attainability*.

11.4 ▪ The concept of attainability

We saw that in the theory of repeated games with vector payoffs, *approachability* deals with the study of the average payoffs and their convergence properties. We also said in the introductory section that *attainability* does the same but with focus on the cumulative payoffs rather than average payoffs [155]. This is formalized in the definition provided below.

Definition 11.5 (Attainability [40, 155]). *A set of payoff vectors A is attainable by player 1 if he has a strategy such that the total payoff up to stage t, x(t), "converges" to A, regardless of the strategy of player 2.*

There is apparently only a subtle distinction between approachability and attainability. However, in what follows we show that such a distinction has deep implications in terms of convergence conditions.

11.4.1 ▪ Attainability in continuous time

In this section we explore attainability in continuous time. In particular, we present the continuous-time repeated game model. After presenting the model, we review the attainability conditions available in the literature. Consider a two-player repeated game $(A_1, A_2, \mathbf{g})$, where we denote by A_i the action space of player i and by $\mathbf{g} : A_1 \times A_2 \to [-1, 1]^d$ the d-dimensional payoff. Define $(a_i^t)_{t \in \mathbb{R}_+}$ as the *nonanticipative behavior strategy* for player i. This strategy has the following characteristics:

- $(a_i^t)_{t \in \mathbb{R}_+}$ takes values in $\Delta(A_i)$;

- there exists an increasing sequence of times $\tau_i^1 < \tau_i^2 < \tau_i^3 < \cdots$ such that a_i^t is measurable with respect to the information available at τ_i^k, $\tau_i^k \leq t < \tau_i^{k+1}$.

An example of nonanticipative behavior strategy is the one depicted in Fig. 11.9. At time τ_i^0 (this corresponds to the origin of the axes), player i chooses the next time τ_i^1 and plays the mixed strategy $(\frac{1}{2}, \frac{1}{2})$. That is to say, player i plays the two actions T or B with uniform probability. He plays this strategy all over the entire interval from τ_i^0 to τ_i^1. At time τ_i^1, player i receives an update on the past play of the opponent. Based on this new information, he selects time τ_i^2 and plays a new mixed strategy, say, for instance, $(1, 0)$. In other words, he plays T all over the interval from τ_i^1 to τ_i^2. At time τ_i^2, there is

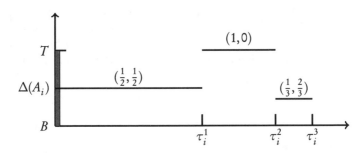

Figure 11.9. *Nonanticipative strategy for player i: $(\frac{1}{2}, \frac{1}{2})$ in the first interval, $(1, 0)$ in the second interval, and $(\frac{1}{3}, \frac{2}{3})$ in the third interval.*

new information available, and therefore player i selects time τ_i^3 and plays a new mixed strategy, say $(\frac{1}{3}, \frac{2}{3})$. This means that he plays T with probability $\frac{1}{3}$ and B with probability $\frac{2}{3}$. He keeps playing this strategy during the entire interval from τ_i^2 to τ_i^3 and so forth.

To complete the model, let us consider the payoff at time t, denoted by g_t, resulting from the mixed actions of the players. The integral describes the cumulative payoff and is given by $x(t) = \int_{\tau=0}^{t} g_\tau$(mixed action pairs at time τ)$d\tau$. With the above in mind, we are in a position to introduce a formal definition of *attainable set*.

Definition 11.6 (Attainable set). *A set A in $\mathbb{R}^d$ is attainable by player 1 if there exists a time $T > 0$ such that for every tolerance $\epsilon > 0$, there exists a strategy σ_1 for player 1 such that*

$$\text{dist}(x(t)[\sigma_1, \sigma_2], A) \leq \epsilon \quad \forall t \geq T, \forall \sigma_2.$$

A geometric illustration of an attainable set is in Fig. 11.10.

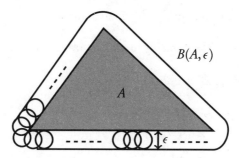

Figure 11.10. *Epsilon ball of attainable set A.*

There, we have a set A, which is the set that player 1 wishes to *attain*. The set $B(A, \epsilon)$ is the set of points whose distance from the set A does not exceed the tolerance ϵ. This corresponds to saying that

$$B(A, \epsilon) := \{z : \text{dist}(z, A) \leq \epsilon\}.$$

After the formal definition of attainable set, in the next section we turn to investigating attainability conditions.

11.4.2 ▪ Main results on attainability

This section illustrates three main attainability conditions. The first one is a condition for the attainability of the origin in the space of the payoffs. The second result deals with attainability of a predefined point different from the origin in the space of the payoffs. Finally, the third result provides conditions for the attainability of any point in the space of the payoffs.

The above results are all centered around the notion of *projected game* and *value of the projected game*.

In particular, the above conditions require that the value of the projected game, which we denote by v_λ, be bounded in sign. That is to say that it must hold that $v_\lambda > 0$ or alternatively $v_\lambda \geq 0$. To see this, consider the game with vector payoffs described by the bimatrix in Fig. 11.11(left). In the matrix the symbol $\langle \cdot, \cdot \rangle$ indicates the inner product in $\mathbb{R}^d$. Having a game with vector payoffs, one can construct a matrix game by simple premultiplication of the entries by a given vector $\lambda \in \mathbb{R}^d$. Let us think of λ as a specific direction in the space of the payoffs. By doing this, one obtains the matrix game on the right. The matrix game represents now a two-player zero-sum game, and for it we can calculate the equilibrium payoff v_λ. Recall that such an equilibrium payoff is the *value* of the game. Obviously, such a *value* is a function of the direction λ. We highlight this by adding the index λ. With in mind the definition of *value of the projected game* mentioned above, and recalling that we can project the game along any direction λ, we can establish the following main theorems.

$$\begin{pmatrix} (\#,\#) & (\#,\#) \\ (\#,\#) & (\#,\#) \end{pmatrix} \Rightarrow \begin{pmatrix} \langle \lambda, (\#,\#) \rangle & \langle \lambda, (\#,\#) \rangle \\ \langle \lambda, (\#,\#) \rangle & \langle \lambda, (\#,\#) \rangle \end{pmatrix}.$$

Figure 11.11. *Game with vector payoffs (left) and its projected game (right).*

The first theorem deals with attainability of the origin $\vec{0}$ in $\mathbb{R}^d$.

Theorem 11.7. *The following conditions are equivalent.*

B1 *vector $\vec{0} \in \mathbb{R}^d$ is attainable by player 1;*

B2 $v_\lambda \geq 0$ *for every $\lambda \in \mathbb{R}^d$.*

Proof (Sketch). The first part of the proof shows that for the origin $\vec{0} \in \mathbb{R}^d$ attainability and approachability are equivalent. In other words, $\vec{0} \in \mathbb{R}^d$ is attainable, namely condition *B1* holds, if and only if the same vector is approachable. We cannot stress enough that such an equivalence is true only for the origin. The second part of the proof makes use of *Blackwell's Approachability Principle*. Actually, from Blackwell's principle, if $\vec{0} \in \mathbb{R}^d$ is approachable, then condition *B2* holds true and vice versa. Consequently, we have the equivalence between conditions *B1* and *B2*.   □

We can interpret the result mentioned above by saying that for the $\vec{0}$ in $\mathbb{R}^d$ to be attainable, the *value of the projected game* along any direction $\lambda \in \mathbb{R}^d$ must be bounded in sign.

Note that the first condition, referred to as condition *B1*, recalls in spirit *Blackwell's Approachability Principle*. This should become clearer by looking at the graph displayed in Fig. 11.12. The graph depicts the above condition when the attainable set coincides with the singleton $\vec{0}$ in $\mathbb{R}^d$. At time τ_1^k, let the cumulative payoff $x(\tau_1^k)$ be given. Let us project $x(\tau_1^k)$ on $\vec{0}$ in $\mathbb{R}^d$. By doing this we find the direction $\lambda = -\frac{1}{\|x(\tau_1^k)\|} x(\tau_1^k)$. Consider any feasible payoff in the set $R_1(p)$. Recall that $R_1(p)$ is the set of payoffs when player 1 plays the mixed action p and for all possible actions of player 2. If the inner product between the payoff and the direction λ turns to be nonnegative, then we know that the payoff and λ are confined within the nonpositive half-space H^-.

The above condition is a necessary condition also if we are interested in the attainability of a given point in the space of payoffs which is not the origin $\vec{0}$ in $\mathbb{R}^d$. This is established in the following theorem.

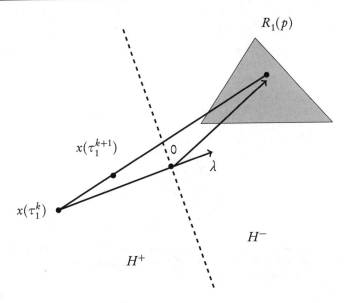

Figure 11.12. *Geometric illustration of the condition for the attainability of $\vec{0}$.*

Theorem 11.8. *Vector $z \in \mathbb{R}^d$ ($\neq \vec{0}$) is attainable by player 1 $\Leftrightarrow$.*

B1 *the vector $\vec{0} \in \mathbb{R}^d$ is attainable by player 1;*

B3 *for every function $f : \Delta(A_1) \to \Delta(A_2)$, vector z is in*

$$Cone(f) := \left\{ y \in \mathbb{R}^d \mid y = \sum_{p \in A_1} \alpha_p g(p, f(p)) : \alpha_p \geq 0 \, \forall p \right\}.$$

Proof **(Sketch).** To understand the role of condition **B3** for the attainability of z, assume that player 1 plays the mixed strategy $p \in \Delta(\{T, B\})$. Here $\Delta(\{T, B\})$ denotes the set of probability distributions over the set of pure actions $\{T, B\}$. Furthermore, assume that player 2 responds with the strategy $f(p)$. Consequently, the payoff $x(\tau_1^1)$ lives in the segment ab, as depicted in Fig. 11.13.

If player 1 plays B all over the interval $0 \leq t \leq \tau_1^1$, the payoff $x(\tau_1^1)$ is the extreme point of the segment in boldface in Fig. 11.14. From this we understand that given all feasible mixed strategies in the second interval the payoff $x(\tau_1^2)$ lies on the segment cd.

Now, assume that player 1 switches to strategy T in the interval $\tau_1^1 \leq t \leq \tau_1^2$. As a consequence, the payoff $x(\tau_1^2)$ coincides with the extreme point of the new segment in boldface in Fig. 11.15. Whatever the mixed strategy of player 2 will be, in the third interval the payoff $x(\tau_1^3)$ lies on the segment ef.

At time τ_1^3, under the assumption that player 1 opts for a mixed strategy $(\frac{1}{2}, \frac{1}{2})$ all over the interval $\tau_1^2 \leq t \leq \tau_1^3$, the corresponding payoff $x(\tau_1^3)$ coincides with the extreme point of the third segment in boldface in Fig. 11.16.

Given that $x(\tau_1^3)$ is approximately close to the point we wish to attain, namely point z, we infer that the cumulative payoff in the interval τ_1^3 to ∞ must necessarily be close to zero. Obviously, this is possible only if $\vec{0}$ is attainable.

It is worth noting that we have constructed the trajectory in such a way that explains also the reason why the feasible trajectories $\{x(\tau_1^k)\}_{k=0,\dots,\infty}$ are contained in the $Cone(f)$. $\quad\square$

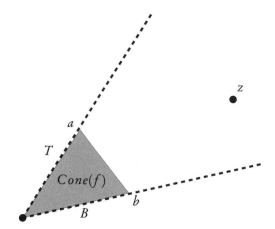

Figure 11.13. *Geometric illustration of Theorem 11.8. Player 1 selects a time τ_1^1 and plays any mixed action p in the set $\Delta(\{T,B\})$. Player 2 plays $f(p)$. At time τ_1^1 the cumulative payoff is a point in the segment ab.*

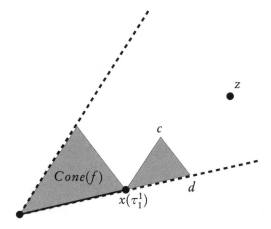

Figure 11.14. *Geometric illustration of Theorem 11.8. Assume that player 1 plays $p(t) = B$ in the interval $[0, \tau_1^1]$. At time τ_1^1 the cumulative payoff $x(\tau_1^1)$ coincides with the extreme point b of the segment ab. Then player 1 selects a new time τ_1^2 and the corresponding $x(\tau_1^2)$ lies on segment cd.*

We conclude this section by stressing that when the value of the projected game is strictly positive for every direction $\lambda \in \mathbb{R}^d$, any vector in the space of payoffs is attainable.

Theorem 11.9. *The following statements are equivalent:*

C1 $v_\lambda > 0$ *for every* $\lambda \in \mathbb{R}^d$;

C2 *every vector $z \in \mathbb{R}^d$ is attainable by player 1.*

Proof (Sketch). One way to conduct this proof is by showing that condition *C1* is equivalent to the existence of a Lyapunov function in the space of the payoffs. Actually, let us take as a candidate Lyapunov function the distance of the current cumulative payoff from the attainable point $z \in \mathbb{R}^d$. It can be proven that condition *C1* holds if the derivative of such a function is strictly negative and vice versa. This corresponds to saying that the distance is a Lyapunov function, and therefore it tends to decrease monotonically to zero.   ☐

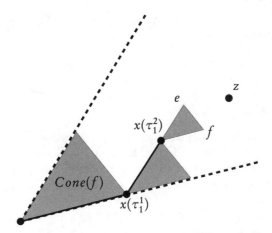

Figure 11.15. *Geometric illustration of Theorem 11.8. Assume that player 1 plays $p(t) = T$ in the interval $[\tau_1^1, \tau_1^2]$. At time τ_1^2 the cumulative payoff $x(\tau_1^2)$ coincides with the extreme point c of the segment cd. Then player 1 selects a new time τ_1^3 and the corresponding $x(\tau_1^3)$ lies on segment ef.*

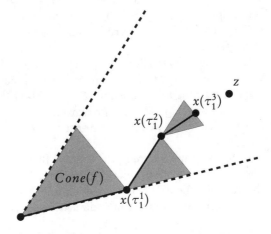

Figure 11.16. *Geometric illustration of Theorem 11.8. At time τ_1^3, under the assumption that player 1 plays the mixed strategy $(\frac{1}{2}, \frac{1}{2})$ all over the interval $\tau_1^2 \le t \le \tau_1^3$, the corresponding payoff $x(\tau_1^3)$ coincides with the extreme point of the third segment in boldface.*

11.5 ▪ Conclusions and future directions

This chapter has developed the theory of repeated games with vector payoffs. After introducing *approachability* and *attainability*, the latter being a new concept developed in [40, 155], we have surveyed the main results available in the literature.

Future directions involve in order the following:

- The study of continuous-time attainability from the point of view of differential game theory [229]. Specifically, one could start from the formal definition of attainability for a continuous-time repeated game available in [40, 155].

- The study of analogies between a main attainability condition (positiveness of the value of any projected game) and the subtangentiality conditions characterizing discriminating sets in viability theory [11, 71], set-valued analysis [12, 13], and set invariance theory [58].

- The investigation of the main attainability condition in connection with the robust stabilizability conditions derived in network flow control [33, 27, 60, 59]. We exploit the analogy with network flow control to characterize attainable sets and associated strategies.

- The study of attainability with infinite horizon discounted payoffs to show that the main condition derived for the undiscounted case (positiveness of the value of any projected game) no longer implies that every point is attainable. Indeed, attainability can be guaranteed only for a small neighborhood of the initial payoff value.

11.6 ▪ Notes and references

The rudiments of *Approachability Theory* are in the seminal work by Blackwell in the 1950s [57]. A detailed description of *Blackwell's Approachability Theorem* and its use in *prediction* and *learning* is also available in the book by Cesa-Bianchi and Lugosi [73, Chap. 7.7]. Examples 11.2 and Fig. 11.3 are borrowed from the book by Mashler, Solan, and Zamir [173]. The topic shares striking similarities with Lyapunov stability (see the Appendix, Chapter C).

Approachability applied to allocation processes in coalitional games is the main focus of [150]. Approachability and regret minimization are discussed in [152, 111]. Approachability in adaptive learning is examined in [73, 98, 109, 110]. Excludability and bounded recall are illustrated in [153]. Weak approachability is presented in [243]. For further reading on approachable sets we refer the reader to [57, 121, 151, 154, 155, 156, 208, 229, 230].

Despite its discrete-time nature in the original Blackwell formulation, approachability has also been extended to continuous-time repeated games, thus showing common elements with Lyapunov theory [111]. Though formalized in a finite-dimensional space, a definition of approachability in an infinite-dimensional space was first proposed by Lehrer in [151].

Approachability can be reframed within differential games and as such can be studied using differential calculus and stability theory [156, 229]. In particular, in [156] the authors show that, beyond the approachability principle being an extension (to a vector space) of the von Neumann minimax theorem, it also has elements in common with differential inclusion [12]. In addition to this, [229] establishes connections with viability theory [11], set-valued analysis [13] (see the comparison of an approachable set with a discriminating set), and set invariance theory [58].

Still within the realm of differential games, it is worth noting that the notion of nonanticipative behavior strategy has a long history [16, 91, 208, 229, 242]. Actually, it turns out that classical feedback strategies in differential games are special nonanticipative strategies.

More recently, approachability conditions have been reviewed and extended to the case where the quantity to regulate is not the average but the cumulative payoff. A new term has been coined to address such a scenario: *attainability* [40, 155].

In [40, 155], attainability conditions are studied for two-player continuous-time repeated games with vector payoffs. The authors show that attainability arises in several application domains, including transportation, distribution, and production networks. For a formal proof of Theorem 11.8 we refer the reader to [40].

Chapter 12

Mean-Field Games

12.1 ▪ Introduction

This chapter provides an overview of the theory of games with many negligible agents. The theory was first developed within the area of Engineering Mathematics, but it has recently attracted the attention of econophysists and sociophysists (see Fig. 12.1). After presenting the main setup we shall discuss a few stylized examples borrowed from [105]. The last part skims through some available results on existence and uniqueness of solutions, linear-quadratic mean-field games, and robustness.

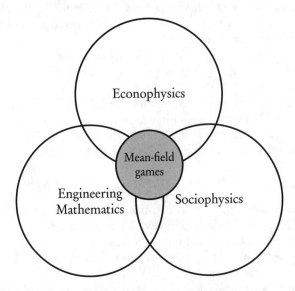

Figure 12.1. *Mean-field games were first formulated within the area of Engineering Mathematics, but the topic shows overlaps with econophysics and sociophysics.*

Section 12.2 introduces first- and second-order mean-field games and highlights different formulations involving finite and infinite horizon cost functionals.

Section 12.3 explores seminal results on existence and uniqueness conditions. Section 12.4 provides examples. Section 12.5 introduces robust mean-field games, examines a general solution, and discusses the new equilibrium concept of robust mean-field equilibrium. Section 12.6 provides conclusions and points out open problems. Finally, Section 12.7 provides notes and references on the topic.

12.2 • Formulating mean-field games

In this section, we shall start with formulating first- and second-order mean-field games. Then we consider finite and infinite horizon models.

12.2.1 • First-order mean-field game

In a mean-field game we have N *homogeneous players*, and we let $N \to \infty$. The term *homogeneous players* means that all the players who share the same state $x \in \mathbb{R}^n$ behave exactly in the same way. That is to say, these players play the same *state-feedback strategy*, denoted by $u(x(t), t)$. Suppose that the state dynamics is given by the first-order differential equation

$$\dot{x}(t) = u(x(t), t), \quad x(0) \in \mathbb{R}^n. \tag{12.1}$$

It is worth noting that, as the right-hand side is a function of x, the state dynamics (12.1) defines a vector field in $\mathbb{R}^n$. To put dynamics (12.1) in context, imagine that the players represent particles of salt which flow on the bed of a river. Then, the state space is the bidimensional Euclidean space, the variable $x(t) \in \mathbb{R}^2$ denotes the position of the particle at a given time t, and $u(x(t), t)$ is the speed of the particle. Needless to say, we can play with our imagination and see the state variable as any abstract entity, such as an opinion in the space of opinions, or the characteristic of an individual (aggressive or nonaggressive) in the space of social behaviors (think, for instance, of the *Hawk and Dove game* in an evolutionary context). The vector field is then a description of how opinions or individual behaviors progress over time.

Let us think again of the salt particles, and let us describe the concentration of particles in a point x at a time t. This requires the use of a density function, denoted by $m(x, t)$, which depends on both space x and time t. From *calculus* and in particular from the definition itself of *divergence operator*, we know that if a scalar function is immersed in a vector field, the time evolution of the scalar function follows the so-called *advection equation*. For a generic n-dimensional vector variable x and a finite horizon $[0, T]$, such an equation is a partial differential equation of the form

$$\partial_t m(x, t) + \operatorname{div}(m(x, t) \cdot u(x, t)) = 0 \quad \text{in } \mathbb{R}^n \times [0, T]. \tag{12.2}$$

The above partial differential equation, which is also referred to as *transport equation*, is essentially a mass conservation law. This law states that if we take the partial derivative of the density with respect to time, the result must be equal to the divergence of the scalar function $m(x, t)$ subjected to the vector field $u(x, t)$. Fig. 12.2 explains the nature of this equation as a mass conservation law. Let us freeze time t and look at point x. If point x is a *source*, then the vector field $u(x, t)$ describes an outgoing flow from x. Recall that the divergence operator yields the flow traversing the spherical surface surrounding point x when, in the limit, the radius of the sphere tends asymptotically to zero. In this context, a mass flow which departs from point x is equivalent to a divergence term $\operatorname{div}(m(x, t) \cdot u(x, t)) > 0$. To counterbalance such a positive term, the first term in (12.2), namely $\partial_t m(x, t)$, is negative. That is to say that the concentration of particles diminishes with time. Conversely, if point x is a *sink*, the second term in (12.2), namely $\operatorname{div}(m(x, t) \cdot u(x, t)) < 0$, and the partial derivative $\partial_t m(x, t)$ must be positive. This means that more and more particles accumulate in point x.

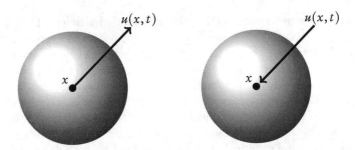

Figure 12.2. *Physical interpretation of the divergence operator used in the advection equation. If the divergence is positive, point x is a source (left); if the divergence is negative, point x is a sink (right).*

After explaining the physical interpretation of the advection equation in (12.2), let us assume that the particles are *rational*. As such they select their velocities $u(x, t)$ in order to minimize a cost functional of the form

$$\int_0^T \left[\underbrace{\frac{1}{2}|u(x(t), t)|^2}_{\text{penalty on control}} + \underbrace{g(x(t), m(\cdot, t))}_{\text{...on state \& distribution}} \right] dt + \underbrace{G(x(T), m(\cdot, T))}_{\text{...on final state}}.$$

Remarkably, the cost functional introduced above presents the same structure of a classical optimal control functional except for the density function $m(.)$ appearing in the integrand and in the terminal penalty. In particular, the first term $\frac{1}{2}|u(x(t), t)|^2$ is a penalty term on control, representing the energy necessary to control the particle. The second term $g(x(t), m(\cdot, t))$ is the running cost and represents a penalty term depending on the current state and distribution. Finally, the third term $G(x(T), m(\cdot, T))$ is the terminal penalty.

From the theory of optimal control, it turns out that the optimal state-feedback control is along the anti-gradient of a well-known function $v(.)$, namely

$$u(x(t), t) = -\nabla_x v(x(t), t). \qquad (12.3)$$

Function $v(.)$ is known in the literature as the *value function*. The value function is nothing but the minimum cost achievable, and such a function will intuitively depend on the initial position of the player. The value function $v(.)$ is the solution of the *Hamilton–Jacobi–Bellman (HJB) equation* (cf. Section 9.2.3). We shall reiterate later how to derive the HJB equation.

In summary, the model of a mean-field game takes the form of two coupled partial differential equations in $\mathbb{R}^n \times [0, T]$:

$$-\partial_t v(x, t) + \frac{1}{2}|\nabla_x v(x, t)|^2 = g(x, m(.)) \qquad \text{(HJB)—backwards}$$

$$u(.) \Big\downarrow \qquad\qquad\qquad \Big\uparrow m(.)$$

$$\partial_t m(x, t) + \text{div}(m(x, t) \cdot u(x, t)) = 0 \qquad \text{(advection)—forwards}$$

The two coupled partial differential equations mentioned above have to be solved imposing the boundary conditions at time 0 for the density function, i.e., $m(\cdot, 0) = m_0$, and at time T for the value function $v(x, T) = G(x, m(\cdot, T))$.

This corresponds to saying that the HJB equation must be solved using dynamic programming backwards. In this equation we can see the density $m(.)$ as a parameter and the *value function* $v(.)$ as the variable. Once we obtain the value function, we get the

optimal control $u(.)$ from (12.3). Put differently, solving the HJB equation means finding
the best response of a single player $u(.)$ to the population behavior, the latter captured by
the density $m(.)$. We say that $u(.)$ is a best response, as it is obtained as optimal control
for a given assumption on $m(.)$; therefore, $u(.)$ is the best response to $m(.)$.

Analogously, in the advection equation, we can interpret the best response $u(.)$ as a
parameter and the density $m(.)$ as the variable. This equation describes the evolution of
the population as a whole under the assumption that all players are rational.

Solving a mean-field game as the one mentioned above means to study existence and
uniqueness and eventually to compute a *fixed point*. The computation may be carried
out iteratively as follows. Let us first assume a given density $m(.)$. Based on the given
$m(.)$, let us solve the HJB equation to obtain a best response $u(.)$. Let us substitute $u(.)$
in the advection equation and compute the density $m(.)$. At a fixed point such a density
coincides with the one we had used in the HJB equation at the beginning of the last cycle.
If a fixed point exists, it is called *mean-field equilibrium*. Remarkably, this equilibrium
is the asymptotic solution of a Nash equilibrium when we take the number of players
tending to infinity.

Let us now turn our attention to the computation of the HJB equation for the example
at hand. The derivation of the aforementioned equation consists in the following steps.

- In step 1, we shall consider the *Bellman Principle* (cf. Section 9.2.3). That is to say
  that the value function $v(x,t)$, which represents today's cost, can be decomposed
  as the sum of a stage cost, denoted by $\min_u [\frac{1}{2}|u|^2 + g(x,m(.))]$, and a future cost-
  to-go, denoted by $v(x+dx,t+dt)$, depending on the future state $x+dx$, which
  we reach by applying the optimal u. In other words we get

$$\underbrace{v(x,t)}_{\text{today's cost}} = \underbrace{\min_u [\frac{1}{2}|u|^2 + g(x,m(.))]dt}_{\text{stage cost}} + \underbrace{v(x+dx,t+dt)}_{\text{future cost}}.$$

- In step 2, we shall perform the Taylor expansion of the future cost. By doing this
  we obtain that

$$v(x+dx,t+dt) = v(x,t) + \partial_t v(x,t)dt + \nabla_x v(x,t)\dot{x}dt.$$

- In step 3, we shall set the gradient of a convex function equal to zero, as it is typical to find the minimizer of a convex function. Actually, after computing for the
  Hamiltonian we have

$$\min_u \underbrace{[\frac{1}{2}|u|^2 + g(x,m(.)) + \partial_t v(x,t) + \nabla_x v(x,t)\overbrace{\dot{x}}^{u}]}_{\text{Hamiltonian}} = 0,$$

where we have dropped the index 0. We note that the optimal control is given by
$u = -\nabla_x v(x,t)$, which in turn yields

$$-\partial_t v(x,t) + \frac{1}{2}|\nabla_x v(x,t)|^2 = g(x,m(.)) \quad \text{(HJB)}.$$

12.2.2 • Second-order mean-field game and chaos

If the particles evolve in a chaotic way, then the state dynamics can be described by a
stochastic differential equation of type

$$dx(t) = u(x(t),t)dt + \sigma dB(t),$$

where $dB(t)$ is the infinitesimal Brownian motion.

As for the deterministic case introduced in the previous sections, the corresponding mean-field game model involves two coupled partial differential equations. The difference is that now we have second-order derivatives of the value function $v(.)$ and of the density $m(.)$ appearing in the equations as displayed below:

$$-\partial_t v(x,t) + \tfrac{1}{2}|\nabla_x v(x,t)|^2 - \tfrac{\sigma^2}{2}\Delta v(x,t) = g(x,m(.)) \qquad \text{(HJB)—backwards}$$

$$u(.)\Bigg\downarrow \qquad\qquad\qquad \Bigg\uparrow m(.)$$

$$\partial_t m(x,t) + \mathrm{div}(m(x,t)\cdot u(x,t)) - \tfrac{\sigma^2}{2}\Delta m(x,t) = 0. \qquad \text{(KFP)—forwards}$$

In the above equations, Δ is the *Laplacian operator*, which is given by

$$\Delta = \sum_{i=1}^{n} \frac{\partial^2}{\partial x_i^2}.$$

Consequently, the above model is called second-order mean-field game. The advection equation is now replaced by the well-known *Kolmogorov–Fokker–Planck (KFP)* equation. This equation usually models diffusion processes and constitutes a fundamental in statistical mechanics.

12.2.3 ▪ Average and discounted infinite horizon formulations

Mean-field games can also be formulated as infinite horizon problems. In this case we have two alternative formulations. If the players are *patient* or *farsighted*, the formulation involves the average infinite horizon cost functional. Differently, if the players are *myopic* or *shortsighted*, the formulation involves a discounted cost functional. We elaborate on the two cases in order.

- **(Average cost)** First, in the case of *shortsighted* players, the cost functional is of the form

$$J = \mathbb{E}\limsup_{T\to\infty} \frac{1}{T}\int_0^T \Big[\tfrac{1}{2}|u(x)|^2 + g(x(t),m(\cdot,t))\Big]dt.$$

Then the mean-field game requires solving in $\mathbb{R}^n$ the system

$$\bar\lambda + \tfrac{1}{2}|\nabla_x \bar v|^2 - \tfrac{\sigma^2}{2}\Delta\bar v = g(x,\bar m) \qquad \text{(HJB)}$$

$$u\Bigg\downarrow \qquad\qquad\qquad \Bigg\uparrow \bar m$$

$$\mathrm{div}(\bar m \cdot u(x)) - \tfrac{\sigma^2}{2}\Delta\bar m = 0 \qquad \text{(KFP)}$$

The formulations mentioned above are such that instantaneous fluctuations of the cost are meaningless. The importance is entirely on the long-term average cost. Note that the problem has the same structure as the other mean-field game formulations with the only difference that now we consider the average stage cost $\bar\lambda$, the long-run average value function $\bar v(.)$, and the long-run average density function $\bar m(.)$.

- **(Discounted cost)** In case of *farsighted* players, the cost functional involves a discount factor as illustrated below:

$$J = \mathbb{E} \int_0^\infty e^{-\rho t} \left[\frac{1}{2} |u(x(t), t)|^2 + g(x(t), m(\cdot, t)) \right] dt.$$

Solving the above mean-field game means to find a fixed point in $\mathbb{R}^n \times [0, T]$ of the following two partial differential equations:

$$-\partial_t v(x, t) + \frac{1}{2} |\nabla_x v(x, t)|^2 - \frac{\sigma^2}{2} \Delta v(x, t) + \rho v = g(x, m(.)) \qquad \text{(HJB)}$$

$$u(.) \Big\downarrow \qquad\qquad\qquad\qquad\qquad \Big\uparrow m(.)$$

$$\partial_t m(x, t) + \mathrm{div}(m(x, t) \cdot u(x, t)) - \frac{\sigma^2}{2} \Delta m(x, t) = 0 \qquad \text{(KFP)}$$

12.3 • Existence and uniqueness

The formulation of mean-field games in the seminal paper by Lasry and Lions [149] is accompanied by some results on existence and uniqueness of mean-field equilibria. These results are also discussed in the lecture notes by Cardaliaguet taken during a course given by Lions at the College de France [70]. For the existence of a solution, we generally refer to the assumptions enumerated below (see Theorem 3.1 in [70]):

- *uniformly boundedness* of running and terminal cost in the space of states and distribution;

- *Lipschitz continuity* of running and terminal cost in the space of states and distribution;

- *absolute continuity* of the initial probability measure with respect to the *Lebesgue measure*.

Under the aforementioned conditions, we have guarantees that the value function and the distribution are "regular." From the third condition, we can also exclude the concentration of masses in specific points. In other words, the distribution cannot have *Dirac impulses*. Remarkably, under the above conditions we have guarantees that a *solution* exists in the *classical sense*. On the contrary, proving existence of *weak* solutions is still a challenging problem. Let us now turn to consider uniqueness conditions.

Uniqueness of a solution is shown to depend on the monotonicity of the cost (see Theorem 3.6 in [70]). Actually, the running cost must satisfy the condition

$$\int_{\mathbb{R}^d} \Big(g(x, m_1) - g(x, m_2) \Big) d(m_1 - m_2)(x) > 0 \quad \forall m_1, m_2 \in \mathscr{P}, \quad m_1 \neq m_2.$$

Likewise, for the terminal penalty it must hold that

$$\int_{\mathbb{R}^d} \Big(G(x, m_1) - G(x, m_2) \Big) d(m_1 - m_2)(x) > 0 \quad \forall m_1, m_2 \in \mathscr{P}.$$

In the above condition, we denote by $\mathscr{P}$ the space of probability distributions. The above inequalities essentially describe situations where a higher density of particles at a given point yields a higher cost for the particles. We use the term *crowd aversion* to mean such a scenario. Crowd aversion is a characteristic of several transportation or pedestrian flow problems [146].

In analogy with the theory of differential games, if the problem is linear-quadratic, then we can compute explicitly the equilibrium strategies. More details on linear-quadratic mean-field games and explicit solutions can be found in the work by Bardi [20].

12.4 • Examples

In this section we develop some examples taken from [105]. These examples are stylized models capable of explaining the generality of the theory and its versatility. These models intersect social science, economics, and production engineering.

Example 12.1 (Mexican wave). This model describes phenomena like *mimicry* and *emulation*. The game has a state, denoted by $x = [y, z]$, where the first component $y \in [0, L)$ represents the horizontal coordinate, and the second component z represents the vertical position, which we henceforth call *posture*. Consider a continuum of players distributed over the interval $[0, L)$, as illustrated in Fig. 12.3. The horizontal coordinate of every player is fixed. The posture lives in the interval from 0 to 1. That is to say,

$$z = \begin{cases} 1 & \text{standing} \\ 0 & \text{seated} \end{cases}, \quad z \in (0, 1) \quad \text{intermediate.}$$

The posture varies in consequence of the input u selected by the players. The input establishes the rate of change of the posture. This is described by the first-order differential equation

$$dz(t) = u(z(t), t)dt.$$

The control u is the variable that the players have to optimize. To obtain the well-known *Mexican wave*, let us introduce a penalty on state and distribution given by

$$g(x, m) = \underbrace{Kz^\alpha(1-z)^\beta}_{\text{comfort}} + \underbrace{\frac{1}{\epsilon^2} \int (z - \tilde{z})^2 m(\tilde{y}; t, \tilde{z}) \frac{1}{\epsilon} s\left(\frac{y - \tilde{y}}{\epsilon}\right) d\tilde{z} d\tilde{y}}_{\text{mimicry}},$$

where K, α, β, and ϵ are given parameters. Note that the above cost includes two terms. The first one accounts for the *comfort* of the player. Evidently, the comfort is maximal at the two extreme values of z, namely $z = 0$, which means that the player is seated, and $z = 1$, which means that the player is standing.

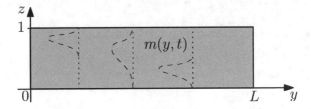

Figure 12.3. *Mexican wave: probability that player in position y takes on posture z.*

To see this, note that the term $z^\alpha(1-z)^\beta$ is concave and that it is null for $z = 0$ and $z = 1$. The second term accounts for the *mimicry*. This term considers the square deviation $(z - \tilde{z})^2$ between the posture of the player and the posture of his neighbors. The penalty decreases with the distance of the player from his neighbor. Actually, note that the term $\frac{1}{\epsilon}s(\frac{y-\tilde{y}}{\epsilon})$ is a Gaussian kernel. This is equivalent to saying that "far neighbors" are less influential than "close neighbors." The penalty is also weighted by the probability $m(\tilde{y}; t, \tilde{z})$. The latter represents the probability that a given neighbor is in a specific posture at a given time.   ∎

Example 12.2 (Meeting starting time). This second example deals with a model of *coordination* among players in response to an *externality*. By externality we mean an exogenous input. Assume that a meeting is scheduled at time t_s. The meeting takes place in a meeting room which is located in the origin of the horizontal axis in Fig. 12.4. Consider a continuum of players who are initially distributed over the negative axis. Players have to choose their speeds when walking to the meeting room. Their speeds depend on their expectations about the time when the meeting will actually start. Suppose the following quorum rule: The meeting starts when θ percent of the participants have reached the room. Thus, θ represents the *quorum*. Evidently, the optimal speed u for a single player depends on the model he uses to predict the other players' behaviors. If the other players are expected to be punctual, then he will have to speed up. Differently, if the other players are expected to be late, then he will slow down. The state evolution of a single player is given by

$$dx(t) = u(x(t), t)dt + \sigma dB(t).$$

In the above dynamics, we use a Brownian motion to introduce a stochastic disturbance in the way in which players approach the meeting room. Let $\tilde{\tau} = \min_s(x(s) = 0)$ be given, which represents the arrival time of the player. Let us also denote by $\bar{t}$ the time when the meeting will actually start, which is in general different from the scheduled time. Note that the actual starting time is a variable of the problem, as it depends on the population behavior.

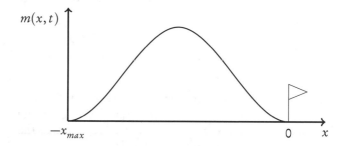

Figure 12.4. *Coordination under externality: the* meeting starting time *example.*

Furthermore, consider a terminal penalty given by

$$G(x(\tilde{\tau}), m(\cdot, \tau)) = \underbrace{c_1[\tilde{\tau} - t_s]_+}_{\text{reputation}} + \underbrace{c_2[\tilde{\tau} - \bar{t}]_+}_{\text{inconvenience}} + \underbrace{c_3[\bar{t} - \tilde{\tau}]_+}_{\text{waiting}},$$

where c_1, c_2, and c_3 are given parameters. Actually, the cost mentioned above shows three different contributions. The first term describes the cost of a bad reputation incurred for arriving after the scheduled time. The second contribution accounts for the inconvenience of arriving late with respect to the actual starting time. Finally, the third contribution is the cost paid if one arrives before the actual starting time and has to wait for the other players to arrive.

After introducing the model, for any time s we can compute how many players are already in the meeting room. This quantity is obtained from the following equation:

$$F(s) = -\int_0^s \partial_x m(0, v) dv.$$

Consequently, the actual starting time is the inverse function, namely

$$\bar{t} = F^{-1}(\theta). \quad \blacksquare$$

Example 12.3 (Herd behavior). In this example we present a model that describes *herd behavior* in social science. To do this, let x be given, which describes the behavior of a single player. For instance, the behavior of the player can describe his political opinion, his social behavior, or his innovation openness. Let us suppose that such a behavior evolves according to the following stochastic differential equation:

$$dx(t) = u(x(t), t) dt + \sigma dB(t).$$

The typical herd behavior arises when we set the running cost as

$$g(x, m) = \beta \left(x - \underbrace{\int y m(y, t) dy}_{\text{average}} \right)^2.$$

The above running cost involves the square difference between the behavior of the player and the average behavior of the individuals in the population. We can use a discounted infinite horizon formulation as the one introduced in the previous section, in which case we have

$$J = \mathbb{E} \int_0^\infty e^{-\rho t} \left[\frac{1}{2} |u(x(t), t)|^2 + g(x(t), m(\cdot, t)) \right] dt.$$

We then arrive at the following mean-field game:

$$-\partial_t v(x, t) + \frac{1}{2} |\nabla_x v(x, t)|^2 - \frac{\sigma^2}{2} \Delta v(x, t) + \rho v(x, t) = g(x, m(.)) \qquad \text{(HJB)}$$

$$u(.) \Bigg\downarrow \qquad\qquad\qquad\qquad \Bigg\uparrow m(.)$$

$$\partial_t m(x, t) + \text{div}(m(x, t) \cdot u(x, t)) - \frac{\sigma^2}{2} \Delta m(x, t) = 0 \qquad \text{(KFP)} \quad \blacksquare$$

Example 12.4 (Oil production). This example deals with a continuum of oil producers. Every producer has an initial reserve of raw material. To model the stock market, we can use the geometric Brownian motion given by

$$dx(t) = [\alpha x(t) + \beta u(x(t), t)] dt + \sigma x(t) d\mathcal{B}(t),$$

where $\beta u(t)$ is the produced quantity. The running cost involves the production costs and the total income, the latter with a negative sign. The cost takes the form

$$g(x, u, m) = -h(\bar{m}) u + \left[\frac{a}{2} u^2 + b u \right],$$

where $h(\bar{m})$ is the sale price of oil. It is reasonable to assume that the sale price decreases in $\bar{m}$. That is to say, the higher the average stock still available among the producers, the lower the current and future sale prices. Furthermore, the terms $[\frac{a}{2}u^2 + bu]$ are quadratic and linear production costs. The terminal penalty penalizes the unexploited reserve at the end of the horizon:

$$G(x(T)) = \phi|x(T)|^2, \quad \phi > 0. \quad \blacksquare$$

12.5 ▪ Robust mean-field games

We shall now consider robust mean-field games. First, we provide the model, and then we analyze a general solution for it which yields a new equilibrium concept called *robust mean-field equilibrium*. We discuss in more detail such a new equilibrium concept at the end of this section.

12.5.1 ▪ The model

Robustness is here related to the presence of a deterministic adversarial disturbance in addition to the classical stochastic disturbance given by the Brownian motion. *Adversarial* means that of all possible realizations, we will consider the worst-case one, in the same spirit as H^∞-optimal control. Fig. 12.5 depicts a classical block system setting up an H^∞-optimal control problem (cf. Section 2.4 and [22]). Here we have a plant G, a feedback controller K, a control u, a disturbance w, and controlled and measured outputs z and y. Inputs and outputs are all measurable in Hilbert spaces, denoted by $\mathcal{H}_u, \mathcal{H}_w, \mathcal{H}_z, \mathcal{H}_y$, respectively. A classical representation of the dynamics of the system is given by

$$\begin{cases} z = G_{11}(w) + G_{12}(u), \\ y = G_{21}(w) + G_{22}(u), \\ u = K(y). \end{cases} \tag{12.4}$$

Here we assume that both the operators G_{ij} and the controller $K \in \mathcal{K}$ are *bounded causal linear operators*, where we denote by $\mathcal{K}$ the controller space. Recall that *causal* means that all subsystems are *nonanticipative*, namely, the output may depend on past and current inputs but not on future inputs.

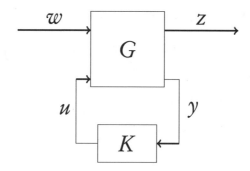

Figure 12.5. *Classical setup of H^∞-optimal control.*

A main issue in robust control is related to the capability of the controlled plant to attenuate the effects of the disturbance, which is called *disturbance attenuation*. Such a

problem can be converted into a zero-sum game between the controller and the disturbance. To see this, for every fixed $K \in \mathcal{K}$, introduce bounded causal linear operators
$T_K : \mathcal{H}_w \to \mathcal{H}_z$:

$$T_K(w) = G_{11}(w) + G_{12}(I - KG_{22})^{-1}(KG_{21})(w).$$

We then look for the worst-case infimum of the operator norm

$$\begin{cases} \inf_{K \in \mathcal{K}} \langle\langle T_K \rangle\rangle =: \gamma^*, \\ \langle\langle T_K \rangle\rangle = \sup_{w \in \mathcal{H}_w} \frac{\|T_K(w)\|_z}{\|w\|_w}. \end{cases} \tag{12.5}$$

This turns the problem into a two-person zero-sum game between the controller and
the disturbance given by

$$\overbrace{\inf_{K \in \mathcal{K}} \sup_{w \in \mathcal{H}_w} \frac{\|T_K(w)\|_z}{\|w\|_w}}^{\text{upper bound}} \geq \overbrace{\sup_{w \in \mathcal{H}_w} \inf_{K \in \mathcal{K}} \frac{\|T_K(w)\|_z}{\|w\|_w}}^{\text{lower bound}}.$$

To move from an H^∞-optimal control problem to a *robust mean-field game*, we need
to consider a large number of copies of the same plant asymptotically tending to infinity.
We then assume that the controlled output depends also on the probability distribution
of the states. This corresponds to saying that each plant plays "against" an adversarial
disturbance (as in H^∞-optimal control) and at the same time "against" the rest of the
population. Such a scenario is illustrated in Fig. 12.6. In the following, we provide a
mathematical formulation of a robust mean-field game.

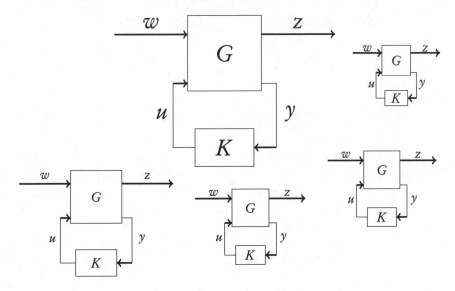

Figure 12.6. *Infinite copies of the plant: the controlled output depends also on the probability
distribution of states.*

Consider N copies of a same plant G. For each copy we have a first player, which is
essentially the controller selecting u, and a second player, namely the disturbance choosing w. Let us denote the set of players by $\mathcal{N} = \{1, \ldots, N\}$. We shall first formulate the

robust game for a finite but fixed N, and then we take $N \to \infty$. To distinguish the formulation with a finite N from the formulation in the asymptotic case $N \to \infty$, we add the index N to the variables in the former case. Given a finite horizon $[0, T]$, for every player $j \in \mathcal{N}$ the state, which we denote by $x_j^N(t) \in \mathbb{R}$, evolves in accordance with the following stochastic differential equation in $\mathbb{R}$:

$$dx_j^N(t) = [\alpha x_j^N(t) + \beta u_j^N(t)]dt + \sigma\left[x_j^N(t)d\mathcal{B}_j(t) + w_j^N(t)dt\right], \qquad (12.6)$$

where α, β, and σ are opportune parameters in $\mathbb{R}$:

- $\mathcal{B}_j(t)$, $t \geq 0$, is a standard Brownian motion; it is independent of the initial state $x_j^N(0)$ and independent across players; notation $\mathcal{B}_j$ is henceforth sporadically used to indicate the Brownian process over the interval $[0, T]$.

- $x_j^N(0)$ represents the initial state of player j. This is randomly extracted from the distribution $m_j^N(0)$. It is assumed that such a distribution converges almost surely to some distribution m_0 for $N \to \infty$ and independently of j.

- $u_j^N(t)$ denotes the control of player j at time t.

- $w_j^N(t) : [0, T] \to \mathbb{R}$ is the disturbance acting on player j at time t.

Dynamics (12.6) represents the dynamics of each player and therefore can be referred to as *microscopic dynamics*. In order to derive the corresponding *collective dynamics*, which we call *macroscopic dynamics*, let us introduce the *Dirac measure* δ, and let $m^N(t) = \frac{1}{N}\sum_{j=1}^N \delta_{x_j^N(t)}$ be the *empirical measure* of the states at time t. We occasionally use m^N to denote the empirical frequency for all time $t \in [0, T]$, i.e., $m^N := (m^N(t))_{t \in [0,T]}$.

After introducing the empirical frequency, consider the cost functional

$$J^N(x_j^N(0), u_j^N, m_j^N, w_j^N) = \mathbb{E}\left(g(x_j^N(T)) + \int_0^T c(x_j^N(t), u_j^N(t), m_j^N(t), t)dt \right.$$
$$\left. -\gamma^2 \int_0^T |w_j^N(t)|^2 dt \right),$$

where $m_j^N = (m_j^N(t))_{t\in[0,T]}$ and $m_j^N(t) := \frac{1}{N-1}\sum_{j'\neq j}\delta_{x_{j'}^N(t)}$, $g(.)$ is the terminal penalty, and $c(.)$ is the running cost.

We assume that each player j plays a control $u_j^N(t)$ which is adapted to the filtration generated by the initial state $x_j^N(0)$ and the Brownian motion $\mathcal{B}_j$, and possibly also adapted to some aggregate filtration associated with other players' dynamics. Specifically, let the following class of individual-and-aggregate state-feedback strategies be given: $u_j^N(t) = \mu_j(t, x_j^N(t), m^N(t))$.

Let us adapt the definition of feedback Nash equilibrium from [23] to the above class of strategies. To do this, ignore the disturbance $w_j(t)$ by setting $w_j^N(t) = 0$ for all $t \in [0, T]$.

Definition 12.5. *A* feedback Nash equilibrium *is a feedback strategy profile*

$$u_j^{N*}(t) = \mu_j^*(t, x_j^N(t), m^N(t)),$$

$j \in \mathcal{N}$, *such that no player has incentive to deviate, i.e., for all* $j \in \mathcal{N}$,

$$J^N(x_j^N(0), u_j^{N*}, m_j^{N*}, 0) = \inf_{\{u_j^N(t)\}_t} J^N(x_j^N(0), u_j^N, m_j^{N*}, 0),$$

where the dynamics of $x_j^N(t)$ are given by

$$dx_j^N(t) = \left[\alpha x_j^N(t) + \beta u_j^N(t)\right] dt + \sigma x_j^N(t) d\mathscr{B}_j(t), \ t \in (0,T], \ x_0 \in \mathbb{R}. \quad (12.7)$$

In the above equation, $m_j^{N*} = (m_j^{N*}(t))_{t\in[0,T]}$, and $m_j^{N*}(t)$ is the empirical measure $\frac{1}{N}\delta_{x_j^N(t)} +$ $\frac{N-1}{N}\frac{1}{N-1}\sum_{j'\neq j}\delta_{y_{j'}^N(t)}$, where $y_{j'}^N(t)$ is the optimal state trajectory for player j', i.e., the state trajectory generated by the feedback best-response control of player j'.

We can modify the above definition in order to consider the worst-case disturbance w_j^N. This leads to the notion of *worst-case disturbance feedback Nash equilibrium*. The problem of computing such an equilibrium is formally stated in the following robust stochastic differential game problem (cf. [21]).

Problem 12.1 (Robust stochastic game). *Let $\mathscr{B}$ be a one-dimensional Brownian motion process defined on $(\Omega, \mathscr{F}, \mathbb{P})$, where $\mathscr{F}$ is the natural filtration generated by $\mathscr{B}$, and $x_j^N(0)$ be any random variable independent of $\mathscr{B}$ having distribution $m_0(x)$. Consider*

$$\inf_{\{u_j^N(t)\}_t} \sup_{\{w_j^N(t)\}_t} J^N(x_j^N(0), u_j^N, m_j^{N*}, w_j^N),$$

where the dynamics of $x_j^N(t)$ are given by

$$dx_j^N(t) = \left[\alpha x_j^N(t) + \beta u_j^N(t) + \sigma w_j^N(t)\right] dt$$
$$+ \sigma x_j^N(t) d\mathscr{B}_j(t), \ t \in (0,T], \ x(0) \in \mathbb{R}. \quad (12.8)$$

In the above equation, $m_j^{N*} = (m_j^{N*}(t))_{t\in[0,T]}$, and $m_j^{N*}(t)$ is the empirical measure

$$\frac{1}{N}\delta_{x_j^N(t)} + \frac{N-1}{N}\frac{1}{N-1}\sum_{j'\neq j}\delta_{y_{j'}^N(t)},$$

where $y_{j'}^N(t)$ is the optimal state trajectory for player j', i.e., the state trajectory generated by the feedback best-response control of player j'.

We are ready to adapt the formulation mentioned above to the asymptotic case, that is to say, when $N \to \infty$. The formulation that we obtain is referred to as a *robust mean-field game* and represents the core of this section.

To this purpose, note that the process $m^N(t)$ has to be replaced by the limiting measure $m(t)$. Analogously, the cost functional J^N has to be replaced by the limiting cost J^∞. Doing this is possible due to the *indistinguishability* of the processes and the convergence results provided by the *de Finetti–Hewitt–Savage* theorem (see the Appendix, Chapter F).

It is worth noting that, in the robust mean-field game, each player responds to the limiting measure of $m_j^{N*}(t)$. That is to say that the players will play their best responses to the mean-field $m^*(t) := (m^*(t))_{j\in\mathcal{N}}$, which is the distribution of the equilibrium state trajectory.

In the asymptotic case, as we are dealing with a continuum of players, the index j can be dropped from all the variables.

Remarkably, for the problem at hand, the convergence of the empirical measure to a limiting measure implies the convergence of the cost functionals and the optimal cost functionals. Furthermore, for the mean of the measure $m(t)$, denoted by $\bar{m}(t)$, it holds that

$$\frac{d}{dt}\bar{m}(t) = \alpha\bar{m}(t) + \beta\mathbb{E}[u(t)] + \sigma\mathbb{E}[w(t)], \quad t \in (0,T], \; \bar{m}_0 \in \mathbb{R}. \tag{12.9}$$

The above equation derives from taking the *expectation* in (12.8). Actually, we have

$$\mathbb{E}\left[\int_0^t \sigma(s)x_j^N(s)d\mathcal{B}_j(s)\right] = 0.$$

Computing the expectation requires that $\mathbb{E}[|x_j^N(t)|] < \infty$ and $\int_0^t \mathbb{E}[|\sigma(s)x_j^N(s)|]ds < \infty$. We show later that the control $u(t)$ and the disturbance $w(t)$ are bounded, and therefore the right-hand side is bounded as well. Then, in order to guarantee $\mathbb{E}[|x_j^N(t)|] < \infty$ and $\int_0^t \mathbb{E}[|\sigma(s)x_j^N(s)|]ds < \infty$, it is sufficient to consider initial distributions with bounded expected value, namely $\mathbb{E}[|x_{j,0}^N|] < \infty$.

We are in a position to give a precise formulation of a robust mean-field game.

Problem 12.2 (Robust mean-field game). *Let $\mathcal{B}$ be a one-dimensional Brownian motion process defined on $(\Omega, \mathcal{F}, \mathbb{P})$, where $\mathcal{F}$ is the natural filtration generated by $\mathcal{B}$. Let $x(0)$ be any random variable independent of $\mathcal{B}$ having distribution $m_0(x)$. We define robust mean-field game by the problem*

$$\inf_{\{u(x(t),t)\}_t} \sup_{\{w(x(t),t)\}_t} J^\infty(x,u,m^*,w),$$

where the dynamics of $x(t)$ are given by

$$dx(t) = [\alpha x(t) + \beta u(x(t),t) + \sigma w(x(t),t)]dt \tag{12.10}$$
$$+ \sigma x(t)d\mathcal{B}(t), \; t \in (0,T], \; x_0 \in \mathbb{R},$$

and $m^*(t)$ is the equilibrium mean-field trajectory obtained when any player at state x implements the control

$$u^*(x(t),t) = \arg\inf_{\{u(x(t),t)\}_t} \sup_{\{w(x(t),t)\}_t} J^\infty(x,u,m^*,w).$$

12.5.2 ▪ A general solution for the robust mean-field game

Under the assumption that $\beta \neq 0, \gamma \neq 0$, let the *robust Hamiltonian* be defined as

$$\tilde{H}(x,p,m,t) = \inf_u \sup_w \{c(x,u,m) - \gamma^2 w^2 + p(\alpha x(t) + \beta u(t) + \sigma\zeta(t))\}.$$

For the supremum part, note that the function $w \longmapsto -\gamma^2 w^2 + p\sigma w$ is strictly concave and has a maximum for

$$w^*(t) = \frac{\sigma}{2\gamma^2}p. \tag{12.11}$$

Let $v(x,t)$ be the *upper value* of the problem with initial time t and initial state x. Consequently, the worst-case disturbance takes the form

$$w^*(t) = \frac{\sigma}{2\gamma^2}\partial_x v(x,t), \tag{12.12}$$

where $v(x,t)$ satisfies the HJB equation

$$\partial_t v(x,t) + \tilde{H}(x, \partial_x v(x,t), m, t) + \frac{\sigma^2 x^2}{2} \partial_{xx}^2 v(x,t) = 0, \qquad (12.13)$$

$$v(x,T) = g(x). \qquad (12.14)$$

Furthermore, the maximum value of the function $-\gamma^2 w^2 + p\sigma w$ is given by $(\frac{\sigma p}{2\gamma})^2$. Then, for the robust Hamiltonian we obtain

$$\tilde{H}(x, p, m, t) = \inf_u \{ c(x, u, m) - \gamma^2 (w^*(t))^2 + p(\alpha x + \beta u + \sigma w^*(t)) \} \qquad (12.15)$$

$$= \inf_u \{ c(x, u, m) + p(\alpha x + \beta u) \} + \left(\frac{\sigma p}{2\gamma} \right)^2. \qquad (12.16)$$

By ignoring the disturbance, we can define the standard Hamiltonian as

$$H(x, p, m, t) = \inf_u \{ c(x, u, m) + p(\alpha x(t) + \beta u(t)) \}.$$

Under the assumption that the cost c is strict convex in u, the derivative of H with respect to p is given by

$$\partial_p H(x, p, m, t) = \alpha x(t) + \beta u^*(t).$$

Consequently, we can formulate the optimal control as a function of the robust Hamiltonian as follows:

$$u^*(x(t), t) = \frac{1}{\beta} \left[\partial_p \tilde{H}(x, p, m, t) - \alpha x(t) - 2 \left(\frac{\sigma}{2\gamma} \right)^2 p \right].$$

Theorem 12.6. *If the cost c is strictly convex in u, the optimal control is given by*

$$u^*(x(t), t) = \frac{1}{\beta} \left[\partial_p H(x(t), \partial_x v(x,t), m(t), t) - \alpha x(t) \right],$$

where an equation generating $v(x,t)$ is yet to be introduced.

Proof. The underlying idea is that under strict convexity the Hamiltonian is well-posed and the derivative of the Hamiltonian with respect to p provides the drift term of the state from which we deduce the feedback optimal control of the player.   □

A direct consequence of the above result is stated in the following corollary.

Corollary 12.7. *The optimal control $u^*(x(t), t)$ depends on the worst-case disturbance $w^*(t)$ as follows:*

$$u^*(x(t), t) = \frac{1}{\beta} \left[\partial_p H(x(t), \frac{2\gamma^2}{\sigma} w^*(x(t), t), m(t), t) - \alpha x(t) \right]. \qquad (12.17)$$

Proof. The result can be obtained from (12.11) by setting $p = \partial_x v(x,t)$.   □

Now, for the drift term of the state at $(u^*(x(t), t), w^*(x(t), t))$, we get

$$\alpha x(t) + \beta u^*(t) + \sigma w^*(t) = \partial_p H + \sigma w^*(t). \qquad (12.18)$$

We are then in a position to give a precise formulation of the robust mean-field game in terms of two coupled partial differential equations.

Theorem 12.8. *The mean-field system of the robust mean-field game is given by*

$$\partial_t v(x,t) + H(x, \partial_x v(x,t), m(t), t) + \left(\frac{\sigma}{2\gamma}\right)^2 |\partial_x v(x,t)|^2$$

$$+ \frac{1}{2}\sigma^2 x^2 \partial_{xx}^2 v(x,t) = 0, \tag{12.19}$$

$$v(x,T) = g(x), \tag{12.20}$$

$$m(x,0) = m_0(x), \tag{12.21}$$

$$\partial_t m(x,t) + \partial_x \left(m(x,t) \partial_p H(x, \partial_x v(x,t), m(t), t) \right)$$

$$+ \frac{\sigma^2}{2\gamma^2}\partial_x(m(x,t)\partial_x v(x,t)) - \frac{1}{2}\sigma^2 \partial_{xx}^2 \left[x^2 m(x,t) \right] = 0, \tag{12.22}$$

where m_0 is the initial population state distribution and g is the terminal penalty.

Proof. The proof is straightforward after noting that the first equation is the HJB equation, which is solved backwards with boundary conditions at final time $T > 0$. The second equation is the KFP equation, which accounts for the distribution evolution. □

In what follows, we analyze sufficiency conditions for the existence of a classical solution. In doing this we use a *fixed point theorem* argument, as in [149].

Consider an initial measure d which is absolutely continuous with a continuous density function with finite second moment, and the terminal function is smooth, bounded, and Lipschitz continuous. Also suppose that the running cost c is convex in u. As the c is concave in the disturbance w, we have that the running cost is a convex-concave function for which the following coercivity condition holds:

$$\frac{c - \gamma^2 \|w\|^2}{\|u\|} \longrightarrow +\infty \quad \text{for } \|u\| \to \infty,$$

$$\tag{12.23}$$

$$\frac{c - \gamma^2 \|w\|^2}{\|w\|} \longrightarrow -\infty \quad \text{for } \|w\| \to \infty.$$

Note that given that the coefficients are bounded, the drift is linear and therefore Lipschitz continuous. Furthermore, let us assume that the Fenchel transform of the running cost c is Lipschitz in (x,m), and that the function $p \longmapsto \frac{\sigma^2}{4\gamma^2}\|p\|^2 + H$ is strictly convex and differentiable and that $\frac{\sigma^2}{4\gamma^2}\|p\|^2 + H$ is Lipschitz continuous. Note that this last condition is weaker than the condition convexity assumption on H. Under the above assumptions, the existence of a solution is established in Theorem 2.6 in [149]. See also Theorems 1 and 2 in [104] and Theorem 3.1 in [70].

Function (12.17) with m^* (solution of (12.22)) in state of the generic m yields the *worst-case disturbance feedback mean-field equilibrium*. The above result simplifies in the deterministic case, as established in the following theorem.

Theorem 12.9. *In the deterministic case, i.e., $\sigma \equiv 0$, the mean-field system reduces to*

$$\partial_t v(x,t) + H(x, \partial_x v(x,t), m(x,t), t) = 0, \tag{12.24}$$

$$v(x,T) = g(x), \tag{12.25}$$

$$\partial_t m(x,t) + \partial_x \left(m(x,t)\partial_p H(x, \partial_x v(x,t), m(x,t), t) \right) = 0, \tag{12.26}$$

$$m(x,0) = m_0(x), \tag{12.27}$$

where $m_0(x)$ is a given initial distribution.

Proof. The proof follows from Theorem 12.8 by letting $\sigma = 0$, which eliminates the disturbance term.   □

We specialize the above result to the oil production application introduced in Example 12.4.

Example 12.10 (Oil production cont'd). For a continuum of oil producers, each one being equipped with a given initial reserve or stock of raw material, consider the geometric Brownian motion stochastic process

$$dx(t) = [\alpha x(t) + \beta u(x(t), t) + \sigma w(x(t), t)]dt + \sigma x(t)d\mathscr{B}(t).$$

The above model describes the time evolution of the reserve. The new term $\sigma w(t)$ represents taxation or inflation on the production.

The penalty involves the total income and the production costs and is given by

$$g(x, u, m, w) = -h(\bar{m}, w)u + \left[\frac{a}{2}u^2 + bu\right],$$

where now the sale price of oil $h(\bar{m}, w)$ depends on the disturbance w. The idea is to tackle the problem considering the worst-case disturbance, as in the book by Başar and Bernhard [22]. This leads to the following inf-sup optimization:

$$\inf_{\{u\}_t} \sup_{\{w\}_t} \mathbb{E}\left(G(x(T)) + \int_0^T g(x, u, m, w)dt - \gamma^2 \int_0^T |w|^2 dt\right).$$

Essentially, we look for the infimum with respect to the control u and the supremum with respect to the disturbance w. A crucial aspect is the selection of an opportune value for γ which makes the problem not ill-posed.   ■

12.5.3 ▪ Discussion on the new equilibrium concept

The considered setup leads to a new equilibrium concept, called *worst-case disturbance feedback mean-field equilibrium* (occasionally also *robust mean-field equilibrium*), which combines two existing concepts. The first one is the worst-case disturbance feedback Nash equilibrium derived in the H^∞ literature [23], while the second one is the mean-field equilibrium. Note that the worst-case disturbance feedback Nash equilibrium accounts for adversarial disturbances but in the case of a finite number of players. On the contrary, the mean-field equilibrium involves an infinite number of players but in the absence of adversarial disturbances. The worst-case disturbance feedback mean-field equilibrium combines both elements: an adversarial disturbance and an infinite number of players.

As for the mean-field equilibrium, also the worst-case disturbance feedback mean-field equilibrium requires the solution of the two coupled partial differential equations displayed in Fig. 12.7. The first block includes the HJI equation, which returns the value function $v(.)$ and with it also the optimal control $u^*(.)$ and the worst-case disturbance $w^*(.)$. Both control and disturbance are then substituted into the KFP equation, as both concur in defining the vector field from which we obtain the new density $m(.)$. Again the worst-case disturbance feedback mean-field equilibrium is the fixed point of such a procedure. Fig. 12.7 sketches the iterative scheme for the computation of fixed points.

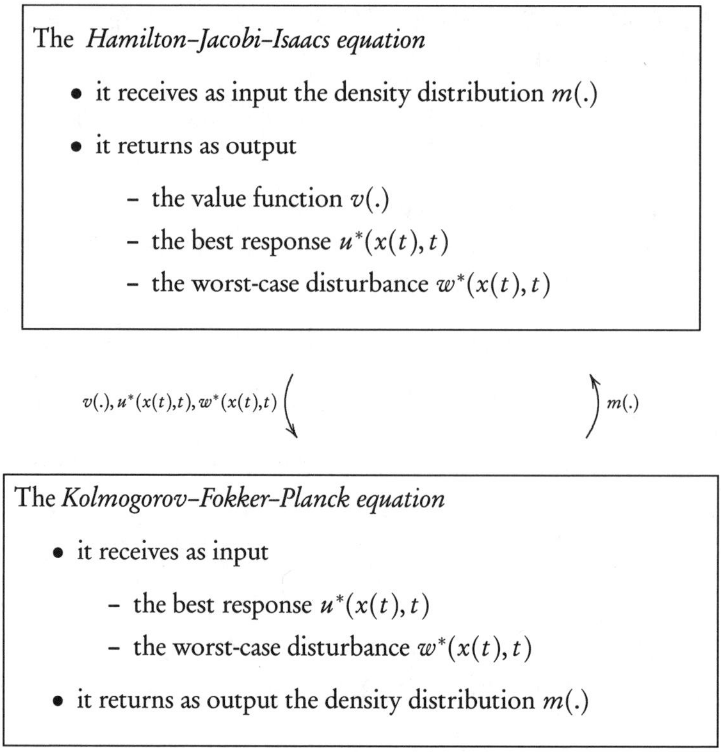

The *Hamilton–Jacobi–Isaacs equation*

- it receives as input the density distribution $m(.)$

- it returns as output

  - the value function $v(.)$
  - the best response $u^*(x(t), t)$
  - the worst-case disturbance $w^*(x(t), t)$

$$v(.), u^*(x(t),t), w^*(x(t),t) \left(\qquad\qquad\qquad\qquad\qquad \right) m(.)$$

The *Kolmogorov–Fokker–Planck equation*

- it receives as input

  - the best response $u^*(x(t), t)$
  - the worst-case disturbance $w^*(x(t), t)$

- it returns as output the density distribution $m(.)$

Figure 12.7. *Iterative scheme for the computation of fixed points in robust mean-field games.*

12.6 ▪ Conclusions and open problems

Mean-field games require solving coupled partial differential equations, the HJB equation and the KFP equation. This chapter illustrates how robustness can be brought into the picture, thus leading to the solution of the HJB equation under the worst-case disturbance. We have called such a new setup *robust mean-field games* and the corresponding equilibrium solution as *worst-case disturbance feedback mean-field equilibrium*.

Key directions for current and future research are

- existence and uniqueness of mean-field equilibria in case of nondifferentiability of the value function, of the probability distribution, and/or of the microscopic state dynamics;

- numerical computation or approximation schemes for mean-field equilibria for non-quadratic nonlinear mean-field games;

- multi-population mean-field games in the presence of heterogeneity of the players;

- applications in other domains, such as engineering, finance, transportation, biology, and social science.

12.7 ▪ Notes and references

The mean-field theory of dynamic games with large but finite populations of asymptotically negligible agents (as the population size goes to infinity) originated in the work of

Huang, Caines, and Malhamé [122, 123, 124, 125] and independently in that of Lasry and Lions [147, 148, 149], where the now standard terminology of mean-field games was introduced. In addition to this, the closely related notion of *oblivious equilibria* for large population dynamic games was introduced by Weintraub, Benkard, and Van Roy in the framework of Markov decision processes [251]. The theory of mean-field games builds upon the notion of *nonatomic player* introduced first by Aumann for a continuum of traders [14] and successively by Jovanovic and Rosenthal for a sequential game [131]. Large robust games are studied in [134]. Mean-field games arise in several application domains, such as economics, physics, biology, and network engineering (see, e.g., [3, 39, 17, 105, 125, 145, 255]).

When the number of players tends to infinity and the players are *homogeneous*, they exhibit identical behavior in a given similar state. In this case the game formulation goes under the name of *anonymous games*. The class of anonymous games has been widely studied in the literature (see [215, 236]). The concept of mass interaction has been used also in evolutionary game theory. Actually, evolutionary games can be reviewed as stationary mean-field games. Preliminary attempts to formulate mean-field games are in [131], where the system involves a value function and a mean-field evolution. The system corresponds to a backward-forward system in the finite horizon case. The equation satisfied by the value is essentially a *Bellman equation*, and the equation satisfied by the mean-field term is a *Kolmogorov equation*. The work [131] has provided sufficiency conditions for the existence of solutions. Mean-field games share striking similarities with consensus problems as highlighted in [125]. A continuous-time version of the mean-field game described in [131] was formulated in [50] in the context of optimal transport, where the backward-forward system consists of a *Hamilton–Jacobi–Bellman equation* and a *Kolmogorov–Fokker–Planck equation*.

Explicit solutions in terms of mean-field equilibria are available for *linear-quadratic mean-field games* [20] and have been recently extended to more general cases in [104]. In addition to explicit solutions, a variety of solution schemes have been recently proposed based on myopic learning, discretization, or numerical approximations (see, e.g., [5, 3, 4, 199]). The idea of extending the state space, which originates in optimal control [212, 213], has also been used to approximate mean-field equilibria in [34] and [35]. In [4], for instance, a fully discrete finite difference approximation scheme of a mean-field game has been proposed and studied.

A mean-field approach in dynamic auctions is discussed in [18, 127]. For a survey on mean-field games and applications we refer the reader to [105]. A first attempt to apply mean-field games to demand side management is in [17].

Finally, based on previous works on H^∞-optimal control [22], the authors in [237] and [238] have established a relation between *risk-sensitive games* and *risk-neutral games* via robust methods in the context of a large number of players. These works together with [42, 43] have led to the formulation of *robust mean-field games*.

Part II

Applications

Chapter 13

Consensus in Multi-agent Systems

13.1 ▪ Introduction

This chapter brings together game theory and *consensus in multi-agent systems*. A multi-agent system involves n *dynamic agents*; these can be vehicles, employees, or computers, each one described by a differential or difference equation. The interaction is modeled through a communication graph. In a *consensus problem* the agents implement a distributed *consensus protocol*, i.e., distributed control policies based on local information. The goal of a consensus problem is to make the agents' reach *consensus*, that is, to converge to a same value, called a *consensus value*.

The core message in this chapter is that the consensus problem can be turned into a *noncooperative differential game*, where the dynamic agents are the players. To do this, we formulate a *mechanism design problem* where a supervisor "designs" the objective functions such that if the agents are rational and use their best-response strategies, then they converge to a consensus value. We illustrate the results by simulating the vertical alignment maneuver of a team of unmanned aerial vehicles (UAVs).

Unfortunately, solving the mechanism design problem is a difficult task, unless the problem can be modeled as an *affine quadratic game*. Given such a game, the main idea is then to translate it into a sequence of more tractable receding horizon problems. At each discrete time t_k, each agent optimizes over an infinite *planning* horizon $T \to \infty$ and executes the controls over a one-step *action* horizon $\delta = t_{k+1} - t_k$. The neighbors' states are kept constant over the planning horizon. At time t_{k+1} each agent reoptimizes its controls based on the new information on neighbors' states which have become available. We then take the limit for $\delta \to 0$.

This chapter is organized as follows. Section 13.2 formulates the consensus problem (Problem 13.1) and the mechanism design problem (Problem 13.2). Section 13.3 provides a solution to the consensus problem. Section 13.4 addresses the mechanism design problem. Section 13.5 illustrates the results on a study case involving a team of UAVs performing a vertical alignment maneuver. Finally, Section 13.6 provides notes and references on the topic.

13.2 ▪ Consensus via mechanism design

Let a set $\Gamma = \{1, \ldots, n\}$ of dynamic agents be given. Let $G = (\Gamma, E)$ be a time-invariant undirected connected network, where Γ is the vertexset and $E \subseteq \Gamma \times \Gamma$ is the edgeset. Such a network describes the interactions between pairs of agents. By *undirected* we mean that if $(i, j) \in E$, then $(j, i) \in E$. By *connected* we mean that for any vertex $i \in \Gamma$ there exists a *path* in E that connects i with any other vertex $j \in \Gamma$. Recall that a path from i to j is a sequence of edges $(i, k_1)(k_1, k_2) \ldots (k_r, j)$ in E. Note that in general, the network G is not complete; that is to say that each vertex i has a direct link only to a subset of other vertices, denoted by $N_i = \{j : (i, j) \in E\}$. This subset is referred to as *neighborhood of i*.

The interpretation of an edge (i, j) in the edgeset E is that the state of vertex j is available to vertex i. As the network is undirected, then communication is bidirectional; namely, the state of agent i is available to agent j.

Let x_i be the state of agent i. The evolution of x_i is determined by the following first-order differential equation driven by a *distributed* and *stationary* control policy:

$$\dot{x}_i = u_i(x_i, x^{(i)}) \quad \forall i \in \Gamma, \tag{13.1}$$

where $x^{(i)}$ represents the vector collecting the states of the only neighbors of i. In other words, for the $j$the component of $x^{(i)}$ we have

$$x_j^{(i)} = \begin{cases} x_j & \text{if } j \in N_i, \\ 0 & \text{otherwise.} \end{cases}$$

The control policy is *distributed*, as the control u_i depends only on x_i and $x^{(i)}$. The control policy is *stationary*, as there is no explicit dependence of u_i on time t. Occasionally, we also call such a control policy *time invariant*. Let the state of the collective system be defined by the vector $x(t) = \{x_i(t), i \in \Gamma\}$, and let the initial state be $x(0)$. Similarly, denote by $u(x) = \{u_i(x_i, x^{(i)}) : i \in \Gamma\}$ the *collective control vector*, which we occasionally call simply *protocol*. Fig. 13.1 depicts a possible network of dynamic agents. In the graph, for some of the vertices, we indicate the corresponding dynamics.

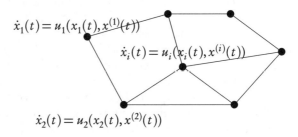

Figure 13.1. *Network of dynamic agents.*

The consensus problem consists in determining how to make the players reach agreement on a so-called consensus value. To give a precise definition of such a value, consider a function $\hat{\chi} : \mathbb{R}^n \to \mathbb{R}$. This function is a generic continuous and differentiable function of n variables $x_1, \ldots, x_n$ which is permutation invariant. In other words, for any permutation $\sigma(.)$ from the set Γ to the set Γ, the function satisfies

$$\hat{\chi}(x_1, x_2, \ldots, x_n) = \hat{\chi}(x_{\sigma(1)}, x_{\sigma(2)}, \ldots, x_{\sigma(n)}).$$

Sporadically, we refer to $\hat{\chi}$ as *agreement function*.

From [194, 206, 252], a protocol $u(.)$ makes the agents reach asymptotically consensus on a *consensus value* $\hat{\chi}(x(0))$ if

$$\|x_i - \hat{\chi}(x(0))\| \longrightarrow 0 \quad \text{for } t \longrightarrow \infty.$$

The above means that the collective system converges to $\hat{\chi}(x(0))\mathbf{1}$, where $\mathbf{1}$ denotes the vector $(1,1,\ldots,1)^T$.

In the rest of this chapter we focus on agreement functions satisfying

$$\min_{i\in\Gamma}\{y_i\} \leq \hat{\chi}(y) \leq \max_{i\in\Gamma}\{y_i\} \quad \forall\, y \in \mathbb{R}^n. \tag{13.2}$$

In other words, the consensus value is a point in the interval from the minimum to the maximum values of the agents' initial states.

In preparation for the formulation of the consensus problem as a game, let us also introduce a *cost functional* for agent i as the one displayed below:

$$J_i(x_i, x^{(i)}, u_i) = \lim_{T\longrightarrow\infty}\int_0^T \left(F(x_i, x^{(i)}) + \rho u_i^2\right)dt, \tag{13.3}$$

where $\rho > 0$ and $F : \mathbb{R} \times \mathbb{R}^n \to \mathbb{R}$ is a nonnegative *penalty function*. This penalty accounts for the deviation of player i from his neighbors. With the above cost functional in mind, a protocol is said to be *optimal* if each control u_i minimizes the corresponding cost functional J_i. Fig. 13.2 depicts a network of dynamic agents and the cost functionals corresponding to different agents.

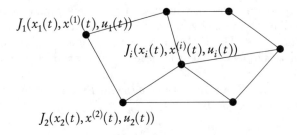

Figure 13.2. *Network of dynamic agents with the cost functionals assigned to the players.*

After this preamble, the problem under study can be stated as follows.

Problem 13.1 (Consensus problem). *Let a network of dynamic agents $G = (\Gamma, E)$ be given. Assume that the agents evolve according to the first-order differential equations (13.1). For any agreement function $\hat{\chi}$ verifying (13.2), design a distributed and stationary protocol as in (13.1) that makes the agents reach asymptotically consensus on $\hat{\chi}(x(0))$ for any initial state $x(0)$.*

We say that a protocol is a *consensus protocol* if it is solution of the above *consensus problem*. Furthermore, we say that a consensus protocol is *optimal* if the controls $u_i(.)$ minimize (13.3). We are in a position to give a precise definition of the mechanism design problem.

Problem 13.2 (Mechanism design problem). *Let a network of dynamic agents $G = (\Gamma, E)$ be given. Assume that the agents evolve according to the first-order differential equations (13.1). For any agreement function $\hat{\chi}(.)$ design a penalty function $F(.)$ such that there exists an optimal consensus protocol $u(.)$ with respect to $\hat{\chi}(x(0))$ for any initial state $x(0)$.*

Note that a pair $(F(.), u(.))$ which is solution to Problem 13.2 must guarantee that all cost functionals in (13.3) converge to a finite value. For this to be true, the integrand in (13.3) must be null in $\chi\mathbf{1}$.

Table 13.1. *Means and corresponding functions f and g.*

Mean	$\hat{\chi}(x)$	$f(y)$	$g(z)$
Arithmetic	$\sum_{i\in\Gamma}\frac{1}{n}x_i$	$\frac{1}{n}y$	z
Geometric	$\sqrt[n]{\prod_{i\in\Gamma}x_i}$	$e^{\frac{1}{n}y}$	$\log z$
Harmonic	$\frac{1}{\sum_{i\in\Gamma}\frac{n}{x_i}}$	$\frac{n}{y}$	$\frac{1}{z}$
Mean of order p	$\sqrt[p]{\sum_{i\in\Gamma}\frac{1}{n}x_i^p}$	$\sqrt[q]{\frac{1}{n}y}$	z^p

13.3 ▪ A solution to the *Consensus Problem*

This section deals with the solution of Problem 13.1, namely the *Consensus Problem*. To this purpose, let us start by considering the following family of agreement function $\hat{\chi}(x)$.

Assumption 13.1 (Structure of $\hat{\chi}(.)$). *Assume that the agreement function $\hat{\chi}(.)$ verifies (13.2) and it is such that $\hat{\chi}(x) = f(\sum_{i\in\Gamma} g(x_i))$ for some $f, g : \mathbb{R} \to \mathbb{R}$ with $\frac{dg(x_i)}{dx_i} \neq 0$ for all x_i.*

It is worth noting that the class of agreement functions contemplated in the above assumption involves any value in the range between the minimum and the maximum of the initial states. This is clear if we look at Table 13.1 and note that to span the whole interval we simply consider the mean of order p and let p vary between $-\infty$ and ∞.

Theorem 13.1 (Solution to the *Consensus Problem*). *The following protocol is solution to the consensus problem:*

$$u_i(x_i, x^{(i)}) = \alpha \frac{1}{\frac{dg}{dx_i}} \sum_{j\in N_i} \hat{\phi}(\vartheta(x_j) - \vartheta(x_i)) \quad \forall i \in \Gamma, \tag{13.4}$$

where

- *the parameter $\alpha > 0$, and the function $\hat{\phi} : \mathbb{R} \to \mathbb{R}$ is continuous, locally Lipschitz, odd, and strictly increasing;*

- *the function $\vartheta : \mathbb{R} \to \mathbb{R}$ is differentiable with $\frac{d\vartheta(x_i)}{dx_i}$ locally Lipschitz and strictly positive;*

- *the function $g(.)$ is strictly increasing, that is, $\frac{dg(y)}{dy} > 0$ for all $y \in \mathbb{R}$.*

Proof. Let us start by observing that from the restrictions imposed on α, $\hat{\phi} : \mathbb{R} \to \mathbb{R}$, and $\vartheta : \mathbb{R} \to \mathbb{R}$, the equilibria are given by $\lambda \mathbf{1}$. We can also infer that if a trajectory $x(t)$ converges to $\lambda_0 \mathbf{1}$, then it holds that $\lambda_0 = \hat{\chi}(x(0))$ for any initial state $x(0)$.

Let us turn to the restrictions on $g(.)$. It is useful to introduce the new variable $\eta = \{\eta_i, i \in \Gamma\}$, where $\eta_i = g(x_i) - g(\hat{\chi}(x(0)))$. Actually, after doing this, consensus implies asymptotic stability of η. Note that η_i is strictly increasing. Furthermore, $\eta = 0$ corresponds to $x = \hat{\chi}(x(0))\mathbf{1}$. Having introduced η, we next prove that the equilibrium point $\eta = 0$ is asymptotically stable in the quotient space $\mathbb{R}^n/\text{span}\{\mathbf{1}\}$. To do this, we consider the following candidate Lyapunov function: $V(\eta) = \frac{1}{2}\sum_{i\in\Gamma}\eta_i^2$. Note that we have $V(\eta) = 0$ if and only if $\eta = 0$. In addition, $V(\eta) > 0$ for all $\eta \neq 0$. Our goal is to show that $\dot{V}(\eta) < 0$ for all $\eta \neq 0$. To this purpose, let us first rewrite $\dot{V}(\eta)$ as follows:

$$\dot{V}(\eta) = \sum_{i \in \Gamma} \eta_i \dot{\eta}_i = \sum_{i \in \Gamma} \eta_i \frac{dg(x_i)}{dx_i} \dot{x}_i. \tag{13.5}$$

Now, from (13.4) we can rewrite (13.5) as

$$\begin{aligned}
\dot{V}(\eta) &= \sum_{i \in \Gamma} \eta_i \frac{dg(x_i)}{dx_i} u_i \\
&= \sum_{i \in \Gamma} \eta_i \frac{dg(x_i)}{dx_i} \alpha \frac{1}{\frac{dg}{dx_i}} \sum_{j \in N_i} \hat{\phi}(\vartheta(x_j) - \vartheta(x_i)) \\
&= \alpha \sum_{i \in \Gamma} \eta_i \sum_{j \in N_i} \hat{\phi}(\vartheta(x_j) - \vartheta(x_i)).
\end{aligned} \tag{13.6}$$

Now, by noting that $j \in N_i$ if and only if $i \in N_j$ for each $i, j \in \Gamma$, from (13.6) we can rewrite

$$\dot{V}(\eta) = -\alpha \sum_{(i,j) \in E} (g(x_j) - g(x_i)) \hat{\phi}(\vartheta(x_j) - \vartheta(x_i)). \tag{13.7}$$

From (13.7) we conclude that $\dot{V}(\eta) \leq 0$ for all η and, more specifically, $\dot{V}(\eta) = 0$ only for $\eta = 0$. To see this, observe that for any $(i, j) \in E$, $x_j > x_i$ implies $g(x_j) - g(x_i) > 0$, $\vartheta(x_j) - \vartheta(x_i) > 0$, and $\hat{\phi}(\vartheta(x_j) - \vartheta(x_i)) > 0$. This is true, as $\alpha > 0$ and $g(.)$, $\hat{\phi}(.)$, and $\vartheta(.)$ are strictly increasing. Therefore we have $\alpha(g(x_j) - g(x_i))\hat{\phi}(\vartheta(x_j) - \vartheta(x_i)) > 0$ if $x_j > x_i$. A similar argument can be used if $x_j < x_i$. □

13.4 ▪ A solution to the *Mechanism Design Problem*

This section deals with Problem 13.2, namely the *Mechanism Design Problem*. We show that the cost functionals can be designed so that the consensus protocol (13.4) is the unique best-response strategy. In other words, consensus is reached when all the agents implement their best-response strategies. This result is significant, as it shows the true nature of consensus as Nash equilibrium and of a consensus protocol as a collection of best-response policies.

However, Problem 13.2 presents some difficulties in that the agents must predict the evolution of their neighbors' states over the horizon. We propose a method that turns Problem 13.2 into a sequence of tractable problems (Problem 13.3). Consider an infinite *planning* horizon, namely $T \to \infty$, and assume that at each discrete-time t_k the agents compute their best-response strategies over this horizon. Remarkably, in doing this, the neighbors' states do not change over the planning horizon. Given the sequence of optimal controls, the agents use only their first controls. In the parlance of *receding horizon* and *Model Predictive Control*, this corresponds to saying that the agents operate on a one-step *action* horizon $\delta = t_{k+1} - t_k$. When new information on the neighbors' states becomes available at time t_{k+1}, the agents use such information to perform a new iteration of the infinite horizon optimization problem. This section concludes by showing that the solution to Problem 13.3 coincides with the solution to Problem 13.2 asymptotically, namely for $\delta \to 0$.

Let the following update times be given: $t_k = t_0 + \delta k$, where $k = 0, 1, \ldots$. Let $\hat{x}_i(\tau, t_k)$ and $\hat{x}^{(i)}(\tau, t_k)$, $\tau \geq t_k$ be the predicted state of agent i and of his neighbors, respectively. The problem we wish to solve is the following one.

Problem 13.3 (Receding horizon). *For all agents $i \in \Gamma$ and times $t_k, k = 0, 1, \ldots$, given the initial state $x_i(t_k)$ and $x^{(i)}(t_k)$, find*

$$\hat{u}_i^\star(\tau, t_k) = \arg\min \mathscr{J}_i(x_i(t_k), x^{(i)}(t_k), \hat{u}_i(\tau, t_k)),$$

where

$$\mathscr{J}_i(x_i(t_k), x^{(i)}(t_k), \hat{u}_i(\tau, t_k)) = \lim_{T \to \infty} \int_{t_k}^{T} \left(\mathscr{F}(\hat{x}_i(\tau, t_k), \hat{x}^{(i)}(\tau, t_k)) + \rho \hat{u}_i^2(\tau, t_k) \right) d\tau \tag{13.8}$$

subject to the following constraints:

$$\dot{\hat{x}}_i(\tau, t_k) = \hat{u}_i(\tau, t_k), \tag{13.9}$$

$$\dot{\hat{x}}_j(\tau, t_k) = \hat{u}_j(\tau, t_k) := 0 \quad \forall j \in N_i, \tag{13.10}$$

$$\hat{x}_i(t_k, t_k) = x_i(t_k), \tag{13.11}$$

$$\hat{x}_j(t_k, t_k) = x_j(t_k) \qquad \forall j \in N_i. \tag{13.12}$$

The above set of constraints involves the predicted state dynamics of agent i and of his neighbors; see (13.9) and (13.10), respectively.

The constraints also involve the boundary conditions at the initial time t_k; see conditions (13.11) and (13.12). Note that, by setting $\hat{x}^{(i)}(\tau, t_k) = x^{(i)}(t_k)$ for all $\tau > t_k$, agent i restrains the states of his neighbors to be constant over the planning horizon.

At t_{k+1} new information on $x^{(i)}(t_{k+1})$ becomes available. Then the agents update their best-response strategies, which we refer to as *receding horizon control policies*. Consequently, for all $i \in \Gamma$, we obtain the *closed-loop* system

$$\dot{x}_i = u_{i_{RH}}(\tau), \quad \tau \geq t_0,$$

where the receding horizon control law $u_{i_{RH}}(\tau)$ satisfies

$$u_{i_{RH}}(\tau) = \hat{u}_i^\star(\tau, t_k), \quad \tau \in [t_k, t_{k+1}).$$

The complexity reduction introduced by the method derives from turning Problem 13.3 into n one-dimensional problems. This is a consequence of constraint (13.10), which forces $\hat{x}^{(i)}$ to be constant in (13.8). Further evidence of this derives from rewriting $\mathscr{F}(.)$, thus highlighting its dependence on the state $\hat{x}_i(\tau, t_k)$. By doing this the cost functional (13.8) takes the form

$$J_i = \lim_{T \to \infty} \int_{t_k}^{T} \left(\mathscr{F}(\hat{x}_i(\tau, t_k)) + \rho \hat{u}_i^2(\tau, t_k) \right) d\tau. \tag{13.13}$$

Consequently, the problem simplifies, as it involves the computation of the optimal control $\hat{u}_i(\tau, t_k)$ that minimizes (13.13).

Fig. 13.3 illustrates the receding horizon formulation. Given a dynamics for $x_j(t)$, for all $j \in N_i$ (solid line), agent i takes for it the value measured at time t_k (small circles) and maintains it constant from t_k on (thin horizontal lines).

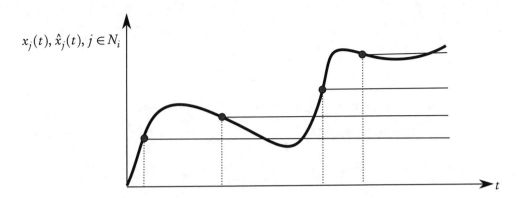

Figure 13.3. *Receding horizon formulation for agent i: at each sampling time (circles) the estimated state of neighbor j, $\hat{x}_j(.)$ is maintained constant over the horizon (thin solid); the actual state $x_j(.)$ changes with time (thick solid).*

Let us now use the Pontryagin Minimum Principle to prove that the control $\hat{u}_i(\tau, t_k)$ is a best-response strategy. Before doing this, let the Hamiltonian function be given by

$$H(\hat{x}_i, \hat{u}_i, p_i) = (\mathscr{F}(\hat{x}_i) + \rho \hat{u}_i^2) + p_i \hat{u}_i, \tag{13.14}$$

where p_i is the co-state. In the above we have dropped dependence on τ and t_k. After doing this, the Pontryagin necessary conditions yield the following set of equalities:

Optimality condition: $\quad \dfrac{\partial H(\hat{x}_i, \hat{u}_i, p_i)}{\partial \hat{u}_i} = 0 \quad \Rightarrow \quad p_i = -2\rho \hat{u}_i. \tag{13.15}$

Multiplier condition: $\quad \dot{p}_i = -\dfrac{\partial H(\hat{x}_i, \hat{u}_i, p_i)}{\partial \hat{x}_i}. \tag{13.16}$

State equation: $\quad \dot{\hat{x}}_i = \dfrac{\partial H(\hat{x}_i, \hat{u}_i, p_i)}{\partial p_i} \quad \Rightarrow \quad \dot{\hat{x}}_i = \hat{u}_i. \tag{13.17}$

Minimality condition: $\quad \dfrac{\partial^2 H(\hat{x}_i, \hat{u}_i, p_i)}{\partial \hat{u}_i^2}\bigg|_{\hat{x}_i = \hat{x}_i^*, \hat{u}_i = \hat{u}_i^*, p_i = p_i^*} \geq 0 \quad \Rightarrow \rho \geq 0. \tag{13.18}$

Boundary condition: $\quad H(\hat{x}_i^*, \hat{u}_i^*, p_i^*) = 0. \tag{13.19}$

The boundary condition (13.19) restrains the Hamiltonian to be null along any optimal path $\{\hat{x}_i^*(t) \forall t \geq 0\}$ (see, e.g., [52, Sect. 3.4.3]).

Recall from Section 9.2.1 that the Pontryagin Minimum Principle yields conditions that are, in general, *necessary* but not sufficient (see also [52]). However, sufficiency is guaranteed under the following additional assumption:

Uniqueness condition: $\quad \mathscr{F}(x_i)$ is convex. $\tag{13.20}$

If we impose further restraints on the structure of $\mathscr{F}(x_i)$, we obtain sufficient conditions that yield a unique optimal control policy $\hat{u}_i(.)$. This is established in the next result.

Theorem 13.2. *Let agent i evolve according to the first-order differential equation (13.1). At times $t_k = 0, 1, \dots$, let the agents be assigned the cost functional (13.8), where the penalty* *is given by*

$$\mathscr{F}(\hat{x}_i(\tau, t_k)) = \rho \left(\frac{1}{\frac{dg}{dx_i}} \sum_{j \in N_i} (\vartheta(x_j(t_k)) - \vartheta(\hat{x}_i(\tau, t_k))) \right)^2, \qquad (13.21)$$

and where $g(.)$ is increasing, $\vartheta(.)$ is concave, and $\frac{1}{\frac{dg(y)}{dy}}$ is convex.

　　Then the control policy

$$\hat{u}_i^\star(\tau, t_k) = u_i(x_i(\tau)) = \alpha \frac{1}{\frac{dg}{dx_i(\tau)}} \sum_{j \in N_i} (\vartheta(x_j(t_k)) - \vartheta(x_i(\tau))), \quad \alpha = 1, \qquad (13.22)$$

is the unique optimal solution to Problem 13.3.

Proof. First, well-posedness of the problem is guaranteed, as for $x_i^* = \vartheta^{-1} \left(\frac{\sum_{j \in N_i} \vartheta(x_j(t_k))}{|N_i|} \right)$ the control policy is null and the cost functional (13.13) converges. This is obtained from the condition that the penalty (13.21) is null in a state $\hat{x}_i^*$, for which it holds that $\sum_{j \in N_i} (\vartheta(x_j(t_k)) - \vartheta(x_i^*)) = 0$.

　　Let us now prove optimality of the control policy (13.22) with $\alpha = 1$. To this purpose, note that it satisfies conditions (13.15)–(13.20). Actually, it is straightforward to see that conditions (13.17) and (13.18) are satisfied. Now, let us compute $\dot{p}_i$ from (13.15) and let us substitute the expression we obtain in (13.16). Thus we get

$$2\rho \dot{\hat{u}}_i = \frac{\partial H(\hat{x}_i, \hat{u}_i, p_i)}{\partial \hat{x}_i}. \qquad (13.23)$$

　　Also, from (13.17), we have $\dot{\hat{u}}_i = \frac{\partial \hat{u}_i}{\partial \hat{x}_i} \dot{\hat{x}}_i = \frac{\partial \hat{u}_i}{\partial \hat{x}_i} \hat{u}_i$. Then, (13.23) yields $2\rho \frac{\partial \hat{u}_i}{\partial \hat{x}_i} \hat{u}_i = \frac{\partial H(\hat{x}_i, \hat{u}_i, p_i)}{\partial \hat{x}_i}$. After integration and from (13.19) we have that the solution of (13.23) must satisfy

$$\rho \hat{u}_i^2 = \mathscr{F}(\hat{x}_i). \qquad (13.24)$$

Then, it suffices to note that $\hat{u}_i(\tau, t_k) = \frac{1}{\frac{dg}{d\hat{x}_i}} \sum_{j \in N_i} (\vartheta(x_j(t_k)) - \vartheta(\hat{x}_i(\tau, t_k)))$ verifies the above condition.

　　To prove uniqueness, let us prove that $\mathscr{F}(\hat{x}_i)$ is convex. To this purpose, we can write $\mathscr{F} = \mathscr{F}_3(F_1(\hat{x}_i), \mathscr{F}_2(\hat{x}_i))$, where function $\mathscr{F}_1(\hat{x}_i) = \left(\frac{\partial g}{\partial x_i} \right)^{-1}$, function $\mathscr{F}_2(\hat{x}_i) = \sum_{j \in N_i} (\vartheta(x_j(t_k)) - \vartheta(\hat{x}_i))$, and $\mathscr{F}_3 = (\mathscr{F}_1(\hat{x}_i) \cdot \mathscr{F}_2(\hat{x}_i))^2$. As $\mathscr{F}_3(.)$ is nondecreasing in each argument, function $\mathscr{F}_3(.)$ is convex if both functions $\mathscr{F}_1(.)$ and $\mathscr{F}_2(.)$ are also convex [64]. Function $\mathscr{F}_1(.)$ is convex, as $\left(\frac{dg}{d\hat{x}_i} \right)^{-1}$ is convex by hypothesis. Analogously, $\mathscr{F}_2(.)$ is convex, as $\vartheta(.)$ is concave, and this concludes the proof. □

　　The above theorem holds also for $\alpha = -1$ if $\frac{dg}{dx_i} < 0$ for all $x_i(0)$.
　　From the above theorem, we can derive the following corollary.

Corollary 13.3. *Let a network of dynamic agents $G = (\Gamma, E)$ be given. Assume that the agents evolve according to the first-order differential equation* (13.1). *At times $t_k = 0, 1, \ldots,$*

let the agents be assigned the cost functional (13.8), where the penalty is given by

$$\mathscr{F}(\hat{x}_i(\tau,t_k)) = \rho\left(\frac{1}{\frac{dg}{dx_i}}\sum_{j\in N_i}(\vartheta(x_j(t_k))-\vartheta(\hat{x}_i(\tau,t_k)))\right)^2, \qquad (13.25)$$

and where $g(.)$ is increasing, $\vartheta(.)$ is concave, and $\frac{1}{\frac{dg(y)}{dy}}$ is convex. If we take $\delta \longrightarrow 0$, then we have

(i) *the penalty function*

$$\mathscr{F}(x_i(\tau,t_k)) \longrightarrow F(x_i,x^{(i)}) = \rho\left(\frac{1}{\frac{dg}{dx_i}}\sum_{j\in N_i}(\vartheta(x_j)-\vartheta(x_i))\right)^2 \qquad (13.26)$$

and

(ii) *the applied receding horizon control law*

$$u_{i_{RH}}^{\star}(\tau) \longrightarrow u_i(x_i,x^{(i)}) = \frac{1}{\frac{dg}{dx_i}}\sum_{j\in N_i}(\vartheta(x_j)-\vartheta(x_i)). \qquad (13.27)$$

The above corollary provides a solution to the mechanism design problem (Problem 13.2). To see this, imagine that a game designer wishes the agents to asymptotically reach consensus on the consensus value $\hat{\chi}(x) = f(\sum_{i\in\Gamma}g(x_i))$. He can accomplish this by assigning the agents the cost functional (13.3), where the penalty is as in (13.26) and where $g(.)$ is increasing, $\frac{1}{\frac{dg(y)}{dy}}$ is convex, and δ is "sufficiently" small.

13.5 ▪ Numerical example: Team of UAVs

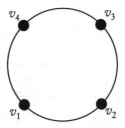

Figure 13.4. *The information flow in a network of four agents.*

Let us now illustrate the results on a team of four UAVs. The UAVs are initially at different heights, and they are performing a vertical alignment maneuver in longitudinal flight. Each vehicle controls the vertical rate on the basis of the neighbors' heights. The UAVs interact as described by the communication network depicted in Fig. 13.4. The goal of the mission is to make the UAVs reach consensus on the *formation center*. We analyze four different vertical alignment maneuvers where the formation center is the (i) *arithmetic mean*, (ii) *geometric mean*, (iii) *harmonic mean*, and (iv) *mean of order* 2 of the initial heights of all UAVs. Set the initial heights as $x(0) = (5,5,10,20)^T$.

Simulations are performed using the following algorithm.

ALGORITHM 13.1. Simulation algorithm for a team of UAVs.

Input: Communication network $G = (V, E)$ and UAVs' initial heights.
Output: UAVs' heights $x(t)$
1 : **Initialize.** Set the initial states equal to the UAVs' initial heights
2 : **for** time $iter = 0, 1, \ldots, T - 1$ **do**
3 :   **for** player $i = 1, \ldots, n$ **do**
4 :   Set $t = iter \cdot dt$ and compute protocol $u_i(.)$ using current $x^{(i)}(t)$
5 :   compute new state $x(t + dt)$ from (13.1)
6 :   **end for**
7 : **end for**
8 : **STOP**

In the first simulation, the UAVs are assigned the cost functional (13.3), where the penalty $F(x_i, x^{(i)}) = \left(\sum_{j \in N_i} (x_j - x_i) \right)^2$. The UAVs use their best responses

$$u(x_i, x^{(i)}) = \sum_{j \in N_i} (x_j - x_i), \tag{13.28}$$

and as a result, they reach asymptotically consensus on the arithmetic mean of $x(0)$. We illustrate this in Fig. 13.5(a).

In the second simulation, the UAVs are assigned a cost functional where the penalty $F(x_i, x^{(i)}) = \left(x_i \sum_{j \in N_i} (x_j - x_i) \right)^2$. By using their best responses

$$u(x_i, x^{(i)}) = x_i \sum_{j \in N_i} (x_j - x_i), \tag{13.29}$$

they reach asymptotically consensus on the geometric mean of $x(0)$. A graphical illustration of this is available in Fig. 13.5(b).

In the third simulation scenario, the UAVs are assigned a cost functional where for the penalty we have $F(x_i, x^{(i)}) = \left(x_i^2 \sum_{j \in N_i} (x_j - x_i) \right)^2$. The implementation of their best responses

$$u(x_i, x^{(i)}) = -x_i^2 \sum_{j \in N_i} (x_j - x_i) \tag{13.30}$$

leads them to reach asymptotically consensus on the harmonic mean of $x(0)$. A sketch of the resulting dynamics is given in Fig. 13.5(c).

In the fourth simulation scenario, the UAVs are assigned cost functionals where the penalty $F(x_i, x^{(i)}) = \left(\frac{1}{2x_i} \sum_{j \in N_i} (x_j - x_i) \right)^2$. The UAVs' best responses

$$u(x_i, x^{(i)}) = \frac{1}{2x_i} \sum_{j \in N_i} (x_j - x_i) \tag{13.31}$$

lead them to reach asymptotically consensus on the mean of order 2 of $x(0)$. This is sketched in Fig. 13.5(d).

Finally, Fig. 13.6 depicts a vertical alignment maneuver when the UAVs use protocol

$$u(x_i, x^{(i)}) = \frac{\max_{i \in \Gamma} \{x_i(0)\}}{2x_i} \sum_{j \in N_i} (x_j - x_i). \tag{13.32}$$

The above protocol is obtained by scaling the protocol (13.31) by twice an upper bound of $\max_{i \in \Gamma} \{x_i(0)\}$.

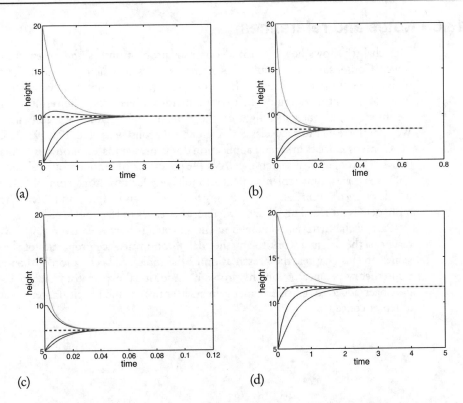

Figure 13.5. *Longitudinal flight dynamics converging to* (a) *the arithmetic mean under protocol* (13.28); (b) *the geometric mean under protocol* (13.29); (c) *the harmonic mean under protocol* (13.30); (d) *the mean of order 2 under protocol* (13.31). *Reprinted with permission from Elsevier* [30].

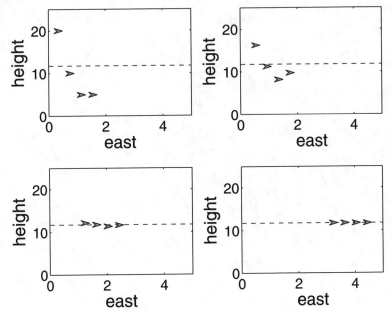

Figure 13.6. *Vertical alignment to the mean of order 2 on the vertical plane. Reprinted with permission from Elsevier* [30].

13.6 ▪ Notes and references

This chapter shows how to turn a consensus problem into a *noncooperative differential game*. Consensus is the result of a mechanism design where a game designer imposes individual objective functions. Then, the agents reach asymptotically consensus as a side effect of the optimization of their own individual objectives. The results of this chapter are important, as they shed light on the game-theoretic nature of a consensus problem. We refer the reader to a few classical references on consensus [128, 194, 193, 205, 206, 252].

Consensus arises in several application domains, such as autonomous formation flight [94, 102], cooperative search of UAVs [46], swarms of autonomous vehicles or robots [100, 161], and joint replenishment in multi-retailer inventory control [31, 32]. More details on *mechanism design* or *inverse game theory* can be found in [196, Chap. 10]. For more details on receding horizon we refer the reader to [89] and [158].

Part of the material contained in this chapter is borrowed from [30]. We refer the reader to the original work for further details on invariance properties of the consensus value. In this chapter, the presentation of the topic has been tailored to emphasize the game theory perspective on the problem. Additional explanatory material and figures have been added to help the reader gain a better insight and physical interpretation of the different concepts.

Chapter 14

Demand Side Management

14.1 ▪ Introduction

This chapter combines mean-field games and *demand side management* for a population of *thermostatically controlled loads* (*TCLs*). Demand side management involves a set of operations aiming at decentralizing the control of loads in power networks. The loads are considered as *fully responsive*. This means that they adjust their consumption patterns in response to the network conditions.

Reframing demand side management within the framework of game theory provides a better understanding of the following aspects in order:

- the *strategic behavior* of the end-use customers and the impact on the collective behavior of the power system;

- the *consistency* of the power system model (by this we mean the study of the relation between microscopic and macroscopic phenomena and how both phenomena can be brought together into a unified framework);

- the *scalability* of the policies adopted by the loads and the corresponding *stability* of the main characteristics of the loads (temperatures, level of charges, and so on) as well as of the system frequency dynamics.

To put the problem in context, Fig. 14.1 depicts a population of plug-in electric vehicles supplied by a renewable energy power plant.

In the same spirit as *prescriptive game theory* and *mechanism design*, we imagine that a game designer assigns cost functionals to the players (the TCLs) in order to penalize those players that are in *on* state in peak hours, as well as those who are in *off* state in off-peak hours. The overall result is a stabilization of the mains frequency.

This chapter is organized as follows. Section 14.2 formulates the problem for a population of thermostatically controlled loads. Section 14.3 turns the problem into a mean-field game. Section 14.4 examines mean-field equilibrium solutions. Section 14.5 provides a numerical example. Finally, Section 14.6 provides notes and references.

14.2 ▪ Population of TCLs

Let a population of TCLs be given, and consider a time horizon window $[0, T]$. At time $t \in [0, T]$, the state of a TCL involves its temperature, denoted by a continuous variable $x(t)$, and its state *on* or *off*, described by a binary variable $\pi_{on}(t) \in \{0, 1\}$.

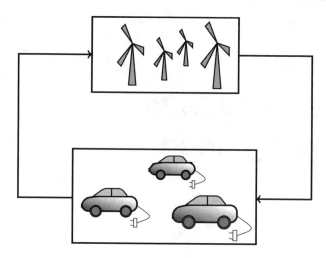

Figure 14.1. *Demand response involves populations of electrical loads (lower block) and energy generators (upper block) intertwined in a feedback-loop scheme.*

Let us denote by x_{on} and x_{off} the lower and upper bounds for the temperature. The temperature decreases exponentially to x_{on} with rate α anytime the TCL is in state *on*. Conversely, the temperature increases exponentially to x_{off} with rate β any time the TCL is in state *off*. Then, the time evolution of the temperature of a TCL with initial value x follows the differential equations

$$\begin{cases} \dot{x}(t) = \begin{cases} -\alpha(x(t) - x_{on}) & \text{if } \pi_{on}(t) = 1, \\ -\beta(x(t) - x_{off}) & \text{if } \pi_{on}(t) = 0, \end{cases} \quad t \in [0, T), \\ x(0) = x, \end{cases} \tag{14.1}$$

where α, β are positive scalar parameters.

In the spirit of [9, 17], let us consider a stochastic model in which the TCLs can be in one of the two states, *on* or *off*, with given probabilities $\pi_{on}(t) \in [0, 1]$ and $\pi_{off}(t) \in [0, 1]$. Let the transition rates be the control variables. In particular, denote by u_{on} the transition rate from *off* to *on*, and by u_{off} the transition rate from *on* to *off*. Let u be a two-component vector including u_{on} and u_{off}. The automata in Fig. 14.2 provide a sketch of the aforementioned model.

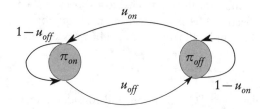

Figure 14.2. *Automata describing transition rates from* on *to* off *and vice versa.*

After introducing the transition rates, the controlled dynamics takes the form

$$\begin{cases} \dot{\pi}_{on}(t) = u_{on}(t) - u_{off}(t), \ t \in [0, T), \\ \dot{\pi}_{off}(t) = u_{off}(t) - u_{on}(t), \ t \in [0, T), \\ 0 \le \pi_{on}(t), \pi_{off}(t) \le 1, \ t \in [0, T). \end{cases} \qquad (14.2)$$

Note that we can take only one of the above dynamics as $\dot{\pi}_{on}(t) + \dot{\pi}_{off}(t) = 0$. Let us introduce the notation $y(t) = \pi_{on}(t)$.

For any x, y in the

"set of feasible states" $\mathscr{S} :=]x_{on}, x_{off}[\times]0$ ̅ ̅

we obtain the set of equations as below:

$$\begin{cases} \dot{x}(t) &= \left(y(t) \left[-\alpha(x(t) - x_{on}) \right] + (1 - y(t)) \left[-\beta(x(t) - x_{off}) \right] \right) \\ &=: f(x(t), y(t)), \ t \in [0, T), \\ x(0) &= x, \\ \dot{y}(t) &= \left(u_{on}(t) - u_{off}(t) \right) \\ &=: g(u(t)), \ t \in [0, T), \\ y(0) &= y. \end{cases} \qquad (14.3)$$

In addition, let us introduce a probability density function $m : [x_{on}, x_{off}] \times [0, 1] \times [t, T] \to [0, +\infty[$, $(x, y, t) \mapsto m(x, y, t)$, which satisfies $\int_{x_{on}}^{x_{off}} \int_{[0,1]} m(x, y, t) dx dy = 1$ for every t. Furthermore, let us define $m_{on}(t) := \int_{x_{on}}^{x_{off}} \int_{[0,1]} y m(x, y, t) dx dy$. Similarly, set $m_{off}(t) = 1 - m_{on}(t)$.

Assume that the mains frequency shows linear dependence on the difference between the proportion of TCLs in the state *on* and a nominal value. Let us refer to such a difference as *error*, and let it be denoted by $e(t) = m_{on}(t) - \bar{m}_{on}$. Here $\bar{m}_{on}$ is the nominal value. The rationale of setting the error as above is that the higher the proportion of TCLs in the *on* state, the lower the mains frequency.

To introduce the cost function of a TCL, let the running cost depend on the distribution $m(x, y, t)$ through the error $e(t)$ as indicated below:

$$c(x(t), y(t), u(t), m(x, y, t)) = \tfrac{1}{2} \left(q x(t)^2 + r_{on} u_{on}(t)^2 + r_{off} u_{off}(t)^2 \right) \\ + y(t)(Se(t) + W), \qquad (14.4)$$

where q, r_{on}, r_{off}, S, and W are opportune positive scalars.

Cost (14.4) includes four terms on which we comment in order. First, $\tfrac{1}{2} q x(t)^2$ penalizes the deviation of the TCL's temperature from the nominal value; we take the nominal value equal to zero. Second, $\tfrac{1}{2} r_{on} u_{on}(t)^2$ is the penalty due to a fast switching; that is to say that this cost is zero when $u_{on}(t) = 0$ (no switching) and is maximal when $u_{on}(t) = 1$ (switching with probability 1). In addition, we have the term $\tfrac{1}{2} r_{off} u_{off}(t)^2$, which admits an interpretation analogous to that of the previous term. Third, $y(t) Se(t)$ is the *incentive*; this term penalizes those appliances that are in state *on* when the demand exceeds the supply, namely $e(t) > 0$. The term $y(t) Se(t)$ turns into a reward if an appliance is *on* when the supply exceeds the demand, that is, $e(t) < 0$. Finally, $y(t) W$ penalizes the power consumption; that is to say that when the TCL is *on*, the power consumption is W.

In addition to the running cost, we also have a terminal penalty $\Psi : \mathbb{R} \rightarrow [0, +\infty[$, $x \mapsto \Psi(x)$ yet to be designed.

We are in a position to give a precise statement of the problem under study.

Problem 14.1 (Population of TCLs). *Consider a population of TCLs. Let a finite horizon $T > 0$ be given, and assume that the initial states of the TCLs are determined by the distribution $m_0 : [x_{on}, x_{off}] \times [0, 1] \rightarrow [0, +\infty[$.*

Minimize over $\mathcal{U}$ subject to the controlled system (14.3) the cost functional

$$ J(x, y, t, u(.)) = \int_0^T (c(x(t), y(t), u(t), m(x, y, t)))dt + \Psi(X(T)), $$

where $\mathcal{U}$ is the set of all measurable state-feedback closed-loop policies $u(.) : [0, +\infty[\rightarrow \mathbb{R}$ and $m(.)$ is the time-dependent function describing the evolution of the distribution of the TCLs' states.

14.3 • Turning the problem into a mean-field game

As a preliminary step, let us develop a mean-field game for the population of TCLs of Problem 14.1.

To this purpose, let us introduce the value function and denote it by $v(x, y, m, t)$. Recall that the value function is the optimal value of $J(x, y, t, u(.))$. In addition, let us set

$$ k(x(t)) = x(t)(\beta - \alpha) + (\alpha x_{on} - \beta x_{off}) $$

and

$$ X(t) = \left[\begin{array}{c} x(t) \\ y(t) \end{array} \right], \quad u(t) = \left[\begin{array}{c} u_{on}(t) \\ u_{off}(t) \end{array} \right]. $$

Henceforth, we occasionally write $v(X, t)$ to mean $v(x, y, m, t)$. The problem under study yields the linear-quadratic problem

$$ \inf_{\{u(t)\}_t} \int_0^T \left[\frac{1}{2} \left(\|X(t)\|_Q^2 + \|u(t)\|_R^2 \right) + L^T X(t) \right] dt, $$

$$ dX(t) = (AX(t) + Bu(t) + C)dt \quad \text{in } \mathcal{S}, $$

(14.5)

where

$$ Q = \left[\begin{array}{cc} q & 0 \\ 0 & 0 \end{array} \right], \qquad R = r = \left[\begin{array}{cc} r_{on} & 0 \\ 0 & r_{off} \end{array} \right], \qquad L(e) = \left[\begin{array}{c} 0 \\ Se(t) + W \end{array} \right], $$

$$ A(x) = \left[\begin{array}{cc} -\beta & k(x(t)) \\ 0 & 0 \end{array} \right], \qquad B = \left[\begin{array}{cc} 0 & 0 \\ 1 & -1 \end{array} \right], \qquad C = \left[\begin{array}{c} \beta x_{off} \\ 0 \end{array} \right]. $$

In (14.5) we simply write A and L to mean $A(x)$ and $L(e)$. From (14.5) we can derive the following mean-field game:

$$\begin{cases} \partial_t v(X,t) + \inf_u \left\{ \partial_X v(X,t)^T (AX + Bu + C) + \frac{1}{2} \left(\|X\|_Q^2 \right. \right. \\ \left. \left. + \|u\|_R^2 \right) + L^T X \right\} = 0 \text{ in } \mathscr{S} \times [0,T[, & \text{(a)} \\ v(X,T) = g(x) \text{ in } \mathscr{S}, \\ \\ u^*(x,t) = \operatorname{argmin}_{u \in \mathbb{R}} \left\{ \partial_X v(X,t)^T (AX + Bu + C) + \frac{1}{2} \|u(t)\|_R^2 \right\} & \text{(b)} \end{cases}$$

(14.6)

and

$$\begin{cases} \partial_t m(x,y,t) + \operatorname{div}[(AX + Bu + C)\, m(x,y,t)] = 0 \text{ in } \mathscr{S} \times]0,T[, \\ m(x_{on},y,t) = m(x_{off},y,t) = 0 \ \forall\, y \in [0,1],\ t \in [0,T], \\ m(x,y,0) = m_0(x,y) \ \forall\, x \in [x_{on},x_{off}],\ y \in [0,1], \\ \int_{x_{on}}^{x_{off}} m(x,t) dx = 1 \ \forall\, t \in [0,T]. \end{cases}$$

(14.7)

Equation (14.6)(a) is the *Hamilton–Jacobi–Bellman equation*. This equation gives the value function $v(x,y,m,t)$ provided the distribution $m(x,y,t)$. The equation has to be solved backwards with a boundary condition at final time T. Such a boundary condition is essentially the equation in the last line of (14.6)(a). Equation (14.6)(b) provides an expression for the optimal closed-loop control $u^*(x,t)$. This is obtained as minimizers of the Hamiltonian function in the right-hand side. Furthermore, equation (14.7) is the advection equation, which gives the distribution $m(x,y,t)$ provided $u^*(x,t)$ and consequently also the vector field $AX + Bu^* + C$. The advection equation has to be solved forwards with boundary condition at the initial time. Such a boundary condition is displayed in the second line of (14.7). Finally, the value for $m(x,y,t)$ which we obtain from (14.7) needs to be substituted in the following expression for the error:

$$\begin{cases} m_{on}(t) := \int_{x_{on}}^{x_{off}} \int_{[0,1]} y\, m(x,y,t) dx dy \ \forall\, t \in [0,T], \\ \\ e(t) = m_{on}(t) - \bar{m}_{on}. \end{cases}$$

(14.8)

The error obtained from the above expression is then substituted in the running cost $c(x,y,m,u)$ in (14.6)(a).

It is worth noting that

$$\bar{X}(t) = \begin{bmatrix} \bar{x}(t) \\ \bar{y}(t) \end{bmatrix} = \begin{bmatrix} \bar{x}(t) \\ m_{on} \end{bmatrix} = \begin{bmatrix} \int_{x_{on}}^{x_{off}} \int_{[0,1]} x\, m(x,y,t) dx dy \\ \int_{x_{on}}^{x_{off}} \int_{[0,1]} y\, m(x,y,t) dx dy \end{bmatrix}.$$

Consequently, we henceforth call *mean-field equilibrium* any pair $(v(X,t), \bar{X}(t))$ which is solution of (14.6)–(14.7).

14.4 • Mean-field equilibrium and stability

After introducing the mean-field game, the solution to Problem 14.1 takes the form of a mean-field equilibrium. This section studies such an equilibrium of the population of TCLs and discusses stability of the TCLs' state dynamics.

As main result we prove that computing a mean-field equilibrium is equivalent to solving three matrix equations.

Theorem 14.1 (Mean-field equilibrium). *A mean-field equilibrium for* (14.6)–(14.7) *is given by*

$$\begin{cases} v(X,t) = \frac{1}{2}X^T P(t)X + \Psi(t)^T X + \chi(t), \\ \dot{\bar{X}}(t) = [A(x) - BR^{-1}B^T P]\bar{X}(t) - BR^{-1}B^T \bar{\Psi}(t) + C, \end{cases} \tag{14.9}$$

where

$$\begin{cases} \dot{P} + PA(x) + A(x)^T P - PBR^{-1}B^T P + Q = 0 \ in \ [0,T[, \ P(T) = \phi, \\ \dot{\Psi} + A(x)^T \Psi + PC - PBR^{-1}B^T \Psi + L = 0 \ in \ [0,T[, \ \Psi(T) = 0, \\ \dot{\chi} + \Psi^T C - \frac{1}{2}\Psi^T BR^{-1}B^T \Psi = 0 \ in \ [0,T[, \ \chi(T) = 0, \end{cases} \tag{14.10}$$

and $\bar{\Psi}(t) = \int_{x_{on}}^{x_{off}} \int_{[0,1]} \Psi(t)m(x,y,t)dxdy$. *Furthermore, the mean-field equilibrium strategy is given by*

$$u^*(X,t) = -R^{-1}B^T[PX + \Psi]. \tag{14.11}$$

Proof. For the first part, let us focus on the *Hamilton–Jacobi–Bellman equation* in (14.6). For given $m(.)$ and for $t \in [0,T]$, it holds that

$$\begin{cases} -\partial_t v(x,y,t) - \left\{ y\left[-\alpha(x - x_{on}) \right] + (1-y)\left[-\beta(x - x_{off}) \right] \right\} \partial_x v(x,y,t) \\ -\sup_{u \in \mathbb{R}} \left\{ -Bu \, \partial_y v(x,y,t) - \frac{1}{2}qx^2 - \frac{1}{2}u^T ru - y(Se + W) \right\} = 0 \\ \quad in \ \mathscr{S} \times]0,T], \\ v(x,y,T) = \Psi(x) \ in \ \mathscr{S}, \\ u^*(x,t) = -r^{-1}B^T \partial_y v(x,y,t). \end{cases} \tag{14.12}$$

The above set of equations can be rewritten in a more convenient way as

$$\begin{cases} -\partial_t v(X,t) - \sup_u \left\{ \partial_X v(X,t)^T (AX + Bu + C) + \frac{1}{2}\left(X^T QX \right. \right. \\ \quad \left. \left. + u^T Ru^T \right) + L^T X \right\} = 0 \ in \ \mathscr{S} \times [0,T[, \\ v(X,T) = g(x) \ in \ \mathscr{S}, \\ u^*(x,t) = -r^{-1}B^T \partial_y v(X,t). \end{cases}$$

Let the value function and optimal strategy be given as follows:

$$\begin{cases} v(X,t) = \frac{1}{2}X^T P(t)X + \Psi(t)^T X + \chi(t), \\ u^* = -R^{-1}B^T[PX + \Psi]. \end{cases}$$

As a result, from (14.12) we have

$$
\begin{cases}
\frac{1}{2}X^T\dot{P}(t)X + \dot{\Psi}(t)X + \dot{\chi}(t) + (P(t)X + \Psi(t))^T\Big[-BR^{-1}B^T\Big](P(t)x + \Psi(t)) \\
\quad + (P(t)x + \Psi(t))^T(AX + C) + \frac{1}{2}\Big(X(t)^TQX(t) + u(t)^TRu(t)^T\Big) \\
\quad\quad\quad\quad\quad\quad\quad + L^TX(t) = 0 \text{ in } \mathscr{S} \times [0,T[, \\
\\
\quad\quad\quad\quad\quad P(T) = \phi, \quad \Psi(T) = 0, \quad \chi(T) = 0.
\end{cases}
\tag{14.13}
$$

For the boundary conditions let us set the following constraint:

$$
v(x,T) = \frac{1}{2}x^TP(T)x + \Psi(T)x + \chi(T) = \frac{1}{2}x^T\phi x.
$$

As (14.13) is an identity in x, the whole procedure culminates in the solution of the following three equations:

$$
\begin{cases}
\dot{P} + PA(x) + A(x)^TP - PBR^{-1}B^TP + Q = 0 \text{ in } [0,T[, \ P(T) = \phi, \\
\\
\dot{\Psi} + A(x)^T\Psi + PC - PBR^{-1}B^T\Psi + L = 0 \text{ in } [0,T[, \ \Psi(T) = 0, \\
\\
\dot{\chi} + \Psi^TC - \frac{1}{2}\Psi^TBR^{-1}B^T\Psi = 0 \text{ in } [0,T[, \ \chi(T) = 0.
\end{cases}
\tag{14.14}
$$

To gain insight on how the incentive term influences the value function, consider the following differential equation for Ψ:

$$
\begin{bmatrix} \dot{\Psi}_1 \\ \dot{\Psi}_2 \end{bmatrix} + \begin{bmatrix} -\beta & 0 \\ k(x(t)) & 0 \end{bmatrix}\begin{bmatrix} \Psi_1 \\ \Psi_2 \end{bmatrix} + \begin{bmatrix} P_{11} & P_{12} \\ P_{21} & P_{22} \end{bmatrix}\begin{bmatrix} \beta x_{off} \\ 0 \end{bmatrix}
$$
$$
- \begin{bmatrix} P_{12}(r_{on}^{-1} + r_{off}^{-1})\Psi_2 \\ P_{22}(r_{on}^{-1} + r_{off}^{-1})\Psi_2 \end{bmatrix} + \begin{bmatrix} 0 \\ Se + W \end{bmatrix}.
\tag{14.15}
$$

The above set of equalities for Ψ yields

$$
\begin{cases}
\dot{\Psi}_1 - \beta\Psi_1 + P_{11}\beta x_{off} - P_{12}(r_{on}^{-1} + r_{off}^{-1})\Psi_2 = 0, \\
\\
\dot{\Psi}_2 + k(x(t))\Psi_1 - P_{22}(r_{on}^{-1} + r_{off}^{-1})\Psi_2 + P_{21}\beta x_{off} + (Se + W) = 0,
\end{cases}
\tag{14.16}
$$

which is of the form

$$
\begin{cases}
\dot{\Psi}_1 + a\Psi_1 + b\Psi_2 + c = 0, \\
\\
\dot{\Psi}_2 + a'\Psi_1 + b'\Psi_2 + c' = 0.
\end{cases}
\tag{14.17}
$$

The set of equalities mentioned above provides the solution $\Psi(x(t), e(t), t)$. Remarkably, the parameters a' and c' depend on x and $e(t)$, respectively.

To get an expression for the closed-loop macroscopic dynamics, let us introduce the mean-field equilibrium strategy given in (14.11) in the open-loop microscopic dynamics provided in (14.5). By averaging both the left-hand side and the right-hand side, we obtain

$$
\dot{\bar{X}}(t) = [A(x) - BR^{-1}B^TP]\bar{X}(t) - BR^{-1}B^T\bar{\Psi}(t) + C,
$$

where $\bar{\Psi}(t) = \int_{x_{on}}^{x_{off}}\int_{[0,1]}\Psi(x,e,t)m(x,y,t)dxdy$, and this concludes our proof. □

It is worth noting that if we introduce the mean-field equilibrium strategy (14.11) in the open-loop microscopic dynamics (14.5), the closed-loop microscopic dynamics takes the form

$$\dot{X}(t) = [A(x) - BR^{-1}B^T P]X(t) - BR^{-1}B^T \Psi(x,e,t) + C. \qquad (14.18)$$

In the rest of this section we analyze stability of the TCLs' microscopic dynamics. To this purpose, let $\mathscr{X}$ be the set of equilibrium points for (14.18). This set is defined as follows:

$$\mathscr{X} = \{(X,e) \in \mathbb{R}^2 \times \mathbb{R} \,|\, [A(x) - BR^{-1}B^T P]X(t) - BR^{-1}B^T \Psi(x,e,t) + C = 0\}.$$

Also, let $V(X(t)) = \mathrm{dist}(X(t), \mathscr{X})$. We are in a position to establish a condition for the asymptotic convergence to the above set of equilibrium points.

Corollary 14.2 (Asymptotic stability). *Let the following inequality hold:*

$$\partial_X V(X,t)^T \Big([A - BR^{-1}B^T P]X(t) - BR^{-1}B^T \Psi^*(x(t),e(t)) + C\Big) \\ < -\|X(t) - \Pi_{\mathscr{X}}[X(t)]\|^2. \qquad (14.19)$$

Then dynamics (14.18) is asymptotically stable, namely, $\lim_{t \to \infty} \mathrm{dist}(X(t), \mathscr{X}) = 0$.

Proof. Consider a solution $X(t)$ of dynamics (14.18) characterized by an initial state $X(0) \notin \mathscr{X}$. Let us set $t = \{\inf t > 0 \,|\, X(t) \in \mathscr{X}\} \leq \infty$. For all $t \in [0, t]$

$$\begin{aligned} V(X(t+dt)) - V(X(t)) &= \|X(t+dt) - \Pi_{\mathscr{X}}[X(t)]\| - \|X(t) - \Pi_{\mathscr{X}}[X(t)]\| \\ &= \|X(t) + dX(t) - \Pi_{\mathscr{X}}[X(t)]\| - \|X(t) - \Pi_{\mathscr{X}}[X(t)]\| \\ &= \tfrac{1}{\|X(t)+dX(t)-\Pi_{\mathscr{X}}[X(t)]\|} \|X(t) + dX(t) - \Pi_{\mathscr{X}}[X(t)]\|^2 \\ &\quad - \tfrac{1}{\|X(t)-\Pi_{\mathscr{X}}[X(t)]\|} \|X(t) - \Pi_{\mathscr{X}}[X(t)]\|^2. \end{aligned}$$

Taking the limit of the difference mentioned above we get

$$\begin{aligned} \dot{V}(X(t)) &= \lim_{dt \to 0} \tfrac{V(X(t+dt)) - V(X(t))}{dt} \\ &= \lim_{dt \to 0} \tfrac{1}{dt}\Big[\tfrac{1}{\|X(t)+dX(t)-\Pi_{\mathscr{X}}[X(t)]\|} \|X(t) + dX(t) - \Pi_{\mathscr{X}}[X(t)]\|^2 \\ &\quad - \tfrac{1}{\|X(t)-\Pi_{\mathscr{X}}[X(t)]\|} \|X(t) - \Pi_{\mathscr{X}}[X(t)]\|^2 \Big] \\ &\leq \tfrac{1}{\|X(t)-\Pi_{\mathscr{X}}[X(t)]\|}\Big[\partial_X V(X,t)^T \Big([A - BR^{-1}B^T P]X(t) \\ &\quad - BR^{-1}B^T \Psi^*(x(t),e(t)) + C\Big) + \|X(t) - \Pi_{\mathscr{X}}[X(t)]\|^2 \Big] < 0, \end{aligned}$$

which implies $\mathscr{L}V(X(t)) < 0$ for all $X(t) \notin \mathscr{X}$, and this concludes our proof. □

In the next section, we illustrate the significance of the above results on a numerical example.

14.5 ▪ Numerical example

This example deals with a population of $n = 10^2$ homogeneous TCLs. The simulations are performed using MATLAB. For the number of iterations, let us set $T = 30$. The time plots are obtained from the following discrete-time version of (14.5):

$$X(t+dt) = X(t) + (A(x(t))X(t) + Bu(t) + C)dt. \qquad (14.20)$$

Table 14.1. *Simulation parameters for a population of TCLs.*

α	β	x_{on}	x_{on}	r_{on}, r_{off}	q	$std(m_0)$	$\bar{m}_0$
1	1	-10	10	1	1	1	0

The parameters are set as in Table 14.1. Specifically, we set $dt = 0.1$ for the step size, $\alpha = \beta = 1$ for the cooling and heating rates, and $x_{on} = -10$ and $x_{off} = 10$ for the lowest and highest temperatures, respectively. Furthermore, we set $r_{on} = r_{off} = 1$ for the penalty coefficients, and $q = 1$. The initial distribution is assumed normal with zero mean and standard deviation $std(m(0)) = 1$.

The algorithm used for the simulations is displayed below.

ALGORITHM 14.1. **Simulation algorithm for a population of TCLs.**

Input: Set of parameters as in Table 14.1.
Output: TCLs' states $X(t)$
  1 : **Initialize.** Generate $X(0)$ given $\bar{m}_0$ and std(m_0)
  2 : **for** time $iter = 0, 1, \ldots, T - 1$ **do**
  3 :   **if** $iter > 0$, **then** compute $m(.)$, $\bar{m}(t)$, and std($m(.)$)
  4 :   **end if**
  5 :   **for** player $i = 1, \ldots, n$ **do**
  6 :   Set $t = iter \cdot dt$ and compute control $u^*(t)$ using current $\bar{m}(t)$
  7 :   compute new state $X(t + dt)$ by executing (14.20)
  8 :   **end for**
  9 : **end for**
 10 : **STOP**

For the optimal control, we set

$$u^* = -R^{-1}B^T[PX + \Psi],$$

where P is obtained from running the MATLAB command [P]=care(A,B,Q,R). The command receives the aforementioned matrices as input and gives as output the solution P to the algebraic Riccati equation. Assuming $BR^{-1}B^T\Psi \approx C$, for the closed-loop dynamics we obtain

$$X(t + dt) = X(t) + [A - BR^{-1}B^T P]X(t)dt.$$

In Fig. 14.3, we plot the time evolution of the states of the TCLs, that is to say, their temperatures $x(t)$ (top) and modes $y(t)$ (bottom). The plot is obtained under the assumption that any 10 seconds the states are subject to an impulsive disturbance. As we can see from the plot, the TCLs respond quickly to the impulsive disturbance and converge again to the equilibrium point. After the TCLs converge a new impulsive disturbance is activated and so forth. This explains the nature of the periodic behavior displayed in the figure.

14.6 ▪ Notes and references

The benefits of demand response in electricity markets is discussed in [192]. Demand side management has been studied in the context of different disciplines, including differential

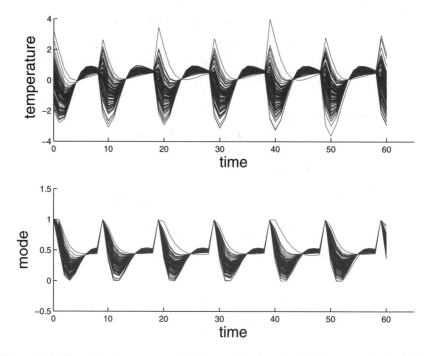

Figure 14.3. *Time plot of temperature $x(t)$ (top row) and mode $y(t)$ (bottom row) of each TCL.*

game theory [17, 82, 198], control and optimization [9, 68, 165, 174, 207], and computer science [228].

Populations of plug-in electric vehicles are dealt with in [9, 165, 174, 198, 228]. Evidence that game theory fits the framework of demand side management is also in [165]. More details on *prescriptive game theory* and *mechanism design* are in [16, 51, 132] and [196, Chap. 10].

Chapter 15

Synchronization of Power Generators

15.1 ▪ Introduction

This chapter illustrates a constructive design of *dynamic demand* for the synchronization of power generators in smart grids. The design is inspired by mechanism design techniques. The game-theoretic nature of the approach provides a better understanding of a series of aspects:

- The responsive loads can be modeled as rational players characterized by *strategic thinking*. Strategic thinking means that the loads respond to the current and the predicted collective behavior of the grid. In order to do this, the loads must show computation capabilities.

- The transient can be regulated by a game designer who designs and assigns cost functionals to the loads. This can be done by considering that the generators have *local interactions* and are subject to disturbances.

- The design must account for the fact that the generators are, in general, *heterogeneous*.

We consider frequency responsive loads which measure the rotor angle deviation between neighbor generators in order to attenuate the *mains frequency* oscillations. These are due to the unbalance between energy demand and supply (see, e.g., [207]).

Oscillations are sketched in Fig. 15.1. The plot shows the time in the x-axis and the rotor angles in the y-axis for a population of 1000 generators. The rotor angles show ample fluctuations around a nominal value. The underlying idea is that such oscillations can be attenuated through an opportune design of the loads' cost functionals.

The transient is modeled as explained next. Let different grids of generators be given. Each grid constitutes a population of generators. Therefore, we have a multi-population scenario. Assume that a virtual load, the player, is connected to each generator. The dynamics of each generator is determined by the *swing equation*. The swing equation involves the mechanical power as input to the generator and the electrical power as output. The electrical power includes the load assigned to the generator and the electrical power in and out from the generator towards other generators. It is a known fact that the swing equation resembles the classical *Kuramoto* oscillators' dynamics [10, 88, 231]. From [193] it is also known that after linearization around zero, the aforementioned oscillator's dynamics turn into a linear consensus dynamics. From a mean-field game perspective, let us

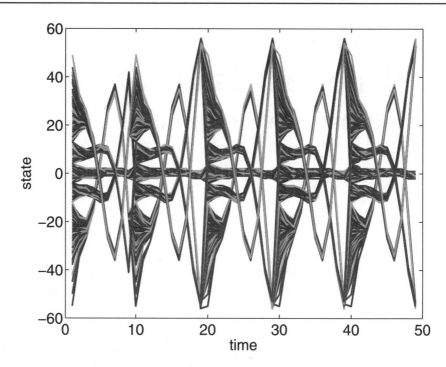

Figure 15.1. *Example of oscillations: qualitative time plot of the state of each TCL, namely temperature (top row) and mode of functioning (bottom row).*

think of the state dynamics of a generator as the *microscopic* dynamics controlled by the virtual load assigned to that generator. The dynamics of the generator describes the time evolution of its rotor angle in the form of a stochastic differential equation. Furthermore, each population of generators is characterized by a *common angle*. This is a measure of the synchronization of the generators. The common angle of each population evolves according to a dynamics, which represents the *macroscopic* dynamics of our mean-field game model. The game-theoretic model involves a finite horizon cost functional for each virtual load. Such a cost functional involves a mean-field term which will incentivize the load to shift the energy consumption from high-peak to off-peak periods.

This chapter is organized as follows. Section 15.2 formulates the synchronization problem. Section 15.3 develops a mean-field game for the problem at hand. Section 15.4 studies mean-field equilibrium solutions. Section 15.5 deals with a numerical example. Finally, Section 15.6 provides notes and references.

15.2 • Multi-machine *transient stability* in power grids

This section illustrates the problem and the mathematical model of the transient stability in multiple interconnected grids of generators. Before addressing multiple grids, let us introduce the model for one grid of generators.

15.2.1 • One grid

Let n generators be given, and let $\delta_i(t)$, $i = 1, 2, \ldots, n$, be their rotor angles. The generators belong to a grid, and therefore their rotor angle dynamics are interconnected as described by the following *swing equations*:

$$\ddot{\delta}_i(t) = \frac{\Pi_{ii}}{\alpha_i} + \frac{1}{\alpha_i}\sum_{j=1}^n \Pi_{ij}\sin(\delta_j(t)-\delta_i(t)), \qquad (15.1)$$

where Π_{ii} is the effective power to generator i, α_i is a damping constant, and Π_{ij} is the maximum power transferred between generators i and j. At an equilibrium, the mechanical power balances the electrical power. At a nonequilibrium condition, the deviation of the mechanical power from the electrical power induces a frequency deviation from the nominal value.

To describe synchronization we use the complex order parameter

$$z = re^{i\Phi} = \frac{1}{n}\sum_{j=1}^n e^{i\delta_j},$$

where Φ is usually referred to as the *common power angle*.

Let us assume that the generators are topologically symmetric. That is to say that $\Pi_{ij} = \Pi_0$ for all $i,j = 1,2,\ldots,n$, $i \neq j$, and $\alpha_i = \alpha_0$ for all $i = 1,2,\ldots,n$. Then, we use the following simplified version of the swing equation:

$$\ddot{\delta}_i(t) = \omega_i + \frac{1}{\alpha_i}\sum_{j=1}^n \frac{\mathscr{K}}{n}\sin(\delta_j(t)-\delta_i(t)). \qquad (15.2)$$

In the above equation, $\omega_i = \frac{\Pi_{ii}}{\alpha_0}$ is the effective power divided by the damping constant, and $\frac{\mathscr{K}}{n} = \frac{\Pi_0}{\alpha_0}$.

Let us now introduce the common power angle in the above dynamics, and let us assume that the players are indistinguishable. The corresponding asymptotic limit for $n \to \infty$ yields the following second-order differential equation:

$$\ddot{\delta}(t) = \omega + \mathscr{K}r\sin(\Phi(t)-\delta(t)). \qquad (15.3)$$

Note that we have dropped index i as we deal with a population of indistinguishable players. By linearizing around zero we obtain

$$\underbrace{\ddot{\delta}}_{\dot{x}(t)} = \underbrace{\omega}_{w(t)} + \underbrace{r(\Phi(t)-\delta(t))}_{u(t)}. \qquad (15.4)$$

Letting $x_1(t) = \delta$, $x_2(t) = \dot{\delta}$, $u(t) = r(\Phi(t)-\delta(t))$, and $w(t) = \omega$, the swing equation takes the form

$$\begin{bmatrix} \dot{x}_1(t) \\ \dot{x}_2(t) \end{bmatrix} = \begin{bmatrix} 0 & 1 \\ 0 & 0 \end{bmatrix}\begin{bmatrix} x_1(t) \\ x_2(t) \end{bmatrix} + \begin{bmatrix} 0 \\ 1 \end{bmatrix}u(t) + \begin{bmatrix} 0 \\ 1 \end{bmatrix}w(t). \qquad (15.5)$$

We can interpret $u(\cdot) \in U$ as the control variable and $w(\cdot) \in W$ as the disturbance, where U and W are the control set and the disturbance set, respectively.

We are in a position to extend the above model to multiple grids.

15.2.2 ▪ Multiple interconnected grids

Let p smart grids be given, each one placed in a different region. Each grid $k \in \{1,2,\ldots,p\}$ involves a population of generators. Let Φ_k be the common power angle of population k. Note that the power angle is now indexed by the population type.

For every population $k \in \{1,\ldots,p\}$, consider a probability density function $m_k :$ $\mathbb{R} \times [0,+\infty[\to \mathbb{R}, (x,t) \mapsto m_k(x,t)$, representing the density of agents of that population in state x at time t, which satisfies $\int_{\mathbb{R}} m_k(x,t)dx = 1$ for every t. Let the initial density be $m_k(.,0) = m_{k0}$. Let the *mean state* of population k at time t be $\bar{m}_k(t) := \int_{\mathbb{R}} x m_k(x,t)dx$. We henceforth denote the common power angle of population k by $\bar{m}_k(t)$.

A network topology is used to model the interconnection between the common power angles of two distinct populations of generators or smart grids; see Fig. 15.2. The topology is essentially represented by a graph $G = (V,E)$, where $V = \{1,\ldots,p\}$ is the set of vertices, one per each population, and $E \subseteq V \times V$ is the set of edges. For sake of simplicity we henceforth assume that $G = (V,E)$ is a balanced graph (or undirected graph). Let $N(k) = \{j \in V | (k,j) \in E\}$ be the neighborhood of k. The rationale of doing this is that the common power angle of neighbor grids is intertwined in a second-order consensus-like form. That is to say that $\bar{m}_k(t)$ evolves based on inputs from $\bar{m}_j(t)$ for all $j \in N(k)$ in order to converge to the local average. The consensus value ρ_k for population k is then expressed by the averaging law:

$$\rho_k = \frac{1}{|N(k)|}\left[\begin{array}{c} \sum_{j \in N(k)} \bar{m}_j(t) \\ \sum_{j \in N(k)} \dot{\bar{m}}_j(t) \end{array}\right], \qquad (15.6)$$

where $\rho_{k,i}$ is the ith component of ρ_k. Let us also denote

$$\bar{m} = (\bar{m}_1,\ldots,\bar{m}_p,\dot{\bar{m}}_1,\ldots,\dot{\bar{m}}_p)^T, \quad \dot{\bar{m}} = (\dot{\bar{m}}_1,\ldots,\dot{\bar{m}}_p,\ddot{\bar{m}}_1,\ldots,\ddot{\bar{m}}_p)^T.$$

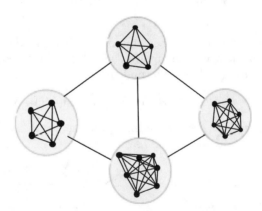

Figure 15.2. *Four distinct populations of generators interconnected.*

Let the virtual loads of population k (recall that we have one virtual load for each generator) be assigned the running cost

$$g(x,\rho_k,u) = \frac{1}{2}\left[a_1(\rho_{k,1} - x_1)^2 + a_2(\rho_{k,2} - x_2)^2 + c u^2\right], \qquad (15.7)$$

where a_1, a_2, and c are given parameters.

Also, let the following terminal cost be given:

$$\Psi(\rho_k,x) = \frac{1}{2}[S_1(\rho_{k,1} - x_1)^2 + S_2(\rho_{k,2} - x_2)^2], \qquad (15.8)$$

where S_1 and S_2 are given parameters.

The synchronization problem for the multiple grid system can be formulated as follows.

Problem 15.1 (Synchronization of generators). *Consider multiple grids $k \in \{1, 2, \ldots, p\}$, and let a finite horizon from 0 to $T > 0$ be given. For each generator in grid k, let the initial state $x(0)$ be obtained from an initial density $m_{k0} : \mathbb{R} \to \mathbb{R}$. Furthermore, let the virtual load linked to the generator be assigned a suitable running cost $g : \mathbb{R}^2 \times \mathbb{R}^2 \times U \to [0, +\infty[$, $(x, \rho_k, u) \mapsto g(x, \rho_k, u)$ as in (15.7); a terminal cost $\Psi : \mathbb{R}^2 \times \mathbb{R}^2 \to [0, +\infty[$ as in (15.8); and $(\rho_k, x) \mapsto \Psi(\rho_k, x)$. Given the linear dynamics $f : \mathbb{R}^2 \times U \times W \to \mathbb{R}$ for x as in (15.5), solve*

$$\min_{u(.)} \max_{w(.)} \int_0^T \left[g(x(t), \rho_k(t), u(t)) - \frac{\gamma^2}{2} w(t)^2 \right] dt + \Psi(\rho_k(T), x(T)), \tag{15.9}$$

where $\gamma > 0$, and $\mathcal{U}$, $\mathcal{W}$ are the sets of all measurable functions $u(.)$ and $w(.)$ from $[0, +\infty[$ to U and W, respectively.

15.3 ▪ Modeling the transient as a mean-field game

This section develops a mean-field game for the problem at hand. After presenting the formulation of the problem in the previous section, and introducing the compact notation

$$x(t) = \left[\begin{array}{c} x_1(t) \\ x_2(t) \end{array} \right] \in \mathbb{R}^2, \quad \rho_k(t) = \left[\begin{array}{c} \rho_{k,1}(t) \\ \rho_{k,2}(t) \end{array} \right] \in \mathbb{R}^2,$$

we understand that for each generator, we have to solve the following linear-quadratic problem:

$$\min_{u(.)} \max_{w(.)} \int_0^T \left[\frac{1}{2} \left(\|x(t)\|_Q^2 + \|\rho_k(t)\|_Q + x(t)^T \hat{Q} \rho_k(t) + R u(t)^2 - \Gamma w(t)^2 \right) \right] dt$$
$$+ \frac{1}{2} \left(\|x(T)\|_S^2 + \|\rho_k(T)\|_S^2 + x(T)^T \hat{S} \rho_k(T) \right)$$

$$\text{subject to} \quad \dot{x}(t) = A x(t) + B u(t) + C w(t),$$

where

$$Q = \left[\begin{array}{cc} a_1 & 0 \\ 0 & a_2 \end{array} \right], \quad \hat{Q} = -2Q = \left[\begin{array}{cc} -2a_1 & 0 \\ 0 & -2a_2 \end{array} \right], \quad R = c, \quad \Gamma = \gamma^2,$$

$$A = \left[\begin{array}{cc} 0 & 1 \\ 0 & 0 \end{array} \right], \quad B = C = \left[\begin{array}{c} 0 \\ 1 \end{array} \right],$$

$$S = \left[\begin{array}{cc} S_1 & 0 \\ 0 & S_2 \end{array} \right], \quad \hat{S} = -2S = \left[\begin{array}{cc} -2S_1 & 0 \\ 0 & -2S_2 \end{array} \right].$$

For every population $k \in \{1, 2, \ldots, p\}$, let us denote by $v_k(x, t)$ the (upper) value of the robust optimization problem under worst-case disturbance starting at time t and at state x. The first step is to show that the problem results in the following multi-population mean-field game system for the scalar functions $v_k(x, t)$ and $m_k(x, t)$ for all $k \in \{1, 2, \ldots, p\}$.

Theorem 15.1. *The synchronization of the generators in a system with multiple grids as introduced in Problem 15.1 can be formulated as a robust mean-field game involving the* *following* Hamilton–Jacobi–Isaacs equation:

$$
\begin{cases}
\partial_t v_k(x,t) + \left(-\frac{1}{2c} BB^T + \frac{1}{2\gamma^2} CC^T \right) |\partial_x v_k(x,t)|^2 + \frac{1}{2} [a_1(\rho_{k,1} - x_1)^2 \\
\qquad\qquad + a_2(\rho_{k,2} - x_2)^2] = 0 \ \ in \ \mathbb{R}^2 \times [0, T[, \\[4pt]
v_k(x,T) = \Psi(\rho_k(T), x) \ in \ \mathbb{R}^2;
\end{cases}
\tag{15.10}
$$

the following advection equation:

$$
\begin{cases}
\partial_t m_k(x,t) + \left(\frac{1}{2\gamma^2} CC^T - \frac{1}{2c} BB^T \right) \partial_x \left(m_k \partial_x v_k \right) = 0 \ \ in \ \mathbb{R}^2 \times [0, T[, \\[4pt]
m_k(x,0) = m_{k0}(x) \ in \ \mathbb{R}, \\
\dot{m}_k(x,0) = \dot{m}_{k0}(x) \ in \ \mathbb{R};
\end{cases}
\tag{15.11}
$$

and the following aggregate dynamics *for the target value:*

$$
\begin{cases}
\bar{m}_k(t) := \int_{\mathbb{R}} x m_k(x,t) dx, \\
\dot{\bar{m}}_k(t) := \int_{\mathbb{R}} x \dot{m}_k(x,t) dx, \\[6pt]
\rho_k = \frac{1}{|N(k)|} \begin{bmatrix} \sum_{j \in N(k)} \bar{m}_j(t) \\ \sum_{j \in N(k)} \dot{\bar{m}}_j(t) \end{bmatrix}.
\end{cases}
\tag{15.12}
$$

Furthermore, the optimal control and worst-case disturbance are

$$
\begin{cases}
u_k^*(x,t) = -\frac{1}{c} B^T \partial_x v_k(x,t), \\
w_k^*(x,t) = \frac{1}{\gamma^2} C^T \partial_x v_k(x,t).
\end{cases}
\tag{15.13}
$$

Proof. Let us start by deriving (15.13). To this purpose, consider the Hamiltonian function

$$
H(x, \partial_x v_k(x,t), \rho_k) = \inf_u \left\{ \frac{1}{2} \left[a_1(\rho_{k,1} - x_1)^2 + a_2(\rho_{k,2} - x_2)^2 + cu^2 \right] \right.
$$
$$
\left. + \partial_x v_k(x,t)^T (Ax + Bu) \right\} = 0. \tag{15.14}
$$

For the robust Hamiltonian we have

$$
\tilde{H}(x, \partial_x v_k(x,t), \bar{m}) = H(x, \partial_x v_k(x,t), \rho_k) + \sup_w \left\{ \partial_x v_k(x,t)^T Cw - \frac{1}{2} \gamma^2 w^2 \right\}.
$$

After differentiation with respect to the control u and the disturbance w we get the following equations:

$$
\begin{cases}
cu + B^T \partial_x v_k(x,t) = 0, \\
-\gamma^2 w + C^T \partial_x v_k(x,t) = 0.
\end{cases}
\tag{15.15}
$$

From the above we then obtain the optimal values for u and w as follows:

$$
\begin{cases}
u_k^*(x,t) = -\frac{1}{c} B^T \partial_x v_k(x,t), \\
w_k^*(x,t) = \frac{1}{\gamma^2} C^T \partial_x v_k(x,t),
\end{cases}
\tag{15.16}
$$

and this yields (15.13).

Let us now turn to prove (15.10)–(15.11). Before doing this, let us note that the second lines of (15.10)–(15.11) are essentially the boundary conditions.

To derive the *Hamilton–Jacobi–Isaacs equation* (15.10), let us replace u_k^* appearing in the Hamiltonian (15.14) by its expression (15.13). After doing this we obtain

$$H(x, \partial_x v_k(x,t), \rho_k) = \frac{1}{2}\left[a_1(\rho_{k,1} - x_1)^2 + a_2(\rho_{k,2} - x_2)^2 + c u_k^{*2}\right] + \partial_x v_k(x,t)^T(Ax + Bu_k^*)$$

$$= \frac{1}{2}\left[a_1(\rho_{k,1} - x_1)^2 + a_2(\rho_{k,2} - x_2)^2\right] - \frac{1}{2c}BB^T\left(\partial_x v_k(x,t)\right)^2.$$

From the above expression, we obtain the following equation for the robust Hamiltonian:

$$\tilde{H}(x, \partial_x v_k(x,t), \bar{m}) = H(x, \partial_x v_k(x,t), \rho_k) + \sup_w \left\{\partial_x v_k(x,t)^T Cw - \frac{1}{2}\gamma^2 w^2\right\}$$

$$= \frac{1}{2}\left[a_1(\rho_{k,1} - x_1)^2 + a_2(\rho_{k,2} - x_2)^2\right] + \partial_x v_k(x,t)^T Ax$$

$$+ \left(-\frac{1}{2c}BB^T + \frac{1}{2\gamma^2}CC^T\right)\left(\partial_x v_k(x,t)\right)^2.$$

Using the above expression of the Hamiltonian in the following *Hamilton–Jacobi–Isaacs equation*,

$$\begin{cases} \partial_t v_k(x,t) + \tilde{H}(x, \partial_x v_k(x,t), \bar{m}) = 0 \text{ in } \mathbb{R}^2 \times [0, T[, \\ v_k(x,T) = \Psi(\rho_k(T), x) \text{ in } \mathbb{R}^2, \end{cases} \tag{15.17}$$

we obtain

$$\begin{cases} \partial_t v_k(x,t) + \left(-\frac{1}{2c}BB^T + \frac{1}{2\gamma^2}CC^T\right)|\partial_x v_k(x,t)|^2 + \partial_x v_k(x,t)^T Ax + \frac{1}{2}[a_1(\rho_{k,1} - x_1)^2 \\ \qquad + a_2(\rho_{k,2} - x_2)^2] = 0 \text{ in } \mathbb{R}^2 \times [0, T[, \\ v_k(x,T) = \Psi(\rho_k(T), x) \text{ in } \mathbb{R}^2, \end{cases}$$

and the *Hamilton–Jacobi–Isaacs equation* (15.10) is proved.

To derive the *advection equation* (15.11), consider the macroscopic dynamics

$$\partial_t m_k(x,t) + \partial_x\left(m_k(x,t)\partial_{\tilde{p}}H(x, \tilde{p}, \rho_k)\right)$$

$$+ \frac{1}{\gamma^2}\partial_x\left(m_k(x,t)\partial_x v_k(x,t)\right) = 0 \text{ in } \mathbb{R}^2 \times [0, T[. \tag{15.18}$$

Introducing $u_k^*(x,t)$ and $w_k^*(x,t)$ as in (15.13) in the above set of equations we obtain

$$\partial_t m_k(x,t) + \partial_x\left[m_k\left(Ax + \left(\frac{1}{2\gamma^2}CC^T - \frac{1}{2c}BB^T\right)\partial_x v_k\right)\right] = 0 \text{ in } \mathbb{R}^2 \times [0, T[.$$

The *advection equation* (15.11) is complemented with the boundary conditions

$$\begin{cases} m_k(x,0) = m_{k0}(x) \text{ in } \mathbb{R}, \\ \dot{m}_k(x,0) = \dot{m}_{k0}(x) \text{ in } \mathbb{R}. \end{cases} \tag{15.19}$$

Finally, for the tracking signal $\rho_k(t)$ we can write

$$
\begin{cases}
\bar{m}_k(t) := \int_{\mathbb{R}} x m_k(x,t)\,dx, \\
\dot{\bar{m}}_k(t) := \int_{\mathbb{R}} x \dot{m}_k(x,t)\,dx, \\[2mm]
\rho_k = \frac{1}{|N(k)|} \left[\begin{array}{c} \sum_{j \in N(k)} \bar{m}_j(t) \\ \sum_{j \in N(k)} \dot{\bar{m}}_j(t) \end{array} \right],
\end{cases}
\tag{15.20}
$$

and this concludes the proof.   □

15.4 ▪ Synchronization explained as stable mean-field equilibrium

In the previous section we have turned the synchronization problem, which we have defined in Problem 15.1, into the robust mean-field game (15.10)–(15.12). This section investigates a solution for the above game in the form of a worst-case disturbance feedback mean-field equilibrium; see Section 12.5.

In particular, the main result states that computing the worst-case disturbance feedback mean-field equilibrium involves solving three matrix equations, provided that the time evolution of the common state is given. In the asymptotic case for $T \to \infty$, we also obtain that the macroscopic dynamics has the form of a second-order consensus dynamics.

Theorem 15.2 (Worst-case mean-field equilibrium). *Let the robust mean-field game be given as in (15.10)–(15.12). A worst-case disturbance feedback mean-field equilibrium can be obtained as follows:*
For all $k \in \{1, 2, \ldots, p\}$

$$
\begin{cases}
v_k(x,t) = \frac{1}{2} x^T \phi(t) x + h(t)^T x + \chi(t), \\
\dot{\bar{m}}_k(\iota) = A\bar{m}_k(\iota) + (-\frac{1}{c} BB^T + \frac{1}{\gamma^2} CC^T)(\phi(\iota)\bar{m}_k(t) + h(t)),
\end{cases}
\tag{15.21}
$$

where

$$
\begin{cases}
\dot{\phi}(t) + \phi^T \left(-\frac{1}{c} BB^T + \frac{1}{\gamma^2} CC^T \right) \phi(t) + \phi^T A + A^T \phi + Q = 0 \text{ in } [0, T[, \\
\phi(T) = S, \\[2mm]
\dot{h}(t) + h(t)^T \left(-\frac{1}{2c} BB^T + \frac{1}{2\gamma^2} CC^T \right) 2\phi(t) + h(t)^T A \\
\qquad\qquad - \rho_k(t)^T Q^T = 0 \text{ in } [0, T[, \\
h(T) = -S\rho_k(T), \\[2mm]
\dot{\chi}(t) + h(t)^T \left(-\frac{1}{2c} BB^T + \frac{1}{2\gamma^2} CC^T \right) h(t) \\
\qquad\qquad + \frac{1}{2} \rho_k(t)^T Q \rho_k(t) = 0 \text{ in } [0, T[, \\
\chi(T) = \frac{1}{2} \rho_k(T)^T S \rho_k(T).
\end{cases}
\tag{15.22}
$$

The corresponding mean-field equilibrium control and disturbance are

$$
\begin{cases}
u^*(x,t) = -\frac{1}{c} B^T (\phi(t) x + h(t)), \\
w^*(x,t) = \frac{1}{\gamma^2} C^T (\phi(t) x + h(t)).
\end{cases}
\tag{15.23}
$$

Furthermore, in the stationary case, namely for $T \to \infty$, set $\bar{m} = (\bar{m}_1, \bar{m}_2, \ldots, \bar{m}_p)^T$. It holds that

$$\dot{\bar{m}}(t) = -\hat{L}\bar{m}(t), \tag{15.24}$$

where

$$\hat{L} = \begin{bmatrix} 0 & I \\ \theta L & \tilde{\theta}(L + \hat{\theta}I) \end{bmatrix},$$

and $L = [L_{kj}]$ is the graph-Laplacian matrix of the network and its elements are defined as follows:

$$L_{kj} = \begin{cases} \phi(\frac{1}{c_1} - \frac{1}{\gamma^2}), & j = k, \\ -\phi(\frac{1}{c_1} - \frac{1}{\gamma^2})\frac{1}{|N(k)|}, & j \in N(k), \ j \neq k, \\ 0 & otherwise. \end{cases} \tag{15.25}$$

Proof. First, we prove (15.22). Extrapolating the *Hamilton–Jacobi–Isaacs equation* in (15.10) for fixed ρ_k, we have

$$\begin{cases} \partial_t v_k(x,t) + \left(-\frac{1}{2c}BB^T + \frac{1}{2\gamma^2}CC^T\right)|\partial_x v_k(x,t)|^2 + \partial_x v_k(x,t)^T Ax + \frac{1}{2}[a_1(\rho_{k,1} - x_1)^2 \\ \quad + a_2(\rho_{k,2} - x_2)^2] = 0 \text{ in } \mathbb{R}^2 \times [0, T[, \\ v_k(x, T) = \Psi(\rho_k(T), x) \text{ in } \mathbb{R}^2. \end{cases} \tag{15.26}$$

Let us consider the value function

$$v_k(x,t) = \frac{1}{2}x^T \phi(t)x + h(t)^T x + \chi(t)$$

so that (15.26) can be rewritten as

$$\begin{cases} \frac{1}{2}x^T \dot{\phi}(t)x + \dot{h}(t)^T x + \dot{\chi}(t) + x^T \phi^T \left(-\frac{1}{2c}BB^T + \frac{1}{2\gamma^2}CC^T\right)\phi x \\ \quad + h(t)^T \left(-\frac{1}{2c}BB^T + \frac{1}{2\gamma^2}CC^T\right)h(t) + h(t)^T \left(-\frac{1}{2c}BB^T + \frac{1}{2\gamma^2}CC^T\right)\phi x \\ \quad + x^T \phi^T \left(-\frac{1}{2c}BB^T + \frac{1}{2\gamma^2}CC^T\right)h(t) + (\phi(t)x + h(t))^T Ax \\ \quad + \frac{1}{2}[a_1(\rho_{k,1} - x_1)^2 + a_2(\rho_{k,2} - x_2)^2] = 0 \text{ in } \mathbb{R}^2 \times [0, T[, \\ \phi(T) = S, \quad h(T) = \frac{1}{2}\hat{S}\rho_k(T) = -S\rho_k(T), \quad \chi(T) = \frac{1}{2}\rho_k(T)^T S\rho_k(T). \end{cases} \tag{15.27}$$

Since this is an identity in x, it reduces to system (15.22).
For the mean-field equilibrium control and worst-case disturbance we have

$$\begin{cases} u^*(x,t) = -\frac{1}{c}B^T(\phi(t)x + h(t)), \\ w^*(x,t) = \frac{1}{\gamma^2}C^T(\phi(t)x + h(t)), \end{cases} \tag{15.28}$$

and this proves (15.23).
By averaging the above expressions and substituting in $\dot{\bar{m}}_k(t) = A\bar{m}_k(t) + B\bar{u}_k(t) + C\bar{w}_k(t)$ we obtain $\dot{\bar{m}}_k(t) = A\bar{m}_k(t) + (-\frac{1}{c}BB^T + \frac{1}{\gamma^2}CC^T)(\phi(t)\bar{m}_k(t) + h(t))$ as in (15.21). In the stationary case, let $t \to \infty$, and set

$$\left(-\frac{1}{c}BB^T + \frac{1}{\gamma^2}CC^T\right) = -\frac{1}{c}\begin{bmatrix} 0 & 0 \\ 0 & 1 \end{bmatrix} + \frac{1}{\gamma^2}\begin{bmatrix} 0 & 0 \\ 0 & 1 \end{bmatrix} = \begin{bmatrix} 0 & 0 \\ 0 & 2x \end{bmatrix}, \tag{15.29}$$

where $x = -\frac{1}{2c} + \frac{1}{2\gamma^2}$. Then for ϕ we have

$$\begin{bmatrix} 2x\phi_{12}\phi_{21} & 2x\phi_{12}\phi_{22} \\ 2x\phi_{22}\phi_{21} & 2x\phi_{22}^2 \end{bmatrix} + \begin{bmatrix} 0 & \phi_{11} \\ 0 & \phi_{12} \end{bmatrix} + \begin{bmatrix} 0 & 0 \\ \phi_{11} & \phi_{12} \end{bmatrix} + \begin{bmatrix} a_1 & 0 \\ 0 & a_2 \end{bmatrix} = 0.$$

(15.30)

Likewise, for h we obtain

$$\begin{bmatrix} h_1 & h_2 \end{bmatrix} \begin{bmatrix} 0 & 0 \\ 2x\phi_{21} & 2x\phi_{22} \end{bmatrix} + \begin{bmatrix} 0 & h_1 \end{bmatrix} - \begin{bmatrix} a_1\rho_{k,1} & a_2\rho_{k,2} \end{bmatrix} = 0.$$

(15.31)

The above equations for h can be rewritten as

$$\begin{cases} 2xh_2\phi_{21} = a_1\rho_1, \\ 2xh_2\phi_{22} = a_2\rho_2 - h_1 \end{cases} \Rightarrow \begin{cases} h_2 = \frac{1}{2x\phi_{21}}a_1\rho_{k,1} = \frac{\phi_{21}}{2x\phi_{21}^2}a_1\rho_{k,1} = -\phi_{21}\rho_{k,1}, \\ h_2 = \frac{1}{2x\phi_{22}}(a_2\rho_{k,2} - h_1) = \frac{\phi_{22}}{2x\phi_{22}^2}(a_2\rho_{k,2} - h_1) \approx -\phi_{22}\rho_{k,2}. \end{cases}$$

(15.32)

In the above we have used $\phi_{21}^2 = -\frac{a_1}{2x}$, which is obtained from (15.29), and $\phi_{22}^2 = -\frac{a_2}{2x}$ and h_1 being negligible.

The closed-loop mean-field equilibrium control is given by

$$\begin{aligned} u^*(x(t), t) &= -\frac{1}{c}[0\ 1]\left(\begin{bmatrix} \phi_{11} & \phi_{12} \\ \phi_{21} & \phi_{22} \end{bmatrix}\begin{bmatrix} x_1(t) \\ x_2(t) \end{bmatrix} + \begin{bmatrix} h_1 \\ h_2 \end{bmatrix}\right) \\ &= -\frac{1}{c}\Big(\phi_{12}(\rho_{k,1} - x_1(t)) + \phi_{22}(\rho_{k,2} - x_2(t))\Big), \end{aligned}$$

(15.33)

and the closed-loop worst-case disturbance is

$$\begin{aligned} w^*(x(t), t) &= \frac{1}{\gamma^2}[0\ 1]\left(\begin{bmatrix} \phi_{11} & \phi_{12} \\ \phi_{21} & \phi_{22} \end{bmatrix}\begin{bmatrix} x_1(t) \\ x_2(t) \end{bmatrix} + \begin{bmatrix} h_1 \\ h_2 \end{bmatrix}\right) \\ &= \frac{1}{\gamma^2}\Big(\phi_{12}(\rho_{k,1} - x_1(t)) + \phi_{22}(\rho_{k,2} - x_2(t))\Big). \end{aligned}$$

(15.34)

Then, the mean states of neighbor populations are related by the following local interaction rule:

$$\begin{aligned} \begin{bmatrix} \dot{\bar{m}}_k(t) \\ \ddot{\bar{m}}_k(t) \end{bmatrix} &= A\bar{m}_k(t) + (-\frac{1}{c}BB^T + \frac{1}{\gamma^2}CC^T)(\phi(t)\bar{m}_k(t) + h(t)) \\ &= A\bar{m}_k(t) + \begin{bmatrix} 0 \\ 1 \end{bmatrix}(\frac{1}{c} - \frac{1}{\gamma^2})\Big(\phi_{12}(\rho_{k,1} - \bar{m}_k(t)) + \phi_{22}(\rho_{k,2} - \dot{\bar{m}}_k(t))\Big) \\ &= A\bar{m}_k(t) + \begin{bmatrix} 0 \\ 1 \end{bmatrix}(\frac{1}{c} - \frac{1}{\gamma^2})\Big(\phi_{12}(\frac{\sum_{j\in N(k)}\bar{m}_j(t)}{|N(k)|} - \bar{m}_k(t)) \\ &\quad + \phi_{22}(\frac{\sum_{j\in N(k)}\dot{\bar{m}}_j(t)}{|N(k)|} - \dot{\bar{m}}_k(t))\Big). \end{aligned}$$

(15.35)

After introducing the compact notation

$$\mu_{\bullet 1}(t) = \big(\bar{m}_1(t), \ldots, \bar{m}_p(t) \big)^T, \quad \mu_{\bullet 2}(t) = \big(\dot{\bar{m}}_1(t), \ldots, \dot{\bar{m}}_p(t) \big)^T,$$

and collecting dynamics (15.35) for all k in a single expression, we obtain the following second-order consensus dynamics:

$$\begin{bmatrix} \dot{\mu}_{\bullet 1}(t) \\ \dot{\mu}_{\bullet 2}(t) \end{bmatrix} = \begin{bmatrix} 0 & I \\ -\theta L & -\tilde{\theta}(L + \hat{\theta}I) \end{bmatrix}\begin{bmatrix} \mu_{\bullet 1}(t) \\ \mu_{\bullet 2}(t) \end{bmatrix} \quad t = 0, 1, \ldots,$$

(15.36)

where the initial condition is

$$\mu_{\bullet 1}(0) = \left(\ \bar{m}_1(0),\ldots,\bar{m}_p(0) \ \right)^T, \quad \mu_{\bullet 2}(t) = \left(\ \dot{\bar{m}}_1(t),\ldots,\dot{\bar{m}}_p(t) \ \right)^T = \left(\ 0,\ldots,0 \ \right)^T,$$

and where L is the normalized (one for the entries in the main diagonal, and the reciprocal of the degree of node i for each adjacent node of i in the ith row) graph-Laplacian matrix of the communication graph $G = (N,E)$. In the above equation, the parameters θ, $\hat{\theta}$, and $\tilde{\theta}$ are the elastic and damping coefficients and are obtained as a by-product from the Riccati equation.

Denoting

$$\hat{L} = \begin{bmatrix} 0 & I \\ \theta L & \tilde{\theta}(L + \hat{\theta}I) \end{bmatrix},$$

where $L = [L_{kj}]$ is the graph-Laplacian matrix of the network as in (15.25), dynamics (15.36) can be rewritten as

$$\dot{\bar{m}}(t) = -\hat{L}\bar{m}(t).$$

This concludes our proof.   □

Remark 1. *Dynamics (15.24) is a consensus dynamics, and as such it guarantees synchronization.*   □

From the above theorem, we derive the following scheme for the computation of an equilibrium:

$$\phi(t), h(t), \chi(t) \text{ as in (15.22)} \qquad \text{Riccati equations—backwards}$$

$$v_k(\cdot) \Big\downarrow \qquad\qquad \Big\uparrow \rho_k(t)$$

$$\dot{\bar{m}}(t) = -L\bar{m}(t) \qquad\qquad \text{Consensus dynamics—forwards}$$

The scheme includes a set of Riccati equations, which involve the variables $\phi(t)$, $h(t)$, $\chi(t)$. The local average $\rho_k(t)$ enters as input in the Riccati equations and plays the role of a target signal for population k. The output is the value function $v_k(x,t)$ for population k.

The second set of equations describe the consensus dynamics. The aforementioned value function enters as input in the consensus dynamics. Recall that the graph-Laplacian matrix L depends on $\phi(t)$, which derives from the Riccati equations. The output of the consensus dynamics is a new trajectory for the target signal ρ_k. At a fixed point, the target signal ρ_k coincides with the one entered in the Riccati equations at the beginning of the iteration.

15.5 ▪ Numerical example

Let us now consider the following numerical example. The example deals with 1000 players, $n = 10^3$, five populations, $p = 5$, and a discretized set of states $\mathscr{X} = \{x_{min}, x_{min} + 1, \ldots, x_{max}\}$, where $x_{min} = -50$ and $x_{max} = 50$. Assume that the graph $G = (V,E)$ is a chain. As for the step size, let us take $dt = 1$. The number of iterations is $T = 50$.

The state dynamics of each generator is approximated by the discrete-time equation

$$\begin{cases} x(t+1) = x(t) + \hat{\xi}(\rho_k - x) + \sigma \ rand[-1,1], & t = 0,1,\ldots,T-1, \\ x(0) = x. \end{cases} \tag{15.37}$$

We also consider a discretized version of the second-order consensus dynamics (15.24) by setting

$$\bar{m} = (\bar{m}_1, \ldots, \bar{m}_5, \dot{\bar{m}}_1, \ldots, \dot{\bar{m}}_5)^T.$$

After introducing the compact notation

$$\mu_{\bullet 1}(t) = \left(\begin{array}{ccc} \bar{m}_1(t), \ldots, \bar{m}_5(t) \end{array} \right)^T, \qquad \mu_{\bullet 2}(t) = \left(\begin{array}{ccc} \dot{\bar{m}}_1(t), \ldots, \dot{\bar{m}}_5(t) \end{array} \right)^T,$$

dynamics (15.24) has the form of the following second-order consensus dynamics:

$$\left[\begin{array}{c} \mu_{\bullet 1}(t) \\ \mu_{\bullet 2}(t) \end{array} \right] = \left[\begin{array}{cc} I & I \\ -\theta L & -\tilde{\theta}(L + \hat{\theta}I) + I \end{array} \right] \left[\begin{array}{c} \mu_{\bullet 1}(t-1) \\ \mu_{\bullet 2}(t-1) \end{array} \right] \quad t = 1, 2, \ldots, T, \qquad (15.38)$$

where the initial condition is

$$\mu_{\bullet 1}(0) = \left(\begin{array}{ccc} \bar{m}_1(0), \ldots, \bar{m}_5(0) \end{array} \right)^T, \qquad \mu_{\bullet 2}(0) = \left(\begin{array}{ccc} \dot{\bar{m}}_1(0), \ldots, \dot{\bar{m}}_5(0) \end{array} \right)^T = \left(\begin{array}{ccc} 0, \ldots, 0 \end{array} \right)^T,$$

and where L is the normalized (one for the entries on the main diagonal, and the reciprocal of the degree of node i for each adjacent node of i in the ith row) graph-Laplacian matrix of the communication graph $G = (V, E)$.

Furthermore, let m_{0k} be Gaussian with mean $\bar{m}_{0k}$ equal to 0 for every population k. Also, let the standard deviation $std(m_{0k})$ be equal to 15 for all k. Then, the initial state x in (15.37) is drawn randomly from m_{0k} for all k. The simulations' plots are obtained from the algorithm displayed below.

ALGORITHM 15.1. **Simulation algorithm for the synchronization of generators.**

Input: Set of parameters
Output: Machine's rotor angle $x(t)$
  1 : **Initialize.** Generate $x(0)$ given $\bar{m}_{0k}$ and $std(m_{0k})$
  2 : **for** time $iter = 0, 1, \ldots, T - 1$ **do**
  3 :   **if** $iter > 0$, **then** compute $m_k(.)$, $\bar{m}_k(t)$, and $std(m_k(.))$ for all k
  4 :   **end if**
  5 :   **for** player $i = 1, \ldots, n$ **do**
  6 :   Set $t = iter \cdot dt$ and compute control $u^*(t)$ using current $\bar{m}(t)$
  7 :   compute $\bar{m}(t) = (\mu_{\bullet 1}(t)\mu_{\bullet 2}(t))^T$ according to (15.38)
  8 :   compute new state $x(t + dt)$ by running (15.37)
  8 :   **end for**
  9 : **end for**
  10 : **STOP**

A first set of simulations examines the impact of different elastic and damping coefficients θ, $\hat{\theta}$, and $\tilde{\theta}$. Specifically, we simulate for increasing damping coefficients $\tilde{\theta} = 0.1, 0.35, 0.55$. The state is reset to the initial value every 10 iterations. Figs. 15.3–15.5 display the time plot of the microscopic dynamics (left) and the time plot of the standard deviation (right). The damping effect increases from Fig. 15.3 to Fig. 15.5.

In a second set of simulations, we investigate the influence of the Brownian motion. Figs. 15.6–15.8 show the time plot of the microscopic dynamics (left) and of the standard deviation (right) for three different values of the parameter $\sigma = 1, 2, 3$. Again, the state is reset to the initial value every 10 iterations. From Fig. 15.6 to Fig. 15.8, note that the higher the coefficient σ, the higher the tolerance in the synchronization dynamics.

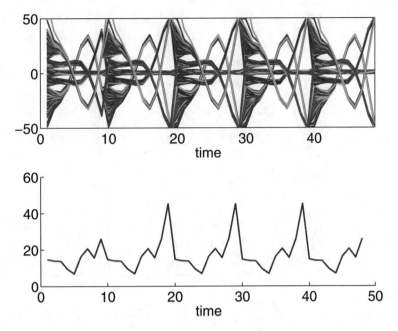

Figure 15.3. *Intercluster oscillation: the influence of the damping coefficient $\tilde{\theta} = 0.1$.*

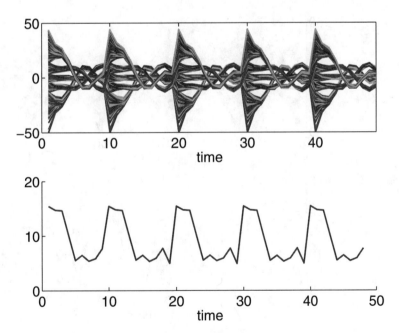

Figure 15.4. *Intercluster oscillation: the influence of the damping coefficient $\tilde{\theta} = 0.35$.*

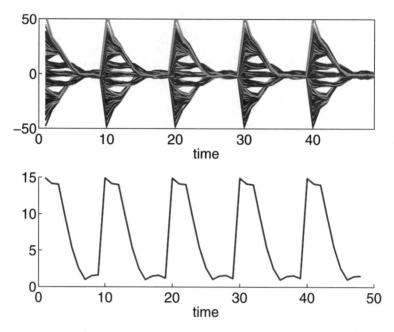

Figure 15.5. *Intercluster oscillation: the influence of the damping coefficient $\tilde{\theta} = 0.55$.*

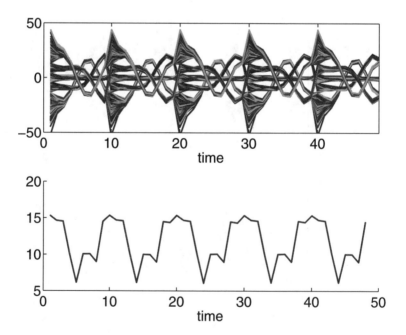

Figure 15.6. *Intercluster oscillation: the influence of the Brownian motion coefficient $\sigma = 1$.*

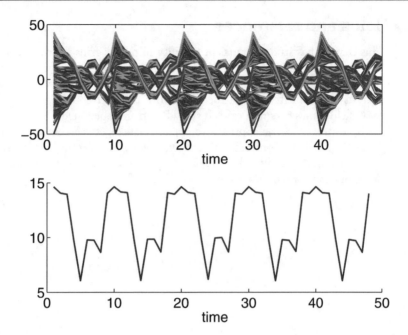

Figure 15.7. *Intercluster oscillation: the influence of the Brownian motion coefficient $\sigma = 2$.*

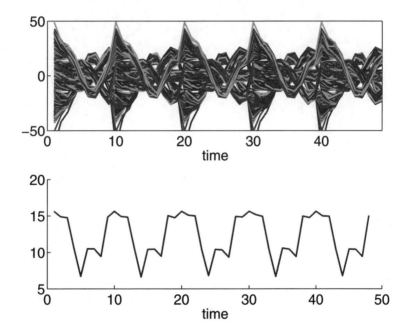

Figure 15.8. *Intercluster oscillation: the influence of the Brownian motion coefficient $\sigma = 3$.*

15.6 ▪ Notes and references

This chapter has studied transient stability in power grids via robust mean-field games. The study has shown multi-scale phenomena involving fast local synchronization and slow intercluster oscillation. Future directions of research involve a detailed stability analysis and the extension of the framework to other coupling effects. The impact of mean-field games on smart grids is still a broad and open field.

A control-theoretic approach to the unbalance between energy demand and supply is in [207]. Connections between the swing equation and the classical *Kuramoto* oscillators' dynamics are explored in [10, 88, 231]. Connections between *Kuramoto* oscillators' dynamics and consensus are pointed out in [193].

Chapter 16

Opinion Dynamics

16.1 ▪ Introduction

This chapter examines opinion dynamics in the context of two-player repeated games with vector payoffs. In general, people's opinions change with time as a consequence of the interactions among individuals.

Sometimes opinions converge to one or multiple values. Social scientists consider *emulation* or *herd behavior* among the root causes of convergence of opinions. It is common practice to distinguish opinion evolutionary patterns in three categories:

- *consensus*, when the opinions converge to a single value;

- *polarization*, when the consensus values are multiple but few in number;

- *plurality*, when the consensus values are multiple and numerous.

Polarization or plurality is often due to *bounded confidence*. Bounded confidence means that the interactions occur only among individuals with "similar" beliefs. Polarization and plurality can also arise in the presence of *stubborn individuals*. Stubborn individuals do not consider their neighbors' opinions. On the contrary, they try to influence the opinions of other individuals.

In the study of opinion dynamics, a common approach uses *Eulerian models*, that is to say models that assimilate the opinion propagation to a mass transport. Eulerian models consider the individuals as homogeneous, in the sense that they have no private identity and are simply identified by their opinions. Fig. 16.1 shows how to turn an Eulerian model into a network model. Consider a density distribution over the space of opinions (top), and imagine discretizing the state space. From left to right we consider increasingly smaller discretization steps. Here rectangles approximate the population in a given state. Now, let us associate to each rectangle a node of a network and let us link the nodes through weighted arcs (different thicknesses correspond to different weights). The resulting networks are depicted in the center row of the figure. The weights model the reciprocal influence between two nodes. Note that the influence between two nodes decreases with the *distance* in the space of opinions. According to the mass transport equation, opinions evolve continuously in space. This means that mass cannot jump. In other words, masses move from one node to an adjacent one, and this is usually described by a chain topology. The chain networks depicted in the bottom row of the figure illustrate this phenomenon.

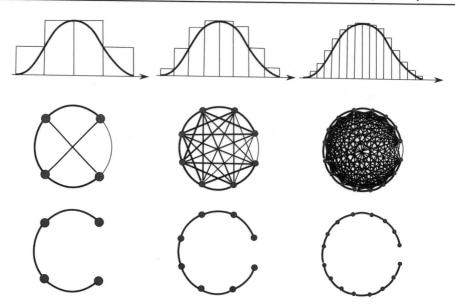

Figure 16.1. *From mean-field models to networks: (top) discretized Eulerian models with increasingly smaller steps from left to right; (center) interaction networks; (bottom) chain networks describing the mass transport between neighbor nodes. Reprinted with kind permission from Springer Science+Business Media* [25].

This chapter models opinion propagation as an *n*-player averaging process with dynamics subject to controls and adversarial disturbances. Adversarial disturbances are opinion leaders trying to influence the opinion of the players. As such, each individual faces an opinion leader in a two-player repeated game with vector payoffs (cf. Chapter 11). Modeling opinion dynamics as a game sheds light on the following aspects:

(i) *Strategic behavior*; namely, individuals are rational and they form their opinions in order to optimize their interests. As an example, they may wish to align their opinions to the mainstream opinion. Strategic behavior requires prediction capabilities, in the sense that the individuals must be able to anticipate the evolution of the mainstream opinion when the rest of the population acts rationally.

(ii) *Heterogeneous stubbornness*; that is to say that the individuals are differently stubborn and have different initial opinions.

(iii) *Local interactions*; namely, the individuals change their opinions or, better, play their strategic game, considering the reactions only of their neighbors, namely, the individuals of the neighbor populations.

We study conditions under which the players achieve robust consensus to some predefined target set. Such conditions build upon the approachability principle in repeated games with vector payoffs. Here the averaging process accounts for social emulation and the input represents the natural opinion changing rate of every individual.

This chapter is organized as follows. Section 16.2 formulates the problem. Section 16.3 discusses *Blackwell's Approachability Principle* (cf. Chapter 11) in connection with opinion dynamics. Section 16.4 gives the main results. Section 16.5 presents a numerical example. Finally, Section 16.6 provides notes and references.

16.2 ▪ Opinion dynamics via local averaging with adversaries

A simple model of opinion dynamics is derived from a classical model of consensus dynamics that also arises in the *Kuramoto oscillator* model [201]. Consider the synchronization of the phase angles of a set of N coupled oscillators, for which the dynamics of the ith oscillator is given by

$$\dot{\Theta}_i = \Omega_i + \frac{\mathscr{K}}{n} \sum_{j \in N} \sin(\Theta_j - \Theta_i),$$

where Θ_i is its phase and Ω_i is its (time-invariant) natural frequency. The coupling term in the right-hand side is responsible for the synchronization in that such a term regulates the angular velocity $\dot{\Theta}_i$ based on the deviation of the ith phase from the average phase computed over the population. The level of synchronization increases with the parameter $\mathscr{K}$ appearing in the global coupling term.

Now, the analogy assimilates oscillators to individuals, phases to opinions, and natural frequencies to natural opinion changing rates. Global coupling is a result of the interactions among the individuals, which depend on the distance between them. Thus the *Kuramoto oscillator* model is changed to

$$\dot{x}_i = \Omega_i + \frac{\mathscr{K}}{n} \sum_{j \in N} \sin(x_j - x_i)e^{|x_j - x_i|}, \quad i \in N,$$

where one can choose to weight the mutual interference between individuals using an exponential damping function.

It is well known from [193] that the *Kuramoto oscillator* model, after linearization around zero, turns into a classical consensus model of type

$$\dot{x}_i = \Omega_i + \sum_{j \in N_i}(x_j - x_i), \quad i \in N,$$

where N_i is the set of neighbors of i. The above model can be rewritten in vector form as

$$\dot{x} = \Omega - Lx,$$

where L is the graph-Laplacian matrix, with entries defined as

$$L_{ij} = \begin{cases} -1, & j \in N_i, \\ |N_i|, & j = i. \end{cases}$$

A discrete-time counterpart of the above model can be obtained as follows. Every player in a set $N = \{1,\ldots,n\}$ is characterized by a vector state $x_i(t) \in \mathbb{R}^{\tilde{n}}$ (its opinion). At every time t this state evolves in accordance with a distributed averaging process that represents the interaction of the player with its neighbors and under the influence of an input variable $u_i(t)$. Let the opinion $x_i(t)$ of player i be determined by the following discrete-time dynamics:

$$x_i(t+1) = \sum_{j=1}^{n} a_j^i(t)x_j(t) + u_i(t), \quad t = 0, 1, \ldots, \tag{16.1}$$

where $a^i = (a_1^i, \ldots, a_n^i)^T$ is a vector of nonnegative weights. These weights are consistent with the sparsity of the *communication graph* $\mathscr{G}(t) = (N, \mathscr{E}(t))$. A link $(j, i) \in \mathscr{E}(t)$

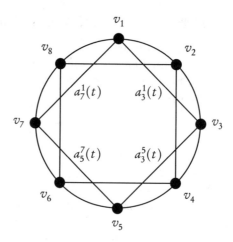

Figure 16.2. *Communication graph.*

implies that $a_j^i(t) \neq 0$, and this means that player j is a neighbor of player i at time t. A graphical illustration of a feasible communication graph is provided in Fig. 16.2.

In the above model, the coupling term accounts for emulation (an individual's opinion is influenced by those of its neighbors) and includes an additional input term (the natural opinion changing rate):

$$\text{"emulation"} = \sum_{j=1}^{n} a_j^i(t) x_j(t), \quad \text{"natural opinion changing rate"} = u_i(t).$$

We assume that the natural opinion changing rate is perturbed by the influence of a persuader. This results in a finite horizon n-player dynamic game in which the input variable for each player is the outcome of another game played against an external persuader.

More formally, for each player $i \in N$, the input $u_i(.)$ is the payoff of a repeated two-player game between player i (player i_1) and an (external) adversary (player i_2). Let S_1 and S_2 be the finite set of actions of players i_1 and i_2, respectively, and let us denote the set of mixed action pairs by $\Delta(S_1) \times \Delta(S_2)$ (set of probability distributions on S_1 and S_2).

For any pair of mixed strategies $(p(t), q(t)) \in \Delta(S_1) \times \Delta(S_2)$ for players i_1 and i_2 at time t, the expected payoff is

$$\begin{cases} u_i(t) = \sum_{j \in S_1, k \in S_2} p_j(t) \phi(j, k) q_k(t), \\ \sum_{j \in S_1} p_j(t) = 1, \quad \sum_{k \in S_2} q_k(t) = 1, \quad p_j, q_k \geq 0. \end{cases} \quad (16.2)$$

Put differently, $\phi(j, k) \in \mathbb{R}^{\tilde{n}}$ is essentially the vector payoff resulting from player i_1 playing the pure strategy $j \in S_1$ and player i_2 playing the pure strategy $k \in S_2$. In Fig. 16.3, we have the continuous action sets for the two players for the case that $S_1 = \{1, 2, 3\}$ and $S_2 = \{1, 2, 3\}$.

In order to enforce one of the aforementioned evolutionary patterns, such as consensus, polarization, or plurality, let us introduce the target set $X \subset \mathbb{R}^{\tilde{n}}$. To keep formalities reasonably simple, let the target set X be a closed convex target set or a convex subset of a nonconvex target set. Let us focus on the case where player i_1 wishes to steer his state $x_i(t)$ towards X, while player i_2 tries to push the same state far from it. The best-response strategy used by both players is the solution of a minimax game involving the distance of the state from X as payoff.

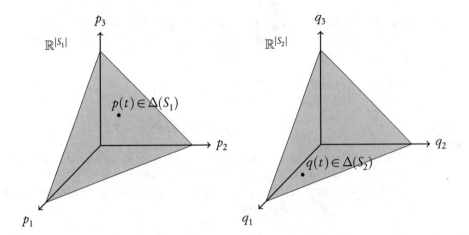

Figure 16.3. *Spaces of mixed strategies for the two players.*

In compact form the problem with finite horizon $[0, T]$ to be solved by player i takes the form

$$\min_{p(0)} \max_{q(0)} \cdots \min_{p(T-1)} \max_{q(T-1)} \sum_{t=0}^{T} \mathrm{dist}(x_i(t), X)^2,$$

$$\left.\begin{array}{l} p(t) \in \Delta(S_1), \quad q(t) \in \Delta(S_2), \\ x_i(t+1) = y_i(t) + u_i(t), \\ u_i(t) = \displaystyle\sum_{j \in S_1, k \in S_2} p_j(t) \phi(j, k) q_k(t), \end{array}\right\} \quad t = 0, \dots, T-1, \tag{16.3}$$

where $y_i(t)$ is the *space average* defined as

$$y_i(t) = \sum_{j=1}^{n} a_j^i(t) x_j(t). \tag{16.4}$$

Now, define by $\xi(t) = (x_1(t), \dots, x_n(t))$ the collective state of all players in the set N at time t and let the value function $V_{i,\tau}(\xi(t), t)$ be given. The value function represents the minimum cost over τ steps starting at $x_i(t)$, where $\tau = T - t$ for $t \in [0, T]$. From dynamic programming and the Bellman principle, we know that the value function satisfies the recursion

$$V_{i,\tau}(\xi(t), t) = \min_{p(t) \in \Delta(S_1)} \max_{q(t) \in \Delta(S_2)} \left\{ \mathrm{dist}(x_i(t), X)^2 + V_{i,\tau-1}(\xi(t+1), t+1) \right\}$$

$$= \mathrm{dist}(x_i(t), X)^2 + \min_{p(t) \in \Delta(S_1)} \max_{q(t) \in \Delta(S_2)} V_{i,\tau-1}(\xi(t+1), t+1),$$

with final value $V_{i,0}(\xi(T), T) = \mathrm{dist}(x_i(T), X)^2$. It is worth noting that the space average in (16.1) implies that the future distance from X of the state x_i depends on the current and future actions of players in N other than player i. As typical of noncooperative games, joint actions are not possible, and therefore the game payoff involves the worst-case cost obtained from maximizing over $u_j(t), j \in N, j \neq i$.

The receding horizon implementation of the optimal strategy for player i defines $p(t)$, and hence $u_i(t)$ in (16.2), as the minimizing argument for the T-stage problem with optimal value function $V_{i,T}(\xi(.), \cdot)$. The stability of a receding horizon control law can be ensured [175] by imposing a terminal constraint such as $x_i(T) \in X$ for all $i \in N$. We therefore impose the local constraint $\text{dist}(x_i(t+1), X) \le \text{dist}(y_i(t), X)$. After doing this, the problem statement can be rewritten as

$$V_{i,\tau}(\xi(t), t) = \text{dist}(x_i(t), X)^2 + \min_{\substack{p(t) \in \Delta(S_1)}} \max_{\substack{q(t) \in \Delta(S_2) \\ u_j(t), j \ne i, j \in N}} V_{i,\tau-1}(\xi(t+1), t+1) \qquad (16.5a)$$

$$\text{subject to } \text{dist}(x_i(t+1), X) \le \text{dist}(y_i(t), X). \qquad (16.5b)$$

Now, we investigate contractivity and invariance of sets for the collective dynamics (16.1)–(16.2). In doing this, we use the collective value function $\sum_{i=1}^{n} V_{i,T}(\xi(t), t)$ assuming that each player $i \in N$ adopts a T-stage receding horizon strategy with the optimal cost $V_{i,T}(\xi(.), \cdot)$ defined in (16.5).

16.3 ▪ Using *Blackwell's Approachability Principle*

This section explains how to bring *Blackwell's Approachability Principle* (cf. Chapter 11) into the framework. In preparation for this, let us make the following assumptions on the information structure of the model (16.1) [188, 185]. Let $A(t)$ be the weight matrix with (i, j)th element $a_j^i(t)$.

Assumption 16.1. *The matrix $A(t)$ is doubly stochastic with positive diagonal. Furthermore, there exists a scalar $\alpha > 0$ such that $a_j^i(t) \ge \alpha$ whenever $a_j^i(t) > 0$ for all t.*

The instantaneous graph $\mathcal{G}(t)$ need not be connected at any given time t; however, the union of the graphs $\mathcal{G}(t)$ over a period of time is assumed to be connected.

Assumption 16.2. *There exists an integer $Q \ge 1$ such that the graph $\left(N, \bigcup_{\tau=tQ}^{(t+1)Q-1} \mathcal{E}(\tau)\right)$ is strongly connected for every nonnegative integer t.*

Now, let us denote by G the one-shot vector-payoff game (S_1, S_2, x_i). Furthermore, consider $\lambda \in \mathbb{R}^{\tilde{n}}$ and let $\langle \lambda, G \rangle$ be the zero-sum game whose set of players and their action sets are as in the game G, and for which the payoff that player i_2 pays to player i_1 is $\lambda^T \phi(j, k)$ for every $(j, k) \in S_1 \times S_2$. We refer to $\langle \lambda, G \rangle$ as the *projected game*. The projected game $\langle \lambda, G \rangle$ is described by the matrix

$$\Phi_\lambda = [\lambda' \phi(j, k)]_{j \in S_1, k \in S_2},$$

and as a zero-sum one-shot game it has a value v_λ, where

$$v_\lambda := \min_{p \in \Delta(S_1)} \max_{q \in \Delta(S_2)} p' \Phi_\lambda q = \max_{q \in \Delta(S_2)} \min_{p \in \Delta(S_1)} p' \Phi_\lambda q.$$

We are in a position to introduce *Blackwell's Approachability Principle* for the opinion dynamics under study [49] (see also [48, Cor. 5.1]).

Assumption 16.3. *The projected game $\langle \lambda, G \rangle$ satisfies*

$$\min_{p \in \Delta(S_1)} \max_{q \in \Delta(S_2)} \left(2p^T \Phi_\lambda q + \left\| \sum_{j \in S_1, k \in S_2} p_j \phi(j, k) q_k \right\|^2 \right) \le 0 \quad \forall \lambda \in \mathbb{R}^{\tilde{n}}.$$

Recall that the condition in Assumption 16.3 is among the foundations of approachability theory since it requires that the value of the projected game satisfies $v_\lambda < 0$ whenever $\lambda \neq 0$. This is sufficient to guarantee that the average vector payoff of a two-player repeated game is locally almost surely convergent to the target set X (see, e.g., [57] and [73, Chap. 7]).

16.4 ▪ *Consensus, polarization,* and *plurality* using contractivity and invariance

In this section we prove contractivity and invariance for the collective dynamics (16.1) under the multi-stage receding horizon strategy defined in (16.5). In preparation for the main result, let us introduce the following lemmas. The first lemma states that the dynamics (16.1) is such that the sum of squared distances of the states x_i, $i \in N$, from the set X decreases with time.

Lemma 16.1. *Let Assumption 16.1 hold. Then the sum of squared distances of the states from the set X decreases; namely, it holds that*

$$\sum_{i=1}^{n} \mathrm{dist}(y_i(t), X)^2 \leq \sum_{i=1}^{n} \mathrm{dist}(x_i(t), X)^2.$$

Proof. Convexity of $\mathrm{dist}(., X)$ implies $\mathrm{dist}(y_i(t), X) \leq \sum_{j=1}^{n} a_j^i(t)\mathrm{dist}(x_j(t), X)$. Hence from convexity of $(.)^2$ we obtain

$$\mathrm{dist}(y_i(t), X)^2 \leq \sum_{j=1}^{n} a_j^i(t)\mathrm{dist}(x_j(t), X)^2.$$

Introducing a sum over $i = 1, \ldots, n$ in the left- and right-hand sides we get

$$\sum_{i=1}^{n} \mathrm{dist}(y_i(t), X)^2 \leq \sum_{i=1}^{n}\sum_{j=1}^{n} a_j^i(t)\mathrm{dist}(x_j(t), X)^2$$

$$= \sum_{j=1}^{n}\left(\sum_{i=1}^{n} a_j^i(t)\right)\mathrm{dist}(x_j(t), X)^2 = \sum_{j=1}^{n} \mathrm{dist}(x_j(t), X)^2,$$

where the last equality follows from the stochasticity of $A(t)$ in Assumption 16.1. □

Before introducing the next lemma, note that from the definition of $\mathrm{dist}(., X)$ and from (16.1) and (16.3), we get

$$\begin{aligned}
\mathrm{dist}(x_i(t+1), X)^2 &= \|x_i(t+1) - \Pi_X[x_i(t+1)]\|^2 \\
&\leq \|x_i(t+1) - \Pi_X[y_i(t)]\|^2 \\
&= \|y_i(t) + u_i(t) - \Pi_X[y_i(t)]\|^2 \\
&= \|y_i(t) - \Pi_X[y_i(t)]\|^2 + \|u_i(t)\|^2 + 2(y_i(t) - \Pi_X[y_i(t)])^T u_i(t).
\end{aligned}$$
$$(16.6)$$

In the following result, we establish that there exists an input $u_i(t)$ given by (16.2) such that the successor state $x_i(t+1)$ is closer to X than the space average $y_i(t)$.

Lemma 16.2. *If Assumptions 16.1–16.3 hold, then, for all $\xi(t) = (x_1(t), \ldots, x_n(t)) \in \mathbb{R}^{\tilde{n}} \times \cdots \times \mathbb{R}^{\tilde{n}}$, there exists $u_i(t)$ satisfying (16.2) and*

$$\text{dist}(x_i(t+1), X)^2 \leq \text{dist}(y_i(t), X)^2 \tag{16.7}$$

for each $i \in N$.

Proof. Rearranging the inequality in (16.6) we obtain

$$\text{dist}(x_i(t+1), X)^2 - \text{dist}(y_i(t), X)^2 \leq \|u_i(t)\|^2 + 2\big(y_i(t) - \Pi_X[y_i(t)]\big)^T u_i(t). \tag{16.8}$$

With $\lambda = y_i(t) - \Pi_X[y_i(t)]$, Assumption 16.3 implies that there exists a mixed strategy $p(t) \in \Delta(S_1)$ for player i_1 such that, for any mixed strategy $q(t) \in \Delta(S_2)$ of player i_2, $u_i(t) = \sum_{j \in S_1} \sum_{k \in S_2} p_j(t)\phi(j,k)q_k(t)$ satisfies

$$\|u_i(t)\|^2 + 2\big(y_i(t) - \Pi_X[y_i(t)]\big)^T u_i(t) \leq 0$$

for all $y_i(t) \in \mathbb{R}^{\tilde{n}}$. Therefore the bound (16.7) follows from (16.8). $\qquad\square$

From Lemma 16.2, we have that the constraint (16.5b) is feasible for all collective states $\xi = (x_1, \ldots, x_n) \in \mathbb{R}^{\tilde{n}} \times \cdots \times \mathbb{R}^{\tilde{n}}$. In the following result we make use of this property to show that the set

$$\Psi(r) = \left\{ (x_1, \ldots, x_n) \in \mathbb{R}^{\tilde{n}} \times \cdots \times \mathbb{R}^{\tilde{n}} \; \Big| \; \sum_{i=1}^{n} \text{dist}(x_i, X)^2 \leq r^2 \right\}$$

is invariant for all $r > 0$.

Lemma 16.3. *If Assumptions 16.1–16.3 hold, then, for any $r > 0$, $\Psi(r)$ is invariant for (16.1) under the receding horizon strategy defined by (16.5) for all $i \in N$.*

Proof. From the constraint in (16.5b) (which, by Lemma 16.2, is necessarily feasible) and Lemma 16.1, we get $\sum_{i=1}^{n} \text{dist}(x_i(t+1), X)^2 \leq \sum_{i=1}^{n} \text{dist}(y_i(t), X)^2 \leq \sum_{i=1}^{n} \text{dist}(x_i(t), X)^2$. Hence $\xi(t+1) \in \Psi(r)$ if $\xi(t) \in \Psi(r)$. $\qquad\square$

In the next lemma we provide bounds on the collective value function $\sum_{i=1}^{n} V_{i,T}(\xi, t)$ in terms of the sum of squared distances of individual players' states from X for all $\xi \in \Psi(R)$.

Lemma 16.4. *Under Assumptions 16.1–16.3, the value functions $V_{i,T}(\xi, \cdot)$, $i \in N$, satisfy, for all $\xi \in \mathbb{R}^{\tilde{n}} \times \cdots \times \mathbb{R}^{\tilde{n}}$,*

$$\sum_{i=1}^{n} \text{dist}(x_i, X)^2 \leq \sum_{i=1}^{n} V_{i,T}(\xi, \cdot) \leq (T+1) \sum_{i=1}^{n} \text{dist}(x_i, X)^2. \tag{16.9}$$

Proof. The lower bounds in (16.9) follow directly from (16.5a) and since $V_{i,T-1}(\xi, \cdot) \geq 0$ for any horizon $T \geq 1$ and all $\xi \in \mathbb{R}^{\tilde{n}} \times \cdots \times \mathbb{R}^{\tilde{n}}$. $\qquad\square$

To introduce the main result of this section, let us denote by $\Psi(r_T)$ a set of initial conditions $\xi(0)$ such that the state $x_i(T)$ of (16.1) is driven into X for all $i \in N$ by the optimal strategy for (16.5) with fixed terminal time $t = T$. Accordingly, let r_T be given by

$$r_T = \max\left\{ r \;\Big|\; \sum_{i=1}^{n} \text{dist}(\hat{x}_i(T), X)^2 = 0 \,\forall \xi(0) \in \Psi(r) \right\}.$$

In the above $\hat{x}_i(t)$ for $t = 0, \ldots, T$ is the trajectory of (16.1) under the minimax strategy with optimal value function $V_{i,T-t}\big((\hat{x}_1(t), \ldots, \hat{x}_n(t)), t\big)$ for all $i \in N$, with $\hat{x}_i(0) = x_i(0)$. Now, from Lemma 16.3 we know that X is invariant under any control law incorporating the constraint (16.5b). Then it follows that r_T is monotonically nondecreasing in T, and hence $\Psi(r_T) \subseteq \Psi(r_{T+1})$ for each $T = 0, 1, \ldots$.

We are now in a position to give a precise statement of the stabilizing properties of the control law defined by (16.5).

Theorem 16.5 (Exponential stability of X). *Let Assumptions 16.1–16.3 hold. For the system (16.1) with the receding horizon strategy with optimal cost $V_{i,T}(\xi(t), t)$ for all $i \in N$, the set X is exponentially stable with a region of attraction that contains $\Psi(r_T)$; namely, for all $\xi(0) \in \Psi(r_T)$ and each $t = 0, 1, \ldots$, we have*

$$\sum_{i=1}^{n} \text{dist}(x_i(t), X)^2 \leq \left(\frac{T}{T+1}\right)^t \sum_{i=1}^{n} V_{i,T}(\xi(0), 0). \tag{16.10}$$

Proof. First, from the definition of r_T and the positive invariance of $\Psi(r_T)$ we obtain the bound in (16.10). As a consequence, under the assumption $\xi(0) \in \Psi(r_T)$, for all $i \in N$ the terminal state of (16.1) verifies $\text{dist}(\hat{x}_i(T), X) = 0$ under the minimax strategy with optimal value function $V_{i,T-t}\big((\hat{x}_1(t), \ldots, \hat{x}_n(t)), t\big)$ for $t = 0, \ldots, T$ and $\hat{x}_i(0) = x_i(0)$. Then

$$V_{i,T}(\xi, \cdot) = V_{i,T-1}(\xi, \cdot) \quad \forall \xi \in \Psi(r_T).$$

In addition to this, from $\xi(0) \in \Psi(r_T)$ we have $\xi(t) \in \Psi(r_T)$ for all $t = 0, 1, \ldots$, and therefore

$$V_{i,T}(\xi(t), t) = \text{dist}(x_i(t), X)^2 + \min_{\substack{p(t) \in \Delta(S_1)}} \max_{\substack{q(t) \in \Delta(S_2) \\ u_j(t), j \neq i, j \in N}} V_{i,T-1}(\xi(t+1), t+1)$$

$$\geq \text{dist}(x_i(0), X)^2 + V_{i,T}(\xi(t+1), t+1)$$

for all $i \in N$. By introducing a sum over $i \in N$ and using the upper bound of (16.9) we obtain

$$\sum_{i=1}^{n} \big[V_{i,T}(\xi(t+1), t+1) - V_{i,T}(\xi(t), t) \big] \leq - \sum_{i=1}^{n} \text{dist}(x_i(t), X)^2$$

$$\leq - \frac{1}{T+1} \sum_{i=1}^{n} V_{i,T}(\xi(t), t)).$$

As a consequence we have $\sum_{i=1}^{n} V_{i,T}(\xi(t), t) \leq \left(\frac{T}{T+1}\right)^t \sum_{i=1}^{n} V_{i,T}(\xi(0), 0)$, and the lower bound of (16.9) yields (16.10). $\quad\square$

Contractivity and invariance are illustrated in Fig. 16.4. A contractive set Φ, including the target set X, exerts an attracting force on the state. Successive state samples $x_i(t)$ (black bullet) and $x_i(t+1)$ (gray bullet) are drawn closer and closer to Φ; see Fig. 16.4(left). Invariance implies that starting from a state inside a region Ψ, the state path remains confined in that region, as illustrated in Fig. 16.4(right).

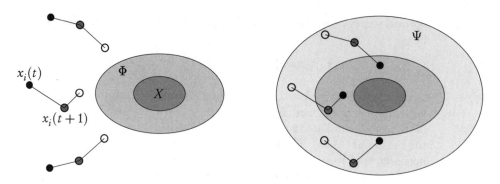

Figure 16.4. *Theorem 16.5: contractivity (left) and invariance (right).*

Table 16.1. *Simulation parameters for the opinion dynamics example.*

ν	x_{min}	x_{max}	σ	$std(m_0)$	T	$\bar{m}_0$	β	θ	$\tilde{\theta}$	$\hat{\theta}$
10^3	0	1	1	10	70	0.5	0.8	0.2	0.32	0.8

16.5 • Numerical example

This section develops a numerical example. The example shows contractivity and invariance, which, in the context of opinion dynamics, correspond to consensus, polarization, or plurality of the opinions. The example deals with $n = 5$ populations and $\nu = 10^3$ individuals. The state space is $\mathscr{X} = \{x_{min}, \ldots, x_{max}\}$, where $x_{min} = 0$ and $x_{max} = 1$. Each population is made by 200 players. The parameters of the example are listed in Table 16.1.

For the number of iterations, we set $T = 70$. The state of each player, namely his opinion, can be $\xi_{ref} \in \{0, 1\}$, but rather than jumps we consider smooth state trajectories in accordance with the following dynamics:

$$\xi(t+1) = \xi(t) + \text{round}\Big(\beta(\xi_{ref} - \xi(t)) + \sigma W(t)\Big), \quad \xi(0) \in \mathscr{X}, \qquad (16.11)$$

where $\sigma = 1$ and $W(t)$ is a random walk.

For every population $i \in N$, set $x_i(t) := (m_i(0, t), m_i(1, t))^T \in [0, 1]^2$, where $m_i(0, t)$ is the probability distribution of individuals that have opinion 0 or are changing their opinions to 0. Similarly, $m_i(1, t)$ is the probability distribution of individuals that have opinion 1 or are changing their opinions to 1. Furthermore, let us denote the average by $\bar{\rho}_i(t) = \sum_{\xi \in \{0,1\}} \xi m_i(\xi, t)$.

Also, let us introduce the compact notation

$$x_{\bullet j}(t) = \Big(\ x_{1j}(t), x_{2j}(t), x_{3j}(t), x_{4j}(t), x_{5j}(t)\ \Big)^T, \quad j = 1, 2,$$
$$\bar{\rho}(t) = \Big(\ \bar{\rho}_1(t), \bar{\rho}_2(t), \bar{\rho}_3(t), \bar{\rho}_4(t), \bar{\rho}_5(t)\ \Big)^T.$$

Then, for given weights a_j^i, $i, j = 1, \ldots, n$, dynamics (16.1) can be written as a second-order consensus dynamics of the form

$$\begin{bmatrix} x_{\bullet 1}(t+1) \\ x_{\bullet 2}(t+1) \end{bmatrix} = \begin{bmatrix} I & I \\ -\theta L & -\tilde{\theta}L + (1-\hat{\theta})I \end{bmatrix} \begin{bmatrix} x_{\bullet 1}(t) \\ x_{\bullet 2}(t) \end{bmatrix} t = 0, 1, \ldots,$$

where the initial condition is given by

$$\left[\begin{array}{c} x_{\bullet 1}(0) \\ x_{\bullet 2}(0) \end{array} \right] = \left[\begin{array}{c} \bar{\rho}(0) \\ 0 \end{array} \right].$$

In the above, L is the normalized graph-Laplacian matrix of the communication graph $\mathcal{G}(t) = (N, \mathcal{E}(t))$ where we have set the coefficients $\theta = 0.2$, $\tilde{\theta} = 0.32$, and $\hat{\theta} = 0.8$. Note that θ is the *elastic coefficient* and that $\tilde{\theta}$ and $\hat{\theta}$ determine the *damping coefficient* of the above second-order consensus dynamics. Recall that for the normalized graph-Laplacian matrix, we have one for the entries on the main diagonal, as well as the reciprocal of the degree of node i for each adjacent node of i in the ith row.

After introducing the above model, we run (16.11) and based on $(x_{11}(t), \ldots, x_{51}(t))^T$ we update the target state for the players at every iteration. For each population, the update consists in setting $\xi_{ref} = 1$ for precisely a number of players equal to the percentage expressed by $x_{i1}(t)$, $i \in N$. So, if $x_{i1}(t) = 0.7$, we set $\xi_{ref} = 1$ for 70% of players (their state approaches 1) and set $\xi_{ref} = 0$ for the remaining 30% of players (their state approaches 0).

Let us consider an initial Gaussian distribution m_0 with mean $\bar{m}_0$ equal to 0.5 and standard deviation $std(m_0)$ equal to 10.

The simulation algorithm is displayed in the following.

ALGORITHM 16.1. **Simulation algorithm for the opinion dynamics example.**

Input: Set of parameters as in Table 16.1.
Output: Player's state trajectories $\xi(t)$, $t \in [0, T]$, and tracked signal $x_i(t)$, $t \in [0, T]$, $i \in N$
  1 : **Initialize.** Generate $\xi(0)$ from Gaussian distribution with
       mean $\bar{m}_0$ and standard deviation $std(m_0)$,
  2 : **for** time $t = 0, 1, \ldots, T - 1$ **do**
  3 :   **if** $t > 0$, **then** compute $m(.)$, $\bar{m}(t)$, and $std(m(.))$,
  4 :   **end if**
  5 :   **for** player $i = 1, 2, \ldots, n$ **do**
  6 :   compute $x_i(t + 1)$ by solving (16.3),
  7 :   **end for**
  8 : **end for;**
  9 : **STOP**

The considered set of simulations analyzes the influence of the communication graph topology on the consensus dynamics.

The microscopic evolution of each agent's state is displayed in Fig. 16.5(left). From top to bottom we have considered different communication graphs, as illustrated in Fig. 16.6. The first topology is a directed chain with the first node, say v_1, which is the one corresponding to the cluster or population with higher average, acting as leader. The second topology is a directed chain with the last node, say v_5, which is the one associated with the cluster having the lower average, acting as leader (middle). The third topology has two connected components (bottom).

In Fig. 16.5 (right), we have the time plot of the average vector $(x_{11}(t), \ldots, x_{51}(t))^T$. In the first two examples, we have consensus. That is to say that the clusters converge to a common value. The reason for this is that the topology has one connected component. We observe intercluster oscillation during the transient. The interpretation of the first

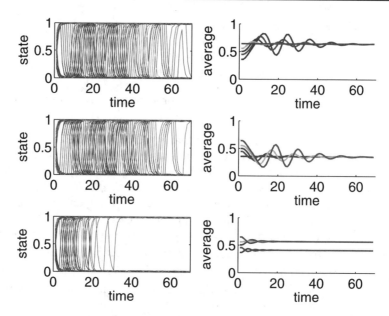

Figure 16.5. *Microscopic time plot (left) and time plot of the average distribution of each population (right).*

two examples is that opinion leaders (influential political parties) may attract the other populations. This leads to a consensus on the leader's opinion value. Differently, the third example shows polarization. In other words, two clusters of opinions arise. The main cause for this is that the topology has multiple connected components.

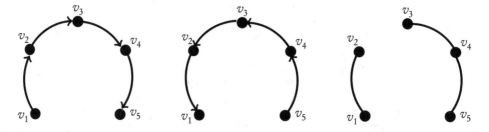

Figure 16.6. *Topologies for the three examples.*

16.6 • Notes and references

This chapter is based on [28] and [29]. In this chapter, we have examined opinion dynamics from a game-theoretic perspective. We have proved invariance and contractivity of the dynamics and have provided physical interpretations in terms of consensus, polarization, or plurality. The considered framework is general and involves also stubborn agents or opinion leaders. Possible future research directions may involve the extension of the analysis to population games with mean-field interactions, as well as averaging algorithms driven by Brownian motions.

Opinion dynamics is studied in [72, Sect. III] and [2]. *Emulation* or *herd behavior* leading to convergence of opinions is the main focus in [6, 19, 61, 72, 140, 115, 201].

Bounded confidence as a cause for polarization or plurality is examined in the well-known model proposed in [140]. *Stubborn agents* as a cause for polarization and plurality are investigated in [1] and [79, 234]. *Eulerian models* are discussed in [180]. *Strategic behavior* in opinion dynamics is studied in [86, 84]. More details on *heterogeneous stubbornness* are available in [2, 1, 79, 160, 181, 193]. *Local interactions* are examined in [69, 84, 115, 140, 163, 180].

Chapter 17

Bargaining

17.1 ▪ Introduction

This chapter discusses bargaining on a *dynamic coalitional game with transferable utility (TU game)* (cf. Chapter 5). The game is a repeated one that produces a stream of different characteristic functions. Our goal is to examine distributed agreement on solutions in the core of the game. *Distributed* means that bargaining involves only *neighbor players*. Neighborhoods are determined by a directed graph, where the vertices are the players and the directed links (i, j) indicate that player i receives a bid from player j at time t. A *bid* from player j is an allocation vector that says how player j would distribute resources among the other players. Then a bid has as many components as the number of players. Let us call such a graph the players' *neighbor-graph*. With the above distributed setup in mind, let us turn to consider the following bargaining mechanism. At every iteration, player i receives bids from some of his neighbors. These bids are combined in a weighted average and based on how such an average player i adjusts his own bid. In a first phase, player i looks for a trade-off between his original bid and the ones of his neighbors. In a second phase, player i verifies whether such an average satisfies the feasibility constraints. These constraints are a subset of the ones characterizing the core of the game, namely, individual rationality and stability with respect to sub-coalitions for the only coalitions involving player i; see Section 6.2. Occasionally we refer to the set of feasible allocations for a player as his *bounding set*. In other words, if a bid lies outside the bounding set for player i, then that same bid receives a veto on the part of player i. Player i then proceeds by projecting the vetoed bid on his bounding set, and the projection is the new bid of player i. In this chapter we study convergence properties of such a bargaining protocol. Results are based on some mild assumptions on the connectivity of the players' neighbor-graph.

This chapter is organized as follows. In Section 17.2, we introduce the game. In Section 17.3, we provide preliminary results. In Section 17.4, we prove the convergence results for the robust game. In Section 17.5, we report some numerical simulations to illustrate our theoretical study, and we conclude with notes and references in Section 17.6.

Given a set X and a scalar $\lambda \in \mathbb{R}$, the set λX is defined by $\lambda X \triangleq \{\lambda x \mid x \in X\}$. Given two sets $X, Y \subseteq \mathbb{R}^n$, the set sum $X + Y$ is defined by $X + Y \triangleq \{x + y \mid x \in X, y \in Y\}$.

Recall from Chapter 5 that a TU game is given by a tuple $\langle N, \eta \rangle$, where N is the set of players and η is the characteristic function. Also, η_S is the value of a coalition S for any nonempty coalition $S \subseteq N$. Furthermore, the *core* is the set of feasible allocation vectors characterized by

$$C(\eta) = \left\{ x \in \mathbb{R}^{|N|} \mid \sum_{i \in N} x_i = \eta_N, \ \sum_{i \in S} x_i \geq \eta_S \ \ \forall \text{ nonempty } S \subset N \right\},$$

where $x_i \in \mathbb{R}$ is an allocation value for player $i \in N$ and $x = (x_1, \ldots, x_{|N|})^T$.

17.2 ▪ Bargaining mechanism

Let $\langle N, \{v(t)\} \rangle$ be a dynamic TU game, where $N = \{1, \ldots, n\}$ is the set of players, and $\{v(t)\}$ for $t = 0, 1, 2, \ldots$ is a sequence of characteristic functions. Put differently, the dynamic TU game $\langle N, \{v(t)\} \rangle$ involves the players in a sequence of *instantaneous TU games* $\langle N, v(t) \rangle$, where $v(t) \in \mathbb{R}^m$ for all $t \geq 0$.

Let $m = 2^n - 1$ be the *number of possible (nonempty) coalitions* $S \subseteq N$, and let $v_S(t)$ be *the value assigned to a nonempty coalition* $S \subseteq N$ in the instantaneous game $\langle N, v(t) \rangle$.

Assumption 17.1. *There exists $v^{\max} \in \mathbb{R}^m$ such that for all $t \geq 0$, we have $v_N(t) = v_N^{\max}$ and $v_S(t) \leq v_S^{\max}$ for all nonempty coalitions $S \subset N$.*

Henceforth, we call $\langle N, v^{\max} \rangle$ a *robust game*.

Assumption 17.2. *We have $C(v^{\max}) \neq \emptyset$.*

From Assumptions 17.1 and 17.2 we have that the core $C(v(t))$ of the instantaneous game is nonempty at any time.

In a generic game $\langle N, \eta \rangle$ the *bounding set of player i* is given by

$$X_i(\eta) \ = \ \{ x \in \mathbb{R}^n \mid e_N^T x = \eta_N, \ e_S^T x \geq \eta_S \ \ \forall S \subset N \text{ with } i \in S \}, \tag{17.1}$$

where $e_S \in \mathbb{R}^n$ is the incidence vector for a nonempty coalition $S \subseteq N$, i.e., the vector with the coordinates given by

$$[e_S]_i = \begin{cases} 1 & \text{if } i \in S, \\ 0 & \text{else.} \end{cases}$$

Remarkably, the intersection of the bounding sets $X_i(\eta)$ of all players $i \in N = \{1, \ldots, n\}$ gives the core $C(\eta)$, i.e.,

$$C(\eta) = \cap_{i=1}^n X_i(\eta). \tag{17.2}$$

Let $x^i(t) \in \mathbb{R}^n$ be the bid of player i at time t, where the jth component $x_j^i(t)$ is the quantity that player i would give to player j. To keep formalities simple let $X_i(t)$ denote the bounding set of player i for the instantaneous game $\langle N, v(t) \rangle$, i.e., for all $i \in N$ and $t \geq 0$,

$$X_i(t) = \left\{ x \in \mathbb{R}^n \mid \sum_{j \in N} x_j = v_N(t), \ \sum_{j \in S} x_j \geq v_S(t) \forall \ S \subset N \text{ s.t. } i \in S \right\}. \tag{17.3}$$

A directed graph $\mathcal{G}(t) = (N, \mathcal{E}(t))$ determines the players and their neighbors at time t. Here, N is the vertex set and $\mathcal{E}(t)$ is the set of directed links. A link $(i, j) \in \mathcal{E}(t)$ exists if player j is a neighbor of player i at time t. We assume that $(i, i) \in \mathcal{E}(t)$ for all t. We refer to graph $\mathcal{G}(t)$ as a *neighbor-graph* at time t. A graphical illustration of a neighbor-graph at two time instances is available in Fig. 17.1.

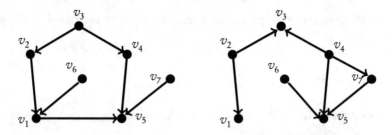

Figure 17.1. *Players' neighbor-graphs for six players and two different time instances. Reprinted with permission from IEEE* [185].

To introduce a formal definition of the bargaining mechanism, let $\mathcal{N}_i(t)$ be the set of neighbors of player i at time t (including himself), i.e., $\mathcal{N}_i(t) = \{j \in N \mid (i,j) \in \mathcal{E}(t)\}$, and let $a_{ij}(t) = 0$ for all $j \notin \mathcal{N}_i(t)$ and all t. The *bargaining mechanism* is then given by, for all $i \in N$ and $t \geq 0$,

$$x^i(t+1) = \Pi_{X_i(t)}\left[\sum_{j=1}^{n} a_{ij}(t)x^j(t)\right], \qquad (17.4)$$

where $\Pi_{X_i(t)}[\cdot]$ is the projection on $X_i(t)$ and $a_{ij}(t) \geq 0$ is a scalar weight that player i assigns to the bid $x^j(t)$ of player $j \in \mathcal{N}_i(t)$. The weights $a_{ij}(t)$, $j \in \mathcal{N}_i(t)$, are assumed to be deterministic scalars chosen by player i (for example, see [187] for some specific possible choices of $a_{ij}(t)$, $j \in \mathcal{N}_i(t)$). The initial allocations $x^i(0)$, $i = 1,\ldots,n$, are drawn randomly and independently of $\{v(t)\}$.

We now discuss the specific assumptions on the weights $a_{ij}(t)$ and the players' neighbor-graph that we use. We let $A(t)$ be the matrix with entries $a_{ij}(t)$.

Assumption 17.3. *Each matrix $A(t)$ is doubly stochastic with positive diagonal, and there exists a scalar $\alpha > 0$ such that $a_{ij}(t) \geq \alpha$ whenever $a_{ij}(t) > 0$.*

Assumption 17.4. *There is an integer $Q \geq 1$ such that the graph $\left(N, \bigcup_{\tau=tQ}^{(t+1)Q-1} \mathcal{E}(\tau)\right)$ is strongly connected for every $t \geq 0$.*

Assumptions 17.3 and 17.4 together guarantee that the players communicate sufficiently often to ensure that the information of each player is persistently diffused over the network in time to reach every other player.

17.3 ▪ Preliminaries: Nonexpansive projection and related bounds

We derive some preliminary results pertinent to the core of the robust game and some error bounds for polyhedral sets applicable to the players' bounding sets $X_i(t)$. We later use these results to establish the convergence of the bargaining mechanism in (17.4).

In our analysis we often use the following relation that is valid for the projection operation on a closed convex set $X \subseteq \mathbb{R}^n$: for any $w \in \mathbb{R}^n$ and any $x \in X$,

$$\|\Pi_X[w] - x\|^2 \leq \|w - x\|^2 - \|\Pi_X[w] - w\|^2. \qquad (17.5)$$

This property of the projection operation is known as a strictly nonexpansive projection property.

We next prove a result that relates the distance $\text{dist}(x, C(\eta))$ between a point x and the core $C(\eta)$ with the distances $\text{dist}(x, X_i(\eta))$ between x and the bounding sets $X_i(\eta)$. This result will be crucial in our later development.

Lemma 17.1. *Let $\langle N, \eta \rangle$ be a TU game with a nonempty core $C(\eta)$. Then, there is a constant $\mu > 0$ such that, for all $x \in \mathbb{R}^n$,*

$$\text{dist}^2(x, C(\eta)) \le \mu \sum_{i=1}^{n} \text{dist}^2(x, X_i(\eta)),$$

where μ depends on the collection of vectors $\{\tilde{e}_S \mid S \subset N, S \neq \emptyset\}$ with each $\tilde{e}_S$ being the projection of e_S on the hyperplane $H = \{x \in \mathbb{R}^n \mid e_N^T x = \eta_N\}$.

Proof **(Sketch).** We here sketch the two main facts that we use in the proof and refer the reader to the original paper [185] for a formal proof. The result essentially relies on the polyhedrality of the bounding sets $X_i(\eta)$ and the core $C(\eta)$ and a special relation for polyhedral sets known as the *Hoffman bound*. The *Hoffman bound* states that for a nonempty polyhedral set $\mathscr{P} = \{x \in \mathbb{R}^n \mid a_\ell^T x \le b_\ell, \ \ell = 1, \dots, r\}$, there exists a scalar $c > 0$ such that

$$\text{dist}(x, \mathscr{P}) \le c \sum_{\ell=1}^{r} \text{dist}(x, H_\ell) \quad \forall x \in \mathbb{R}^n, \tag{17.6}$$

where $H_\ell = \{x \in \mathbb{R}^n \mid a_\ell^T x \le b_\ell\}$ and the scalar c depends on the vectors $a_\ell, \ell = 1, \dots, r$, only.

Another fact that we use in the proof is that the square distance from a point x to a closed convex set X contained in an affine set H is given by

$$\text{dist}^2(x, X) = \|x - \Pi_H[x]\|^2 + \text{dist}^2(\Pi_H[x], X). \tag{17.7}$$

An illustration of the above equation is provided in Fig. 17.2. $\square$

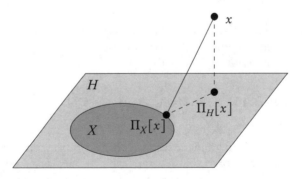

Figure 17.2. *Projection on a set X contained in an affine set H. Reprinted with permission from IEEE [185].*

From Lemma 17.1, we obtain the following result for the instantaneous game $\langle N, v(t) \rangle$.

Lemma 17.2. *Let Assumptions 17.1 and 17.2 hold. We then have for all $t \geq 0$, $x \in \mathbb{R}^n$ that*

$$\text{dist}^2(x, C(v(t))) \leq \mu \sum_{i=1}^{n} \text{dist}^2(x, X_i(t)),$$

where $C(v(t))$ is the core of the game $\langle N, v(t) \rangle$, $X_i(t)$ is the bounding set of player i, and μ is the constant from Lemma 17.1.

Proof. By Assumption 17.2, the core $C(v^{\max})$ is nonempty. Furthermore, under Assumption 17.1, we have $C(v^{\max}) \subseteq C(v(t))$ for all $t \geq 0$, implying that the core $C(v(t))$ is nonempty for all $t \geq 0$. Under Assumption 17.1, each core $C(v(t))$ is defined by the same affine equality corresponding to the grand coalition value, $e_N^T x = v_N^{\max}$. Moreover, each core $C(v(t))$ is defined through the set of hyperplanes $H_S(t) = \{x \in \mathbb{R}^n \mid e_S^T x \geq v_S(t)\}$, $S \subset N$, which have time-invariant normal vectors e_S, $S \subseteq N$. Thus, the result follows from Lemma 17.1. □

17.4 ▪ Convergence of the bargaining mechanism

In this section, we prove convergence of the bargaining mechanism (17.4) to a random allocation vector in the core of the robust game with probability 1. To this purpose, let us rewrite the bargaining mechanism (17.4) as

$$x^i(t+1) = w^i(t) + e^i(t) \quad \forall i \in N, t \geq 0, \tag{17.8}$$

where the linear term is the vector $w^i(t)$ defined as

$$w^i(t) = \sum_{j=1}^{n} a_{ij}(t) x^j(t) \quad \forall i \in N, t \geq 0, \tag{17.9}$$

and the nonlinear term is the error

$$e^i(t) = \Pi_{X_i(t)}[w^i(t)] - w^i(t). \tag{17.10}$$

In preparation for the main result, let us introduce two lemmas. The first lemma shows that the errors $e^i(t)$ decrease with time.

Lemma 17.3. *Let Assumptions 17.1 and 17.2 hold. Also, assume that each matrix $A(t)$ is doubly stochastic. Then, for the bargaining protocol (17.9)–(17.8), we have the following:*

(a) *The sequence $\left\{ \sum_{i=1}^{n} \|x^i(t+1) - x\|^2 \right\}$ converges for every $x \in C(v^{\max})$.*

(b) *The errors $e^i(t)$ in (17.10) are such that $\sum_{t=0}^{\infty} \sum_{j=1}^{n} \|e^i(t)\|^2 < \infty$. In particular, $\lim_{t \to \infty} \|e^i(t)\| = 0$ for all $i \in N$.*

Proof. To prove (a) it suffices to show that

$$\sum_{i=1}^{n} \|x^i(t+1) - x\|^2 \leq \sum_{i=1}^{n} \|x^i(t) - x\|^2 - \sum_{i=1}^{n} \|e^i(t)\|^2. \tag{17.11}$$

Actually, from (17.11) we have that the scalar sequence $\{\sum_{i=1}^{n} \|x^i(t+1) - x\|^2\}$ is nonincreasing for any given $x \in C(v^{\max})$, and therefore the sequence must be convergent.

To prove (17.11), from $x^i(t+1) = \Pi_{X_i(t)}[w^i(t)]$ and from (17.5) we have that for any $i \in N$, $t \geq 0$, and $x \in X_i(t)$,

$$\|x^i(t+1) - x\|^2 \leq \|w^i(t) - x\|^2 - \|e^i(t)\|^2. \tag{17.12}$$

Under Assumptions 17.1 and 17.2, and from the fact that relation (17.12) holds for all $x \in C(v^{\max})$, we can sum both sides over $i \in N$. Then we obtain for all $t \geq 0$ and $x \in C(v^{\max})$ that

$$\sum_{i=1}^{n} \|x^i(t+1) - x\|^2 \leq \sum_{i=1}^{n} \|w^i(t) - x\|^2 - \sum_{i=1}^{n} \|e^i(t)\|^2). \tag{17.13}$$

By the definition of $w^i(t)$ in (17.9), using the stochasticity of $A(t)$ and the convexity of the squared norm, we obtain $\sum_{i=1}^{n} \|w^i(t) - x\|^2 \leq \sum_{j=1}^{n} \left(\sum_{i=1}^{n} a_{ij}(t)\right) \|x^j(t) - x\|^2$. Since $A(t)$ is doubly stochastic, we have $\sum_{i=1}^{n} a_{ij}(t) = 1$ for all j, implying $\sum_{i=1}^{n} \|w^i(t) - x\|^2 \leq \sum_{i=1}^{n} \|x^i(t) - x\|^2$. By substituting this relation in (17.13), we arrive at (17.11), and this concludes the proof of condition (a).

To prove (b) let us sum both sides in (17.11) over $t = 0, \ldots, s$. By taking the limit as $s \to \infty$, we obtain $\sum_{t=0}^{\infty} \sum_{i=1}^{n} \|e^i(t)\|^2 \leq \sum_{i=1}^{n} \|x^i(0) - x\|^2$, and this implies that $\lim_{t \to \infty} e^i(t) = 0$ for all $i \in N$. □

The next lemma involves the instantaneous average of players' allocations defined as

$$y(t) = \frac{1}{n} \sum_{j=1}^{n} x^j(t) \qquad \forall t \geq 0.$$

In the lemma we show that the deviation of $x^i(t)$ for any player i from the average $y(t)$ converges to 0 as time goes to infinity.

Lemma 17.4. *Let Assumptions 17.3 and 17.4 hold. Suppose that for the bargaining protocol* (17.9)–(17.8) *we have* $\lim_{t \to \infty} \|e^i(t)\| = 0$ *for all* $i \in N$. *Then, for every player* $i \in N$ *we have*

$$\lim_{t \to \infty} \|x^i(t) - y(t)\| = 0, \qquad \lim_{t \to \infty} \|w^i(t) - y(t)\| = 0.$$

Proof (**Sketch**). The proof essentially uses the line of analysis that has been employed in [188], where the sets $X_i(t)$ are static in time, i.e., $X_i(t) = X_i$ for all t. In addition, we also use the rate result for doubly stochastic matrices that has been established in [186]. We refer the reader to the original paper [185] for a formal proof. □

Lemma 17.4 captures the effects of the matrices $A(t)$ that represent players' neighborgraphs. At the same time, Lemma 17.3 is basically a consequence of the projection property only. So far, the polyhedrality of the sets $X_i(t)$ has not been used at all. We now put all pieces together.

Bringing together Lemmas 17.2–17.4, we get the following result.

Theorem 17.5. *Consider a robust TU game* $\langle N, v^{\max} \rangle$, *and let Assumptions 17.1–17.4 hold. Also, assume that* Prob $\{v(t) = v^{\max} \ i.o.\} = 1$, *where i.o. stands for infinitely often. Then,* *the players' allocations $x^i(t)$ generated by the bargaining protocol (17.9)–(17.8) converge with probability 1 to an allocation in the core $C(v^{\max})$; i.e., there is a random vector $\tilde{x} \in C(v^{\max})$ such that $\lim_{t\to\infty} \|x^i(t) - \tilde{x}\| = 0$ for all $i \in N$ with probability 1.*

Proof. By Lemma 17.3, for each player $i \in N$, the sequence $\{\sum_{i=1}^{n} \|x^i(t) - x\|^2\}$ is convergent for every $x \in C(v^{\max})$ and $\|e^i(t)\| \to 0$. Then, by Lemma 17.4, we have $\|x^i(t) - y(t)\| \to 0$ for every i. Hence, for every $x \in C(v^{\max})$,

$$\{\|y(t) - x\|\} \text{ is convergent.} \tag{17.14}$$

We want to show that $\{y(t)\}$ is convergent and that its limit is in the core $C(v^{\max})$ with probability 1. For this, we note that since $x^i(t+1) \in X_i(t)$, it holds that, for all $t \geq 0$,

$$\sum_{i=1}^{n} \operatorname{dist}^2(y(t+1), X_i(t)) \leq \sum_{i=1}^{n} \|y(t+1) - x^i(t+1)\|^2.$$

The preceding relation and $\|x^i(t) - y(t)\| \to 0$ for all $i \in N$ (cf. Lemma 17.4) imply that

$$\lim_{t\to\infty} \sum_{i=1}^{n} \operatorname{dist}^2(y(t+1), X_i(t)) = 0.$$

Under Assumptions 17.1 and 17.2, by Lemma 17.2, we obtain, for all $t \geq 0$,

$$\operatorname{dist}^2(y(t+1), C(v(t))) \leq \mu \sum_{i=1}^{n} \operatorname{dist}^2(y(t+1), X_i(t)).$$

By combining the preceding two relations, we see that

$$\lim_{t\to\infty} \operatorname{dist}^2(y(t+1), C(v(t))) = 0. \tag{17.15}$$

By our assumption, the event $\{v(t) = v^{\max} \text{ infinitely often}\}$ happens with probability 1. We now fix a realization $\{v_\omega(t)\}$ of the sequence $\{v(t)\}$ such that $v_\omega(t) = v^{\max}$ holds infinitely often (for infinitely many t's). Let $\{t_k\}$ be a sequence such that $v_\omega(t_k) = v^{\max}$ for all k. All the variables corresponding to the realization $\{v_\omega(t)\}$ are denoted by a subscript ω. By relation (17.14), the sequence $\{y_\omega(t)\}$ is bounded, and therefore $\{y_\omega(t_k)\}$ is bounded. Without loss of generality (by passing to a subsequence of $\{t_k\}$ if necessary), we assume that $\{y_\omega(t_k)\}$ converges to some vector $\tilde{y}_\omega$, i.e., $\lim_{k\to\infty} y_\omega(t_k) = \tilde{y}_\omega$. This and (17.15) imply that $\tilde{y}_\omega \in C(v^{\max})$. Then, by relation (17.14), we have that $\{\|y_\omega(t) - \tilde{y}_\omega\|\}$ is convergent, from which we conclude that $\tilde{y}_\omega$ must be the unique accumulation point of the sequence $\{y_\omega(t)\}$, i.e.,

$$\lim_{t\to\infty} y_\omega(t) = \tilde{y}_\omega, \qquad \tilde{y}_\omega \in C(v^{\max}). \tag{17.16}$$

Since (17.16) is true for every realization ω such that $v_\omega(t) = v^{\max}$ holds infinitely often and since $\operatorname{Prob}\{v(t) = v^{\max} \text{ i.o.}\} = 1$, it follows that the sequence $\{y(t)\}$ converges with probability 1 to a random point $\tilde{y} \in C(v^{\max})$. By Lemma 17.4, we have $\|x^i(t) - y(t)\| \to 0$ for every i. Thus, the sequences $\{x^i(t)\}$, $i = 1, \ldots, n$, converge with probability 1 to a common random point in the core $C(v^{\max})$. $\quad\square$

Table 17.1. *Coalitions' values for the two simulations scenarios.*

	$v_{\{1\}}$	$v_{\{2\}}$	$v_{\{3\}}$	$v_{\{i,j\}}$ for all i, j	$v_{\{1,2,3\}}$
I	$[4,7]$	$[0,3]$	0	0	10

17.5 ▪ Numerical example

In this section we provide a numerical example that illustrates the convergence behavior of the bargaining mechanism (17.9)–(17.8).

Consider a three-player dynamic TU game, where the characteristic function is as in Table 17.1. Set the number of coalitions $m = 7$. The characteristic functions $v_S(t)$ for coalitions $\{1\}$ and $\{2\}$ are drawn independently with identical uniform distribution over an interval. All the other coalition values are zero except for the grand coalition, which has value 10.

The algorithm used for the simulation is the one displayed below.

ALGORITHM 17.1. **Simulation algorithm for the bargaining example.**

Input: Game $\langle N, \{v(t)\}\rangle$, neighbor-graph $\mathscr{G}(t) = (N, \mathscr{E}(t))$
Output: Allocation vectors $x^i(t)$ for all $i \in N$
 1 : **Initialize.** Set the initial allocations $x^i(0)$
 2 : **for** time $t = 0, 1, \ldots, T-1$ **do**
 3 :   **for** player $i = 1, \ldots, n$ **do**
 4 :   run the bargaining mechanism (17.4)
 5 :   **end for**
 6 : **end for**
 7 : **STOP**

We run 50 different Monte Carlo trajectories, each one having 100 iterations. All plots include the sampled average and sampled variance for the 50 trajectories that were simulated. The initial allocations are set to $x^1(0) = [10\ 0\ 0]^T$, $x^2(0) = [0\ 10\ 0]^T$, and $x^3(0) = [0\ 0\ 10]^T$.

The graphs for the times $t = 0, 1, 2$ are as follows: players 2 and 3 connected at time $t = 0$ (see Fig. 17.3(a)), then players 3 and 1 connected at time $t = 1$ (Fig. 17.3(b)), and finally players 1 and 2 connected at time $t = 2$ (Fig. 17.3(c)). These graphs are then repeated consecutively in the same order. In this way, the players' neighbor-graph is connected every 3 time units (Assumption 17.4 is satisfied with $Q = 2$).

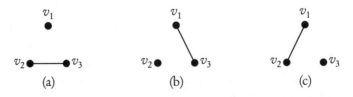

Figure 17.3. *Topology of players' neighbor-graph at three distinct times: $t = 0, 1,$ and 2. Reprinted with permission from IEEE* [185].

The matrices that we associate with these three graphs, are given by, respectively,

$$A(0) = \begin{bmatrix} 1 & 0 & 0 \\ 0 & \frac{1}{2} & \frac{1}{2} \\ 0 & \frac{1}{2} & \frac{1}{2} \end{bmatrix}, \qquad A(1) = \begin{bmatrix} \frac{1}{2} & 0 & \frac{1}{2} \\ 0 & 1 & 0 \\ \frac{1}{2} & 0 & \frac{1}{2} \end{bmatrix}, \qquad A(2) = \begin{bmatrix} \frac{1}{2} & \frac{1}{2} & 0 \\ \frac{1}{2} & \frac{1}{2} & 0 \\ 0 & 0 & 1 \end{bmatrix}.$$

At any time t, the matrix $A(t)$ is doubly stochastic, with positive diagonal, and every positive entry bounded below by $\frac{1}{2}$, so Assumption 17.3 is satisfied with $\alpha = \frac{1}{2}$. All simulations are carried out with MATLAB. The run time of each simulation is around 90 seconds.

The characteristic function $v^{\max}$ for the robust game is $v^{\max} = [7\,3\,0\,0\,0\,0\,10]^T$, and the resulting core of the robust game is given by

$$C(v^{\max}) = \{x \in \mathbb{R}^3 : x_1 \geq 7, x_2 \geq 3, x_3 \geq 0, x_1 + x_2 \geq 0, x_1 + x_3 \geq 0,$$
$$x_2 + x_3 \geq 0, x_1 + x_2 + x_3 = 10\}.$$

This core contains a single point, namely $[7\,3\,0]^T$. To ensure that $v(t) = v^{\max}$ infinitely often, as required by Theorem 17.5 for the convergence of the protocol, we adopt the following randomization mechanism. At each time $t = 1, \ldots, 100$, we flip a coin, and if the outcome is "heads" (probability $1/2$), the coalition values $v_{\{1\}}(t)$ and $v_{\{2\}}(t)$ are extracted from the intervals $[4,7]$ and $[0,3]$, respectively, with uniform probability independently of the other times. If the outcome of the coin flip is "tails," then we assume that the robust game realizes and take $v(t) = v^{\max}$.

We next present the results obtained by the Monte Carlo runs for the bargaining protocol in (17.9)–(17.8).

At time $t = 1$, bargaining involves players 2 and 3, who update the allocations, respectively, as $x^2(1) = [0\ 5\ 5]^T$ and $x^3(1) = [0\ 5\ 5]^T$. These allocations are feasible for their bounding sets, so the projections on these sets are not performed. At time $t = 2$, the bargaining involves players 1 and 3, who update their allocations, respectively, as $x^1(2) = [5\ 2.5\ 2.5]^T$ and $x^3(2) = [5\ 2.5\ 2.5]^T$. Again, these allocations are feasible for their bounding sets and the projections are not performed. Finally, at time $t = 3$, the bargaining involves players 1 and 2, who update their allocations, resulting in $x^1(3) = [7\ 1.5\ 1.5]^T$ and $x^2(3) = [2.5\ 3.75\ 3.75]^T$. Notice that $x^1(3)$ is obtained after player 1 projects onto his bounding set.

In Figs. 17.4 and 17.5, we report our simulation results for the average of the sample trajectories obtained by Monte Carlo runs. Fig. 17.4 shows the sampled average and variance of the allocations $x^i(t)$, $i = 1, 2, 3$, per iteration t. In accordance with the convergence result of Theorem 17.5, the sampled averages of the players' allocations $x^i(t)$ converge to the same point, namely $x = [7\,3\,0]^T$, which is in the core of the robust game $C(v^{\max})$. Fig. 17.5 shows that the sample average and sampled variance of the errors $e^i(t)$ converge to 0, as expected in view of Lemma 17.3(b).

17.6 ▪ Notes and references

This chapter is heavily based on [185]. For a sequence of TU games, each with a random characteristic function, we have designed a decentralized allocation process defined over a communication graph of players. The proposed bargaining mechanism is proven to converge with probability 1 to the robust game under mild assumptions on the communication topology and the stochastic properties of the random characteristic function. This bargaining application is an opportunity to introduce novel aspects, including (i) the

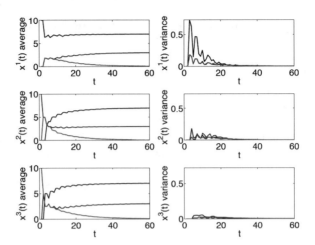

Figure 17.4. *Sampled average (left) and variance (right) of players' allocations $x^i(t)$, $i =$ 1, 2, 3, for the bargaining protocol (17.9)–(17.8) and the robust game associated with the data in Table 17.1. Sampled averages of the allocations $x^i(t)$ converge to the same point $\tilde{x} = [7\,3\,0]^T \in C(v^{\max})$, while sampled variances decrease to zero. Reprinted with permission from IEEE [185].*

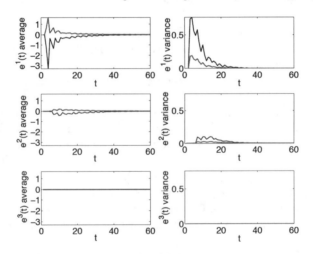

Figure 17.5. *Sampled average (left) and sampled variance (right) of the errors $e^i(t)$, $i = 1, 2, 3$, for the bargaining protocol (17.9)–(17.8) and the robust game associated with the data in Table 17.1. Sampled averages and the variances of the errors $e^i(t)$ converge to zero. Reprinted with permission from IEEE [185].*

formalization of a dynamic coalitional game with transferable utility, (ii) the definition of a robust game, and (iii) the use of a time-varying communication graph over which the bargaining mechanism takes place.

More details on the choice of the edge weights in a consensus problem are in [187]. The strictly nonexpansive projection property is discussed in [93, volume II, 12.1.13 lemma on page 1120]. The *Hoffman bound* is established by Hoffman [120].

In this chapter we present only a sketch of the proof of Lemma 17.4. We refer the reader to the original paper [185] for a formal proof.

Chapter 18

Pedestrian Flow

18.1 ▪ Introduction

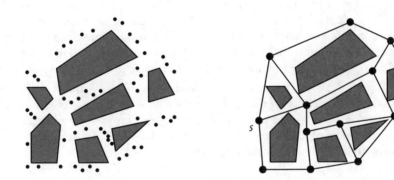

Figure 18.1. *Pedestrian flow (left) and the corresponding network model (right) with a source node s and a destination node d.*

This chapter deals with a pedestrian flow problem. The problem is initially modeled as an *optimal planning problem* over a discrete state space. The optimal planning problem is then turned into a special mean-field game where the players share a same common cost functional. After introducing the mean-field game, we solve it and provide stability conditions which mirror classical convergence conditions in repeated games with vector payoffs. Such conditions can be reviewed also as set inclusion conditions.

The problem involves a continuum of pedestrians walking through the center of a city. Once at a crossroads, a routing policy tells the pedestrians along which way to proceed. For such a scenario, we formulate an optimal planning problem over a network. Actually, we assume that the pedestrians are the players traversing the edges of a network in an attempt to reach a destination node starting from a source node (see, e.g., Fig. 18.1). The microscopic part of the model describes the players jumping from one edge to an adjacent one according to a continuous-time Markov model. The transition rates represent the controls. The macroscopic part of the model involves the dynamics of the density in each edge. Such dynamics are forward *Kolmogorov ordinary differential equations* subject to adversarial disturbances.

In Section 18.2 we formulate the pedestrian flow problem as an optimal planning problem. In Section 18.3 we turn the problem into a mean-field game with common cost functional for the players. In Section 18.4 we illustrate an extended state space solution approach. In Section 18.5 we solve the mean-field game and provide stability conditions. In Section 18.6 we provide simulations. In Section 18.7 we provide conclusions, notes, and references.

18.2 ▪ Model and problem setup

Consider a graph $G = (V, E)$, where $V = \{1, \dots, n\}$ is the set of vertices and $E = \{1, \dots, m\}$ is the set of edges. Vertices correspond to crossroads, and edges correspond to streets. Let $\varepsilon^+(i)$ and $\varepsilon^-(i)$ be the sets of edges departing from vertex i and edges arriving at vertex i, respectively, for all vertices $i \in V$. Occasionally, we will call the edges in $\varepsilon^+(i)$ *outgoing edges from i* and those in $\varepsilon^-(i)$ *incoming edges to i*. Let a time horizon window $[0, T]$ be given, and consider a continuum of pedestrians. Pedestrians are modeled as particles, each one characterized by state $X(t) \in E$ at time $t \in [0, T]$. The state of a particle indicates the edge where the particle currently lies. A particle in a given edge means a pedestrian walking along the corresponding street. When the pedestrians reach a crossroad a routing policy decides which way they proceed based on the congestion configuration of the network. The routing policy is given by a vector-valued function $\alpha(.) : \mathbb{R}_+ \to [0, 1]^m$, $t \mapsto \alpha(t)$, where $[0, 1]^m$ denotes the m-dimensional column vector whose entries are within the interval $[0, 1]$. There is no accumulation of pedestrians at the crossroads. Therefore, for all vertices $i \in V$ from the conservation of the mass, it must hold that $\sum_{e \in \varepsilon^+(i)} \alpha_e = 1$, where α_e is the eth entry of $\alpha(t)$. In other words, $\alpha(t)$ lives in $\Delta^{|\varepsilon^+(1)|} \times \dots \times \Delta^{|\varepsilon^+(n)|}$, where $\Delta^{|\varepsilon^+(i)|}$ is the simplex in $\mathbb{R}^{|\varepsilon^+(i)|}$ and $|\varepsilon^+(i)|$ is the cardinality of set $\varepsilon^+(i)$. Recall that the cardinality of a set is the number of elements in that set, and therefore $|\varepsilon^+(i)|$ is the number of outgoing edges from i for all vertices $i \in V$. Let the current state of a pedestrian be $k \in E$. The dynamics of that pedestrian follows the continuous-time Markov stochastic process:

$$\{X(t),\, t \geq 0\},$$

$$q_{kj}(h, \phi_k, \alpha_j) = \begin{cases} \alpha_j \phi_k h, & j \in Adj(k), \\ 1 - \phi_k h, & j = k, \\ 0 & \text{otherwise}, \end{cases} \tag{18.1}$$

where

- $q_{kj}(h, \phi_k, \alpha_j)$ (q_{kj}) are the infinitesimal transition probabilities from k to j;

- h is the infinitesimal time interval;

- $\phi_k \in \mathbb{R}_+$ is the transition rate in state $k \in E$;

- $Adj(k) = \{j \in E \mid j \in \varepsilon^+(i), k \in \varepsilon^-(i)\}$ is the set of adjacent edges to k.

From previous consideration on mass conservation, the routing policy α appearing in (18.1) lives in

$$\mathcal{U} = \{\alpha \in [0, 1]^m \mid \{\alpha_j\}_{j \in \varepsilon^+(i)} \in \Delta^{|\varepsilon^+(i)|} \,\forall i = 1, \dots, n\}.$$

The above is essentially equivalent to saying that $\sum_{j \in \varepsilon^+(i)} \alpha_j = 1$ for all $i = 1, \dots, n$.

We model congestion using a density function on the edges. To do this, denote by ρ the vector of densities on edges, which means that the sum of the components is equal to one. Thus we have $\rho \in \mathscr{D} := \{\hat{\rho} \in [0, 1]^m : \sum_{e \in E} \hat{\rho}_e = 1\}$, where $\hat{\rho}_e$ is the eth entry of $\hat{\rho}$. Let the flow function $f(\cdot) : \mathcal{D} \to \mathbb{R}_+^m$ be given by $f_e(\rho) = \phi_e \rho_e$, where $f_e(\rho)$ is the eth entry of $f(\rho)$. In other words, the flow function maps densities to flows across the edges. To maintain notation reasonably simple, we have assumed that the flow is linear in the density. With the above in mind, we are in a position to provide a precise formulation of the dynamics of the density, which is given by the following *Kolmogorov ordinary differential equation*:

$$\begin{cases} \dot{\rho}(t) = \left(\tilde{B}^T(\alpha)\hat{B} - I\right)f(\rho), \\ \rho(0) = \rho_0, \end{cases} \tag{18.2}$$

where the following hold:

- $\tilde{B}(.)$ is a matrix-valued function linking vertices to outgoing edges. In particular, $\tilde{B}(.) : \mathcal{U} \to [0,1]^{n \times m}$, $\alpha \mapsto \tilde{B}(\alpha)$. Furthermore, $\tilde{B}_{ij}(\alpha) = \alpha_j$ if $j \in \varepsilon^+(i)$ and $\tilde{B}_{ij}(\alpha) = 0$ otherwise. Here we use $[0,1]^{n \times m}$ to mean the $n \times m$-dimensional matrix whose entries are within the interval $[0,1]$. Also, $\tilde{B}_{ij}(\alpha)$ is the entry in the ith row and jth column of $\tilde{B}(\alpha)$.

- $\hat{B} \in \{0,1\}^{n \times m}$ is a matrix relating nodes to incoming edges. In particular, $\hat{B}_{ij} = 1$ if $j \in \varepsilon^-(i)$ and $\hat{B}_{ij} = 0$ otherwise. Here we use $\{0,1\}^{n \times m}$ to mean the $n \times m$-dimensional matrix whose entries are either 0 or 1. Also, $\hat{B}_{ij}$ is the entry in the ith row and jth column of $\hat{B}$.

- ρ_0 is the initial density, and it is assigned.

Equation (18.2) states that the change in the density in each edge is determined by the difference between the outgoing flow from and the incoming flow to that edge. Actually, the term $f(\rho)$ models the outgoing flows, while the term $\tilde{B}^T(\alpha)\hat{B}f(\rho)$ models the incoming flows. It is worth noting that $\tilde{B}^T(\alpha)$ is a column (left) stochastic matrix, i.e., $\sum_{i=1,\ldots,m}(\tilde{B}^T(\alpha))_{ij} = 1$ for all $j = 1,\ldots,n$.

For the model to be amenable to analysis and design, let us assume that the graph is directed and acyclic, and has one source node, call it s, and one destination node, call it d. Let $\mathcal{P}$ be a set of paths from s to d; namely, each element of $\mathcal{P}$ is an $s - d$ path $\{s,\ldots,i,\ldots,d\}$. Let us introduce the matrix $C \in \{0,1\}^{|\mathcal{P}| \times m}$, which links paths to edges. The rows of C contain only ones and zeros, depending on which edges are included in which paths. Let the output vector-valued function $y(.) : \mathbb{R}_+ \to \mathbb{R}^{|\mathcal{P}|}$ be given, where $t \mapsto y(t)$. This function represents the collective density in each path and is given by $y(t) = C\rho(t)$.

Recall that a Wardrop equilibrium is characterized by uniform density over all available paths. Then, consider the pedestrians as players with a common cost functional. In particular, for each player, let the running cost $g(.) : E \times [0,1]^m \to [0,+\infty[$, $(x,\rho) \mapsto g(x,\rho)$ be given as follows:

$$g(x,\rho) = \text{dist}(\rho,\mathcal{M}), \tag{18.3}$$

$$\mathcal{M} = \{\rho \in \mathcal{D} : y = C\rho = 1p \text{ for any } p \in [0,1]\}, \tag{18.4}$$

where $\mathcal{M}$ is the consensus manifold/Wardrop equilibrium set.

In the above, we write $\text{dist}(\rho,\mathcal{M})$ to mean the distance of the vector ρ from the manifold $\mathcal{M}$. Furthermore, $\mathbf{1}$ denotes the $|\mathcal{P}|$-dimensional column vector of ones.

Finally, to account for model misspecifications, consider an additional adversarial disturbance perturbing the evolution of the density. More formally, let us rewrite dynamics (18.2) as follows:

$$\dot{\rho}(t) = \left(\tilde{B}^T(\alpha, \omega)\hat{B} - I \right) f(\rho), \tag{18.5}$$

where ω is the disturbance. We assume that the disturbance ω is bounded and belongs to the polytope

$$\mathcal{W} = \{ \omega \in [-1,1]^m \,|\, \{\omega_j\}_{j \in \varepsilon^+(i)} \in \Delta_0^{|\varepsilon^+(i)|} \forall i = 1, \ldots, n \},$$

where $\Delta_0^{|\varepsilon^+(i)|}$ is the simplex translated to the origin in $\mathbb{R}^{|\varepsilon^+(i)|}$. In other words, the above corresponds to the constraint $\sum_{j \in \varepsilon^+(i)} \omega_j = 0$ for all $i = 1, \ldots, n$.

The problem of interest is then the following.

Problem 18.1 (Pedestrian flow problem). *Design a routing policy to minimize the output disagreement; i.e., each player solves the following problem:*

$$\begin{cases} \inf_{\alpha(.)} \sup_{\omega(.)} J(x, \alpha(.), \rho[\cdot](.)), \\ J(.) = \mathbb{E}\left[\int_0^T g(X(\tau), \rho(\tau))d\tau + g(X(T), \rho(T)) \right], \\ \{X(t), t \geq 0\} \text{ as in (18.1)}, \\ \dot{\rho}(t) = \left(\tilde{B}^T(\alpha, \omega)\hat{B} - I \right) f(\rho), \end{cases} \tag{18.6}$$

where $\alpha(.)$ and $\omega(.)$ are measurable functions taking values in $\mathcal{U}$ and $\mathcal{W}$.

18.3 ▪ Mean-field formulation with common cost functional

In the above problem, every player minimizes a common cost functional which depends on the density function of the whole population. The density in turn depends on both the control $\alpha(.)$ and the disturbance $\omega(.)$. Let us denote by $v(x, t)$ the value of the optimization problem starting from time t at state x. Furthermore, let $\mathcal{H}(x, \Delta(v), t)$ be the robust Hamiltonian function defined as

$$\mathcal{H}(x, \Delta(v), t) = \inf_{\alpha \in \mathcal{U}} \sup_{\omega \in \mathcal{W}} \left\{ \sum_{z \in E} q_{xz}(v(z,t) - v(x,t)) + g(x, \rho) \right\}. \tag{18.7}$$

In the Hamiltonian function, the symbol $\Delta(v)$ stands for the difference of the value function computed in two successive states, and q_{xz} is the transition rate given in (18.1). We can derive the following mean-field game.

Theorem 18.1. *The mean-field game with common cost functional for the pedestrian flow problem formulated in Problem 18.1 is given by*

$$\begin{cases} \dot{v}(x,t) + \mathcal{H}(x, \Delta(v), t) = 0 \text{ in } E \times [0, T[, \\ v(x, T) = g(x, \rho(T)) \forall x \in E, \\ \dot{\rho}(t) = \left(\tilde{B}^T(\alpha^*, \omega^*)\hat{B} - I \right) f(\rho) \text{ in } [0, T[, \\ \rho(0) = \rho_0, \ \rho_0 \text{ given.} \end{cases} \tag{18.8}$$

Furthermore, the optimal time-varying control $\alpha^*(x,t)$ and worst-case disturbance $\omega^*(x,t)$ are given by

$$\begin{aligned} \alpha^*(x,t) &\in \arg\min_{\alpha\in\mathscr{U}}\left\{\sum_{z\in E} q_{xz}(v(z,t)-v(x,t))+g(x,\rho)\right\}, \\ \omega^*(x,t) &\in \arg\max_{\omega\in\mathscr{W}}\left\{\sum_{z\in E} q_{xz}(v(z,t)-v(x,t))+g(x,\rho)\right\}. \end{aligned} \tag{18.9}$$

Proof. The third and fourth equations of (18.8) are the *forward Kolmogorov equation* and the corresponding boundary condition on the initial distribution law. To derive the first equation of (18.8), we know that from dynamic programming it holds that

$$\dot{v}(x,t)+\inf_{\alpha\in\mathscr{U}}\sup_{\omega\in\mathscr{W}}\left\{\sum_{z\in E} q_{xz}(v(z,t)-v(x,t))+g(x,\rho)\right\}=0 \quad \text{in } E\times[0,T[.$$

By introducing the robust Hamiltonian $\mathscr{H}(x,\Delta(v),t)$ given in (18.7), we obtain the first equation. Note that the transition rates depend on the routing policy/control α. This is then obtained as the minimizer in the computation of the robust Hamiltonian as expressed by (18.9). Note that the second equation in (18.8) is the boundary condition on the terminal penalty. $\square$

18.4 ▪ State space extension

After introducing the mean-field game with a common cost functional, the idea is now to extend the state space in order to include the density, henceforth referred to also as *common state*. By doing this, the robust Hamiltonian in the extended state space can be written as

$$\tilde{\mathscr{H}}(x,\rho,\Delta(v),\partial_\rho V,t)=\inf_{\alpha\in U}\sup_{\omega\in\mathscr{W}}\left\{\sum_{z\in E} q_{xz}(V(z,\rho,t)-V(x,\rho,t))\right.$$

$$\left.+\partial_\rho V(x,\rho,t)^T\left[\left(\tilde{B}^T(\alpha,\omega)\hat{B}-I\right)f(\rho)\right]+g(x,\rho)\right\}.$$

Then the mean-field system turns into the system of equations below in the value function $V(x,\rho,t)$ in $E\times[0,1]^m\times[0,T[$:

$$\begin{cases} \partial_t V(x,\rho,t)+\tilde{\mathscr{H}}(x,\rho,\Delta(v),\partial_\rho V,t)=0 \text{ in } E\times[0,1]^m\times[0,T[, \\ V(x,\rho,T)=g(x,\rho(T)) \;\forall\,(x,\rho)\in E\times[0,1]^m, \end{cases} \tag{18.10}$$

where the optimal time-varying state-feedback control $\alpha^*(x,t)$ is obtained as

$$\alpha^*(x,t)\in\arg\min_{\alpha\in\mathscr{U}}\left\{\sum_{z\in E} q_{xz}(V(z,\rho,t)-V(x,\rho,t))\right.$$

$$\left.+\partial_\rho V(x,\rho,t)^T\left[\left(\tilde{B}^T(\alpha,\omega^*)\hat{B}-I\right)f(\rho)\right]+g(x,\rho)\right\},$$

and the worst-case adversarial disturbance is given by

$$\omega^*(x,t)\in\arg\sup_{\omega\in\mathscr{W}}\left\{\sum_{z\in E} q_{xz}(V(z,\rho,t)-V(x,\rho,t))\right.$$

$$\left.+\partial_\rho V(x,\rho,t)^T\left[\left(\tilde{B}^T(\alpha^*,\omega)\hat{B}-I\right)f(\rho)\right]+g(x,\rho)\right\}.$$

Let $Ext\{\mathcal{U}\}$ and $Ext\{\mathcal{W}\}$ be the sets of the indices of all the vertices of $\mathcal{U}$ and $\mathcal{W}$, respectively. Also, let $a^{(k)}$ and $w^{(k)}$ be the generic vertices of $\mathcal{U}$ and $\mathcal{W}$, respectively. Set $|Ext\{\mathcal{U}\}| = p$ and $|Ext\{\mathcal{W}\}| = q$, where $|Ext\{\mathcal{U}\}|$ and $|Ext\{\mathcal{W}\}|$ are the cardinalities of $Ext\{\mathcal{U}\}$ and $Ext\{\mathcal{W}\}$, respectively. As a consequence, $\mathcal{U} = hull\{a^{(k)}, k \in Ext\{\mathcal{U}\}\}$ and $\mathcal{W} = hull\{w^{(k)}, k \in Ext\{\mathcal{W}\}\}$, where we use $hull$ to denote the convex hull.

From the Carathéodory theorem, we can describe any point in the polytope as a convex combination of a subset of vertices, i.e.,

$$\alpha = \sum_{k \in Ext\{\mathcal{U}\}} a_k a^{(k)}, \qquad \sum_{k \in Ext\{\mathcal{U}\}} a_k = 1,$$

$$\omega = \sum_{k \in Ext\{\mathcal{W}\}} w_k w^{(k)}, \qquad \sum_{k \in Ext\{\mathcal{W}\}} w_k = 1.$$

That is to say that a_k is the weight of the convex combination $\sum_{k \in Ext\{\mathcal{U}\}} a_k a^{(k)}$. Analogously w_k is the weight of the convex combination $\sum_{k \in Ext\{\mathcal{W}\}} w_k w^{(k)}$.

By doing this, α and ω can be reviewed as mixed strategies of the two-player game with vector payoffs displayed in Table 18.1.

Table 18.1. *Two-player game with vector payoffs.*

$a^{(i)}/w^{(j)}$	$w^{(1)}$	...	$w^{(q)}$
$a^{(1)}$	$\left(\tilde{B}^T(a^{(1)}, w^{(1)})\hat{B} - I\right)f(\rho)$	...	$\left(\tilde{B}^T(a^{(1)}, w^{(q)})\hat{B} - I\right)f(\rho)$
$\vdots$	$\vdots$	$\ddots$	$\vdots$
$a^{(p)}$	$\left(\tilde{B}^T(a^{(p)}, w^{(1)})\hat{B} - I\right)f(\rho)$	...	$\left(\tilde{B}^T(a^{(p)}, w^{(q)})\hat{B} - I\right)f(\rho)$

The cumulative payoff up to time t gives the density at time t, i.e.,

$$\begin{cases} \rho(t) = \int_0^t \dot{\rho}(\tau)d\tau = \int_0^t \left(\tilde{B}^T(\alpha, \omega)\hat{B} - I\right)f(\rho)d\tau, \\ \rho(0) = \rho_0. \end{cases} \qquad (18.11)$$

We show next that conditions for the convergence of the cumulative payoffs depend on the *value of the projected game* (cf. Chapter 11). Recall that the *projected game*, which is characterized by the payoff matrix displayed in Table 18.2, is obtained by premultiplying each entry by a given vector $\lambda \in \mathbb{R}^m$.

Table 18.2. *Two-player projected game.*

$a^{(i)}/w^{(j)}$	$w^{(1)}$	...	$w^{(q)}$
$a^{(1)}$	$\lambda^T\left(\tilde{B}^T(a^{(1)}, w^{(1)})\hat{B} - I\right)f(\rho)$	...	$\lambda^T\left(\tilde{B}^T(a^{(1)}, w^{(q)})\hat{B} - I\right)f(\rho)$
$\vdots$	$\vdots$	$\ddots$	$\vdots$
$a^{(p)}$	$\lambda^T\left(\tilde{B}^T(a^{(p)}, w^{(1)})\hat{B} - I\right)f(\rho)$	...	$\lambda^T\left(\tilde{B}^T(a^{(p)}, w^{(q)})\hat{B} - I\right)f(\rho)$

As a result, we obtain a two-player game with scalar payoffs. For this game we can compute the *value*, which is given by

$$val[\lambda] := \inf_{\alpha} \sup_{\omega} \left\{\lambda^T\left(\tilde{B}^T(\alpha, \omega)\hat{B} - I\right)f(\rho)\right\}.$$

Assumption 18.1 (Attainability of set $\mathcal{M}$). *Let $\mathcal{M}$ be given as in (18.4), let $r > 0$, and let $U = \{\rho \in \mathbb{R}^m : \mathrm{dist}(\rho, \mathcal{M}) < r\}$. For all $\rho \in U \setminus \mathcal{M}$ there exists $y \in \Pi_{\mathcal{M}}[\rho]$ such that the value of the projected game, $val[\lambda]$, is negative for every $\lambda = \rho - y$, i.e.,*

$$val[\lambda] := \inf_{\alpha} \sup_{\omega} \left\{ (\rho - y)^T \left(\tilde{B}^T(\alpha, \omega)\hat{B} - I \right) f(\rho) \right\} < 0 \quad \forall \lambda = \rho - y. \quad (18.12)$$

Condition (18.12) in Assumption 18.1 is as in [40, 155]. A graphical illustration of the above condition is depicted in Fig. 18.2.

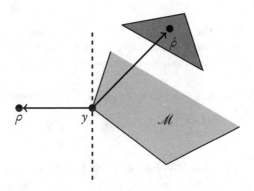

Figure 18.2. *Geometric illustration of the attainability condition.*

The condition stated in the aforementioned assumption guarantees that, given a target manifold, there always exists a routing policy $\alpha(t)$ that steers the density ρ to the manifold independently of the disturbance. Here, λ is the vector from y (the current density projection point on the target manifold) to the current density point $\rho(t)$, with the direction pointing out from the target manifold. The dark gray triangle indicates all feasible values of vector $\dot{\rho}$. We can then establish the following result.

Theorem 18.2. *Let Assumption 18.1 hold true. Then the mean-field game for the routing problem in Problem 18.1 is given by*

$$\begin{cases} \partial_t V(x, \rho, t) + val[\partial_\rho V(x, \rho, t)] + g(x, \rho) = 0 & in\ E \times [0,1]^m \times [0, T[, \\ V(x, \rho, T) = g(x, \rho(T)) & \forall (x, \rho) \in E \times [0,1]^m. \end{cases} \quad (18.13)$$

Furthermore, the optimal control and worst-case disturbance are

$$\alpha^*(x, \rho, t) = \arg\min_{\alpha} \left\{ \partial_\rho V(x, \rho, t)^T \cdot \left[\left(\tilde{B}^T(\alpha, \omega^*)\hat{B} - I \right) f(\rho) \right] \right\},$$
$$\omega^*(x, \rho, t) = \arg\min_{\alpha} \left\{ \partial_\rho V(x, \rho, t)^T \cdot \left[\left(\tilde{B}^T(\alpha^*, \omega)\hat{B} - I \right) f(\rho) \right] \right\}. \quad (18.14)$$

Proof. From (18.12) we have

$$val[\partial_\rho V(x, \rho, t)] = \inf_{\alpha \in \mathcal{U}} \sup_{\omega \in \mathcal{W}} \left\{ \partial_\rho V(x, \rho, t)^T \left[q_{x\bullet} + \dot{\rho} \right] \right\}$$
$$= \inf_{\alpha \in \mathcal{U}} \sup_{\omega \in \mathcal{W}} \left\{ \partial_\rho V(x, \rho, t)^T \left[q_{x\bullet} + \left(\tilde{B}^T(\alpha, \omega)\hat{B} - I \right) f(\rho) \right] \right\}$$
$$= \tilde{\mathcal{H}}(x, \rho, \Delta(v), \partial_\rho V, t) - g(x, \rho).$$

Invoking (18.10), we obtain the first equation in (18.13). The second equation in (18.13) is again the boundary condition on the terminal penalty. It remains to note that the optimal control is the minimizer in the computation of the extended Hamiltonian and thus is obtained from (18.14). Analogously, the worst-case disturbance is the maximizer in the computation of the extended Hamiltonian and thus is obtained from the second equation in (18.14).   □

18.5 ▪ Stability

We now consider the infinite horizon problem obtained by taking $T \to \infty$. The value function is now interpreted as a Lyapunov function. Such a function is now stationary, and therefore we drop explicit dependence on time and write simply $V(\rho(t))$.

Theorem 18.3. *Let Assumption* 18.1 *hold. Then, dynamics* (18.5) *converges asymptotically to $\mathcal{M}$, namely*

$$\lim_{t \to \infty} \text{dist}(\rho(t), \mathcal{M}) = 0.$$

Proof. Let ρ be a solution of dynamics (18.5) with initial value $\rho(0) \in U \setminus \mathcal{M}$. Set $\tau = \{\inf t > 0 | \rho(t) \in \mathcal{M}\} \leq \infty$, and let $V(\rho(t)) = \text{dist}(\rho(t), \mathcal{M})$. For all $t \in [0, \tau]$ and $y \in \Pi_{\mathcal{M}}[\rho(t)]$,

$$\begin{aligned}
V(\rho(t + dt)) - V(\rho(t)) &= \|\rho(t + dt) - y\| - \|\rho(t) - y\| \\[2mm]
&= \|\rho(t) + \dot{\rho}(t)dt - y\| - \|\rho(t) - y\| + |dt|\epsilon(dt) \\[2mm]
&= \frac{\|\rho(t) + \dot{\rho}(t)dt - y\|^2}{\|\rho(t) + \dot{\rho}(t)dt - y\|} - \frac{\|\rho(t) - y\|^2}{\|\rho(t) - y\|} + |dt|\epsilon(dt),
\end{aligned}$$

where $\lim_{dt \to 0} \epsilon(dt) = 0$. Hence

$$\begin{aligned}
\dot{V}(\rho(t)) &= \lim_{dt \to 0} \frac{1}{dt} \left(\frac{\|\rho(t) + \dot{\rho}(t)dt - y\|^2}{\|\rho(t) + \dot{\rho}(t)dt - y\|} - \frac{\|\rho(t) - y\|^2}{\|\rho(t) - y\|} + |dt|\epsilon(dt) \right) \\[2mm]
&= \lim_{dt \to 0} \frac{1}{dt} \left(\frac{\|\rho(t) + \dot{\rho}(t)dt - y\|^2}{\|\rho(t) - y\| + O(\sqrt{dt})} - \frac{\|\rho(t) - y\|^2}{\|\rho(t) - y\|} + |dt|\epsilon(dt) \right) \\[2mm]
&= \frac{1}{\|\rho(t) - y\|} \lim_{dt \to 0} \frac{1}{dt} \left(\|\rho(t) + \dot{\rho}(t)dt - y\|^2 - \|\rho(t) - y\|^2 \right) \\[2mm]
&= \frac{1}{\|\rho(t) - y\|} \frac{d}{dt} \left(\|\rho(t) - y\|^2 \right) \leq \frac{2}{\|\rho(t) - y\|} (\rho(t) - y)^T \dot{\rho}(t).
\end{aligned}$$

Now, as $\mathcal{M}$ is a compact set, from Assumption 18.1 we have that for all $\rho \in U \setminus \mathcal{M}$ there exists $y \in \Pi_{\mathcal{M}}[\rho]$ such that the affine hyperplane orthogonal to $[\rho(t), y]$ at y separates $\rho(t) - y$ from $\dot{\rho}(t)$, namely

$$\begin{aligned}
val[\rho(t) - y] &:= \inf_\alpha \sup_\omega \left\{ (\rho - y)^T \left(\tilde{B}^T(\alpha, \omega)\hat{B} - I \right) f(\rho) \right\} \\[2mm]
&= \inf_\alpha \sup_\omega \left\{ (\rho - y)^T \dot{\rho}(t) \right\} < 0,
\end{aligned} \tag{18.15}$$

from which we have

$$\dot{V}(\rho(t)) \leq \frac{2}{\|\rho(t)-y\|}(\rho(t)-y)^T \dot{\rho}(t) < 0,$$

and this concludes our proof. □

Assumption 18.1 ensures that a specific manifold $\mathcal{M}$ is attainable. However, any manifold can be attainable under a stronger condition, which we copy and adapt from [40, 155].

Assumption 18.2 (Exponential attainability of set $\mathcal{M}$). *Let $\mathcal{M}$ be given as in (18.4), let $r > 0$, and let $U = \{\rho \in \mathbb{R}^m : dist(\rho, \mathcal{M}) < r\}$. For all $\rho \in U \setminus \mathcal{M}$ there exists $y \in \Pi_{\mathcal{M}}[\rho]$ such that the value of the projected game, $val[\lambda]$, is upper bounded by $-\lambda^T \lambda$ for every $\lambda = \rho - y$, i.e.,*

$$val[\lambda] := \inf_{\alpha} \sup_{\omega} \left\{ \lambda^T \left(\tilde{B}^T(\alpha, \omega)\hat{B} - I \right) f(\rho) \right\} < -\lambda^T \lambda \quad \forall \lambda = \rho - y.$$

Theorem 18.4. *Let Assumption 18.2 hold true. Then, dynamics (18.5) converges exponentially to $\mathcal{M}$, namely*

$$dist(\rho(t), \mathcal{M}) = e^{-t} dist(\rho(0), \mathcal{M}).$$

Proof. Let ρ be a solution of dynamics (18.5) with initial value $\rho(0) \in U \setminus \mathcal{M}$. Set $\tau = \{\inf t > 0 | \rho(t) \in \mathcal{M}\} \leq \infty$, and let $V(\rho(t)) = dist(\rho(t), \mathcal{M})$. For all $t \in [0, \tau]$ and $y \in \Pi_{\mathcal{M}}[\rho(t)]$,

$$V(\rho(t+dt)) - V(\rho(t)) = \|\rho(t+dt)-y\| - \|\rho(t)-y\|$$

$$= \|\rho(t)+\dot{\rho}(t)dt-y\| - \|\rho(t)-y\| + |dt|\epsilon(dt)$$

$$= \frac{\|\rho(t)+\dot{\rho}(t)dt-y\|^2}{\|\rho(t)+\dot{\rho}(t)dt-y\|} - \frac{\|\rho(t)-y\|^2}{\|\rho(t)-y\|} + |dt|\epsilon(dt),$$

where $\lim_{dt \to 0} \epsilon(dt) = 0$. Hence

$$\dot{V}(\rho(t)) = \lim_{dt \to 0} \frac{1}{dt} \left(\frac{\|\rho(t)+\dot{\rho}(t)dt-y\|^2}{\|\rho(t)+\dot{\rho}(t)dt-y\|} - \frac{\|\rho(t)-y\|^2}{\|\rho(t)-y\|} + |dt|\epsilon(dt)| \right)$$

$$= \frac{1}{\|\rho(t)-y\|} \lim_{dt \to 0} \frac{1}{dt} \left(\|\rho(t)+\dot{\rho}(t)dt-y\|^2 - \|\rho(t)-y\|^2 \right)$$

$$= \frac{1}{\|\rho(t)-y\|} \frac{d}{dt} \left(\|\rho(t)-y\|^2 \right) \leq \frac{2}{\|\rho(t)-y\|} (\rho(t)-y)^T \dot{\rho}(t).$$

Now, as $\mathcal{M}$ is a compact set, from Assumption 18.2 we have that for all $\rho \in U \setminus \mathcal{M}$ there exists $y \in \Pi_{\mathcal{M}}[\rho]$ such that the affine hyperplane orthogonal to $[\rho(t), y]$ at y separates $\rho(t)-y$ from $\dot{\rho}(t)$, namely

$$val[\rho(t)-y] := \inf_{\alpha} \sup_{\omega} \left\{ (\rho(t)-y)^T \left(\tilde{B}^T(\alpha, \omega)\hat{B} - I \right) f(\rho) \right\}$$

$$= \inf_{\alpha} \sup_{\omega} \left\{ (\rho(t)-y)^T \dot{\rho}(t) \right\} < -(\rho(t)-y)^T (\rho(t)-y),$$

from which we have

$$\dot{V}(\rho(t)) \leq \frac{2}{\|\rho(t)-y\|}(\rho(t)-y)^T\dot{\rho}(t)$$

$$= -2V(\rho(t)) + \frac{2}{\|\rho(t)-y\|}(\rho(t)-y)^T(\rho(t)-y+\dot{\rho}(t)),$$

and this concludes our proof.   □

The above result mirrors the convergence conditions in set inclusion theory discussed in [48, 49].

18.6 ▪ Numerical example

This example involves a network with four vertices and five edges, as depicted in Fig. 18.3. The vertex indicated by s is the source, and the vertex indicated by d is the destination. The flow on edge e is marked with f_e, and the incoming flow f_0 is equal to the outgoing flow $f_6 = f_4 + f_5$.

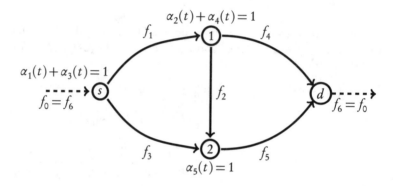

Figure 18.3. *Network system.*

The matrices introduced in the sections above are

$$\tilde{B}^T(\alpha) = \begin{bmatrix} \alpha_1 & 0 & 0 \\ 0 & \alpha_2 & 0 \\ \alpha_3 & 0 & 0 \\ 0 & \alpha_4 & 0 \\ 0 & 0 & \alpha_5 \end{bmatrix}, \qquad \hat{B} = \begin{bmatrix} 0 & 0 & 0 & 1 & 1 \\ 1 & 0 & 0 & 0 & 0 \\ 0 & 1 & 1 & 0 & 0 \end{bmatrix}.$$

The density evolution expressed by (18.2) takes on the following form, where we use $f_e(\rho_e(t)) = \phi\rho_e(t)$:

$$\begin{cases} \dot{\rho}_1(t) &= \alpha_1(t)(\phi\rho_4(t)+\phi\rho_5(t))-\phi\rho_1(t), \\ \dot{\rho}_2(t) &= \alpha_2(t)\phi\rho_1(t)-\phi\rho_2(t), \\ \dot{\rho}_3(t) &= \alpha_3(t)(\phi\rho_4(t)+\phi\rho_5(t))-\phi\rho_3(t), \\ \dot{\rho}_4(t) &= \alpha_4(t)\phi\rho_1(t)-\phi\rho_4(t), \\ \dot{\rho}_5(t) &= \alpha_5(t)(\phi\rho_2(t)+\phi\rho_3(t))-\phi\rho_5(t) \end{cases} \qquad (18.16)$$

Table 18.3. *Parameters of the overall system.*

Parameter	Value	Variable	Initial Value
ϕ	0.8	$\rho(t)$	$(0.3, 0.5, 0.2, 0, 0)$
Time step h	0.01	$\alpha(t)$	$(0.6, 0.5, 0.4, 0.5, 1)$
Time span T	20		

and

$$\begin{cases} \alpha_1(t) + \alpha_3(t) = 1, \\ \alpha_2(t) + \alpha_4(t) = 1, \\ \alpha_5(t) \qquad\quad = 1. \end{cases} \tag{18.17}$$

Let us consider the paths $\{1,4\}$, $\{1,2,5\}$, and $\{3,5\}$. In other words, $\mathscr{P} = \big\{\{1,4\},$ $\{1,2,5\},\{3,5\}\big\}$, which corresponds to defining an output

$$\begin{bmatrix} y_1(t) \\ y_2(t) \\ y_3(t) \end{bmatrix} = \underbrace{\begin{bmatrix} 1 & 0 & 0 & 1 & 0 \\ 1 & 1 & 0 & 0 & 1 \\ 0 & 0 & 1 & 0 & 1 \end{bmatrix}}_{C} \begin{bmatrix} \rho_1(t) \\ \rho_2(t) \\ \rho_3(t) \\ \rho_4(t) \\ \rho_5(t) \end{bmatrix}.$$

Table 18.3 displays the parameters' values. The algorithm used for the simulations is displayed below.

ALGORITHM 18.1. **Simulation algorithm for the pedestrian flow example.**

Input: Set of parameters as in Table 18.3.
Output: Density $\rho(t)$, policy $\alpha(t)$, and dist$(\rho(t), \mathcal{M})$
  1 : **Initialize:** Set of initial values as in Table 18.3.
  2 : **for** time $t = 0, h, 2h, \ldots, T - h$ **do**
  3 :   compute projected point of $\rho(t)$ on $\mathcal{M}$
  4 :   compute the optimal control $\alpha^*(t)$ using
         Theorem 18.2, and the distance dist$(\rho(t), \mathcal{M})$
  5 :   set $\beta(0) = \alpha(t)$
        for $k = 0, 1, \ldots, 100$ **do**
           compute $\beta(k+1) = \beta(k) + \frac{h}{100}(\alpha^*(t) - \beta(k))$
        end for
        set $\alpha(t) = (\beta_1(101), \beta_2(101), 1 - \beta_1(101), 1 - \beta_2(101), 1)$
  6 :   compute $\rho(t + h)$
  7 : **end for**
  8 : **STOP**

The simulations are carried out with MATLAB, and the results are illustrated in Figs. 18.4–18.6. From the conservation law we have $\sum_e \dot{\rho}_e(t) = 0$, and therefore $\sum_e \rho_e(t) = \sum_e \rho_e(0) = 1$, which is shown in Fig. 18.4 (gray solid line). When achieving consensus, $\rho_2(t) = 0$ holds (dashed line), indicating that all players choose either the path involving edge 1 and edge 4, or the path involving edge 3 and edge 5. Moreover, the

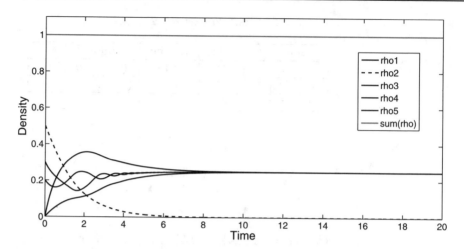

Figure 18.4. *Simulation results: density.*

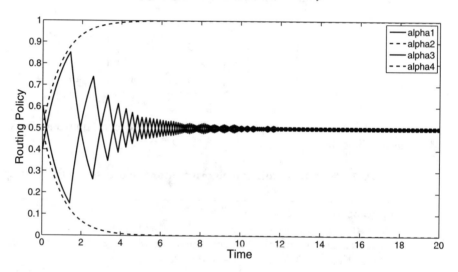

Figure 18.5. *Simulation results: routing policy ($\alpha_5(t) = 1$ holds all the time).*

players choose these two routes almost equiprobably, i.e., $\alpha_1 \approx \alpha_3 \approx 0.5$, as illustrated
in Fig. 18.5. The distance from the consensus manifold converges to zero, as illustrated
in Fig. 18.6. Note that in order to avoid chirping in $\alpha(t)$, we have introduced lowpass
dynamics $\dot{\beta}(t) = \alpha^*(t) - \beta(t)$ (the relevant transfer function is $\beta(s) = \frac{1}{s+1}\alpha^*(s)$ which is
actually a lowpass filter for $\alpha(t)$), corresponding to step 5 in the algorithm.

18.7 ▪ Notes and references

In this chapter, we have examined a mean-field game formulation with common cost functional of a distributed routing problem. The problem overlaps recent research on optimal
planning [3]. The problem setup has been motivated by an idea in [81, 80] that develops
a dynamic model for the density at network edges in a locally responsive traffic network.
This chapter has provided several contributions. Beyond the mean-field game formulation, we have illustrated an extended state space solution approach applied to the worst-

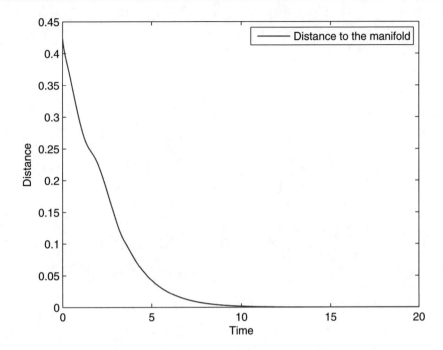

Figure 18.6. *Simulation results: distance to the consensus manifold.*

case scenario. Such an extended state space solution approach was first developed in [34] and [35].

The study has analyzed convergence conditions of the density to a preassigned manifold. Connections with repeated games with vector payoffs and set inclusion theory have been highlighted. The simulations have been conducted by Dr. Xuan Zhang from the Department of Engineering Science of the University of Oxford.

Future directions involve

- alternative models of adversarial disturbances including energy bounded disturbances in the spirit of H^∞-optimal control;

- a detailed analysis of the impact of the graph properties (graph connectivity, degree of nodes, eigenvalues of graph-Laplacian matrices) on the speed of convergence;

- the analysis of nonconservative flows (actually, a more complex scenario would consider the case where the number of players changes with time, perhaps dependently on the level of congestion in the network).

The stability conditions provided in this chapter mirror classical convergence conditions in repeated games with vector payoffs [40, 155]. Such conditions can be reviewed also as set inclusion conditions [27, 59, 60].

Chapter 19

Supply Chain

19.1 ▪ Introduction

In joint replenishment applications, retailers can benefit from placing joint orders and consequently sharing the transportation costs. When they do this, we say that the retailers form a coalition. This casts joint replenishment applications within the framework of coalitional games with transferable utilities (cf. Chapters 5–6). This chapter shows that in the above application, where the values of coalitions are not known with certainty, one needs to develop a model which is more sophisticated than those provided by classical coalitional games with transferable utilities.

We consider a sequence of games where the *average* coalitions' values (over time) are known with certainty but the instantaneous values are *unknown but bounded* by a polyhedron. This model may be seen as a dynamic extension of cooperative interval games where a coalition value is a closed interval on the real line.

At each point in time a certain revenue is allocated to each player. In general, these revenues will not meet the actual instantaneous value of the coalitions. To keep track of this allocation error, an excess vector stores the difference between the instantaneous value of each coalition and the sum of the allocated revenues to all its players. We may interpret this excess vector as the state variable describing the history of our dynamic system. Under the assumption that the only information available at each time is the excess of the coalitions, our goal is to design *robust* allocation rules, i.e., allocation rules that (i) keep the excess vector bounded within a predefined threshold ϵ at each time (we will refer to such rules as ϵ-stabilizing), while (ii) guaranteeing a certain average allocation vector over time. Justification for keeping the excess vector bounded follows from the observation that a fair allocation should not allocate the maximum excess to the same coalition each time.

One may notice that our problem is similar in spirit to classical problems in machine learning.

This chapter is organized as follows. Section 19.2 introduces the supply-chain model with multiple retailers. The model builds on the concept of family of balanced games which we formalize in Section 19.3. Section 19.4 turns the family of games into a dynamic system. Section 19.5 designs the allocation rule. Section 19.6 considers allocation rules based on the Shapley value. Section 19.7 shows a numerical example. Finally, Section 19.8 draws some conclusions and provide notes and references for this chapter.

We denote by $\langle N, v \rangle$ a coalitional TU game, where $N = \{1, \ldots, n\}$ is the set of n players and v is the characteristic function returning the value of each nonempty coalition $S \subseteq N$. We define by $m = 2^n - 1$ the number of nonempty coalitions. Furthermore, occasionally we also interpret v as a vector in $\mathbb{R}^m$, namely, $v = [v(S)]_{S \subseteq N}$. For $\xi \in \mathbb{R}^m$, let ξ_i denote the ith component of ξ, and define

$$|\xi| = \max_i |\xi_i|.$$

Let $\mathbb{Z}$ denote the set of integers and $\mathbb{Z}^+$ the set of nonnegative integers. Also, let $f = \{f(0), f(1), f(2), \ldots\}$ be any bounded one-sided sequence in $\mathbb{R}^m$, and define

$$\|f(k)\| = \sup_{k \in \mathbb{Z}^+} |f(k)|.$$

19.2 • Supply chain with multiple retailers and uncertain demand

Let us consider a supply chain consisting of a single-period one-warehouse multi-retailer inventory system such as the one depicted in Fig. 19.1.

A central warehouse W serves a number of retailers R_i, $i = 1, \ldots, n$, each one facing a demand d_i unknown but bounded by preassigned values $d_i^- \in \mathbb{R}$ and $d_i^+ \in \mathbb{R}$. After demand d_i is known to retailer R_i, he must decide whether to fulfill the demand or not. Fulfilling the demand implies reordering just in time from the warehouse, as the retailers hold no inventory. Retailers that place joint orders share the transportation cost K. In particular, if only retailer R_i places an order—in the parlance of coalitional TU games we say that he does not play in coalition with other players—he pays the full transportation cost K. This is illustrated by the dashed path (W, R_1, W) in the network of Fig. 19.1(a), which describes a single truck that serves only R_1 and goes back to the warehouse. The cost of not reordering is the cost of the unfulfilled demand d_i.

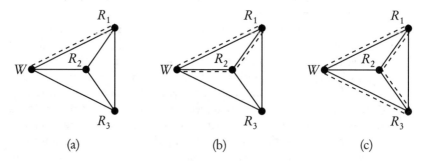

Figure 19.1. *Example of one warehouse W and three retailers R_1, R_2, and R_3: (a) Truck leaving W, serving R_1, and returning to W; (b) Truck leaving W, serving R_1 and R_2, and returning to W; (c) Truck leaving W, serving R_1, R_2, and R_3, and returning to W. Reprinted with permission from IEEE [185].*

If two or more retailers "play" in a coalition, they agree on a joint decision ("everyone reorders" or "no one reorders"). The cost of reordering for the coalition also equals the total transportation cost that must be shared among the retailers. This is illustrated, with reference to coalition $\{R_1, R_2\}$, by the dashed path (W, R_1, R_2, W) in Fig. 19.1(b).

When necessary, a single truck will serve all retailers in the coalition and get back to the warehouse. This is illustrated, with reference to coalition $\{R_1, R_2, R_3\}$, by the dashed path (W, R_1, R_2, R_3, W) in Fig. 19.1(c). The cost of not reordering is the sum of the un-fulfilled demands of all retailers.

How the players will share the cost is part of the solution generated by the warehouse manager, which plays here the role of a game designer.

The cost scheme can be captured by a game with the set $N = \{R_1, \ldots, R_n\}$ of players where the cost of a nonempty coalition $S \subseteq N$ is given by (assume a unitary cost for the unfulfilled demand)

$$c(S) = \min\left\{K, \sum_{i \in S} d_i\right\}.$$

Note that the bounds on the demand d_i reflect into the bounds on the cost as follows: for all nonempty $S \subseteq N$,

$$\min\left\{K, \sum_{i \in S} d_i^-\right\} \leq c(S) \leq \min\left\{K, \sum_{i \in S} d_i^+\right\}. \tag{19.1}$$

The coalition values represent the cost savings of the coalitions. The cost savings $v(S)$ of a coalition S are the difference between the sum of the costs of the coalitions of the individual players in S and the cost of the coalition itself, namely,

$$v(S) = \sum_{i \in S} c(\{i\}) - c(S). \tag{19.2}$$

Given the bound for $c(S)$ in (19.1), the value $v(S)$ is also bounded, as given: for any $S \subset N$,

$$v(S) \leq \sum_{i \in S} \min\left\{K, d_i^+\right\} - \min\left\{K, \sum_{i \in S} d_i^-\right\}.$$

Thus, the cost savings (value) of each coalition is bounded uniformly by a maximum value.

Letting $v_{min}(S)$ and $v_{max}(S)$ be lower and upper bounds for the value of coalition $S \subseteq N$, the game model turns into a family of cost-savings games $\langle N, \mathcal{V} \rangle$, where

$$\mathcal{V} = \{v \in \mathbb{R}^m : v_{min}(S) \leq v(S) \leq v_{max}(S) \forall S \subseteq N\}, \tag{19.3}$$

and $v(S)$ is as in (19.2).

In a dynamic context, the same situation is repeated in time. That is to say that at each time (day, week) $k = 0, 1, \ldots$, the warehouse manager allocates the costs and a new demand is realized.

In the next section, we introduce a formal definition of family of balanced games with coalition values lying on preassigned closed intervals.

19.3 ▪ Family of balanced games

Consider a coalitional TU game $\langle N, v \rangle$, where $N = \{1, \ldots, n\}$ is a set of n players and v is the characteristic function returning the value of each nonempty coalition $S \subseteq N$. Let $m = 2^n - 1$ be the number of nonempty coalitions, and, with a little abuse of notation, let us interpret v as a vector in $\mathbb{R}^m$, namely, $v = [v(S)]_{S \subseteq N}$.

Definition 19.1 (Family of TU games). *A family of games $\langle N, \mathcal{V} \rangle$ is the set of games $\langle N, v \rangle$ obtained when v varies within a polyhedron $\mathcal{V} = \{v \in \mathbb{R}^m : v_{min} \leq v \leq v_{max}\}$, where the bounds v_{min} and v_{max} are given.*

For the sake of simplicity, let us take $v \geq 0$. Furthermore, denote by 2^N the family of subsets of N. Let us recall the definitions of balanced map and balanced game.

Definition 19.2 (Balanced map). *A map* $\lambda : 2^N \setminus \{\emptyset\} \to \mathbb{R}^+$ *is called a* balanced map *if* $\sum_{S \subseteq N} \lambda(S) e^S = e^N$.

Here, $e^S \in \mathbb{R}^n$ is the *characteristic vector* of coalition S with $e_i^S = 1$ if $i \in S$ and $e_i^S = 0$ if $i \in N \setminus S$.

Definition 19.3 (Balanced game). *An n-person game $\langle N, v \rangle$ is called a* balanced game *if, for each balanced map* $\lambda : 2^N \setminus \{\emptyset\} \to \mathbb{R}^+$,

$$\sum_{S \subseteq N} \lambda(S) v(S) \leq v(N). \tag{19.4}$$

The polyhedron $\mathscr{V}$ represents a *family of balanced games* if the above condition is satisfied for each game $v \in \mathscr{V}$. This is formalized in the next definition.

Definition 19.4 (Family of balanced games). *A family of balanced games $\langle N, \mathscr{V}_b \rangle$ is the set of games $\langle N, v \rangle$ obtained when v varies within a polyhedron*

$$\mathscr{V}_b = \{ v \in \mathscr{V} : condition \ (19.4) \ holds \},$$

where the bounds v_{min} and v_{max} are given.

Recall the notions of core and allocation rules from Chapters 5–6. Also, recall that a game is balanced if and only if the core is nonempty [63, 221]. By definition each game $\langle N, v \rangle$ with $v \in \mathscr{V}_b$ is balanced, and so the core $C(v)$,

$$C(v) = \left\{ a \in \mathbb{R}^n : \frac{a}{v(N)} \in \Delta^n, \ \sum_{i \in S} a_i \geq v(S) \forall S \subseteq N \right\},$$

is nonempty. This is equivalent to saying that there exists an allocation $a \in C(v)$ such that the grand coalition is stable with respect to any sub-coalition.

For the aforementioned family of balanced games, the problem turns into finding an allocation rule $a(v)$ such that $a(v) \in C(v)$ for all games $v \in \mathscr{V}_b$. To this purpose, note that the core is a convex set described by linear equations and inequalities.

By introducing a vector of nonnegative surplus variables $s = [s_1, \dots, s_{m-1}]^T$, finding an allocation rule a in the core $C(v)$ corresponds to finding an *allocation vector* $u \in \mathbb{R}^{n+m-1}$ in the set

$$\mathscr{U}(v) = \{ u : Au = v, \ u \geq 0 \}, \tag{19.5}$$

where

$$A = \left[\begin{array}{c|c} B & \begin{array}{c} -I \\ \hline 0 \dots 0 \end{array} \end{array} \right], \tag{19.6}$$

and where $B \in \mathbb{R}^{m \times n}$ is an incidence matrix with the characteristic vectors e^S as rows and I is the $(m-1)$-dimensional identity matrix.

Note that if $u \in \mathscr{U}(v)$, then $u = \begin{bmatrix} a \\ s \end{bmatrix}$ for some $a \in C(v)$. Observe that, in general, $\mathscr{U}(v)$ is a polyhedron of dimension $n-1$.

Furthermore, it is worth noting that for each coalition, the surplus variable indicates the deviation of the allocated value to that coalition and the value of the coalition itself, namely $\sum_{i \in S} a_i - v(S)$. Notice that we only need $m - 1$ surplus variables because $\sum_{i \in N} a_i = v(N)$ due to the efficiency condition of the core.

19.4 ▪ Turning the repeated TU game into a dynamic system

After introducing a family of balanced games in the previous section, let us turn to consider a stream of games with characteristic function bounded within the polyhedron $\mathcal{V}_b$:

$$\mathbf{v}(t), \; t = 1, 2, \ldots, \text{ with } \mathbf{v}(t) \in \mathcal{V}_b \, \forall t. \tag{19.7}$$

In the above expression, $\mathbf{v}(t) = [v(t, S)]_{S \subseteq N}$ is the vector of coalition values.
Let the *average vector of coalition values* $\bar{\mathbf{v}}$ be defined as

$$\bar{\mathbf{v}} = \lim_{T \to \infty} \frac{1}{T} \sum_{k=0}^{T} \mathbf{v}(t), \tag{19.8}$$

and assume that such a value is given.

Furthermore, assume that allocations to players are made at a higher rate than the rate of change of the coalitional values, which equals 1. In particular, let Θ be the time between two successive allocations. Consequently, the integer number $1/\Theta$ is the rate of allocations. Now, by stretching the time scale by the rate $1/\Theta$, we obtain the new sequence of games

$$v(k) = \mathbf{v}(t)\Theta, \quad k = \frac{t-1}{\Theta} + 1, \; \ldots, \; \frac{t}{\Theta}, \quad t = 1, 2, \ldots. \tag{19.9}$$

We can interpret the above sequence of games as follows. In the original time interval $(t - 1, t]$ the vector of coalitional values equals $\mathbf{v}(t)$. We distribute these values equally over the $1/\Theta$ allocations that occur in this time period, so this results in values $\mathbf{v}(t)\Theta$ for each point in time where allocations are made. This way we can ensure that the total amount allocated to the players in the new interval $((t - 1)/\Theta, t/\Theta]$ does not exceed the available amount $v(t, N)$.

If we use the notation $\mathcal{V}_b^\Theta = \Theta \cdot \mathcal{V}_b$, the sequence of games (19.7)–(19.8) is equivalent to the sequence of games

$$v(k), \, k = 1, 2, \ldots, \text{ with } v(k) \in \mathcal{V}_b^\Theta \text{ for each } k = 1, 2, \ldots,$$
$$\bar{v} = \lim_{T \to \infty} \frac{1}{T} \sum_{k=0}^{T} v(k), \tag{19.10}$$

where $\bar{v} = \Theta\bar{\mathbf{v}}$. In the remainder of this chapter, we will refer to the sequence of games in (19.10).

Now, denote by $x(k+1) \in \mathbb{R}^m$ a vector of variables describing the aggregate coalition excesses over all previous games $v(1), \ldots, v(k)$ (the value $x(0)$ is the excess at time 0), i.e.,

$$x(k + 1, S) = x(k, S) + \sum_{i \in S} a_i(k) - (s_S(k) + v(k, S)) \quad \forall S \subseteq N, \tag{19.11}$$

where $a_i(k)$ is the revenue allocated to player i and $s_S(k)$ is a desired surplus for coalition S.

The aggregate coalition excess $x(k + 1, S)$ is the coalition excess summed over all previous games $v(1), \ldots, v(k)$ and therefore represents the *state* of the system. We rewrite (19.11) in the following matrix form:

$$x(k + 1) = x(k) + Au(k) - v(k), \quad v(k) \in \mathcal{V}_b^\Theta, \quad k = 1, 2, \ldots, \tag{19.12}$$

where $u(k) = \begin{bmatrix} a(k) \\ s(k) \end{bmatrix}$, $a(k) = [a_i(k)]_{i \in N}$, and $s(k) = [s_S(k)]_{S \subset N}$. The condition $u(k) \geq 0$ is omitted for the sake of notation. Now, let the vector $\bar{u} \in \mathcal{U}(\bar{v})$ be arbitrarily chosen, where $\bar{v}$ is assigned once given the sequence of games (19.10).

Lemma 19.5 (Average constraint). *Let the sequence of games (19.10) be given. There exists an allocation rule $f : \mathbb{R}^m \longrightarrow \mathbb{R}^{n+m-1}$ such that, for $u(k) = f(v(k))$,*

$$Au(k) = v(k), \tag{19.13}$$

$$\lim_{T \longrightarrow \infty} \frac{1}{T} \sum_{k=0}^{T} u(k) = \bar{u} \tag{19.14}$$

if and only if there exists a matrix $D \in \mathbb{R}^{(n+m-1) \times m}$ that satisfies

$$AD = I \in \mathbb{R}^{m \times m}, \tag{19.15}$$

$$D(v - \bar{v}) + \bar{u} \geq 0 \quad \forall v \in \mathcal{V}_b^\Theta. \tag{19.16}$$

The allocation rule is linear on $v(k)$, that is,

$$u(k) = \bar{u} + D(v(k) - \bar{v}). \tag{19.17}$$

In the following we call $\lim_{T \longrightarrow \infty} \frac{1}{T} \sum_{k=0}^{T} u(k)$ the *average allocation* (vector).

Note that condition (19.13) implies that $u(k) = \begin{bmatrix} a(k) \\ s(k) \end{bmatrix} \in \mathcal{U}(v(k))$ at each time k. This in turn means that $a(k)$ is an element of the core $C(v(k))$ of the game $\langle N, v(k) \rangle$ obtained from freezing the coalition values at time k.

The linear allocation rule (19.17) builds on the hypothesis that the coalition values are known at each sample time. We can turn the above rule into a *feedback rule* which allocates revenues at time k based on the aggregate coalition excesses $x(k)$.

Our goal is to find dynamic allocation rules that keep the excess vector bounded and such that the average allocation is $\bar{u}$. For this we need the following definition of feasible dynamic allocation rule.

Definition 19.6 (ϵ-stabilizing allocation rule). *Given $\epsilon > 0$ and a reference value x_{ref} for system (19.12), an ϵ-stabilizing allocation rule is a feedback rule for which there exists a continuous positive function $\phi(k)$, monotonically decreasing and converging to 0 as $k \longrightarrow \infty$ such that for all $x(0)$, the following condition holds true:*

$$\|x(k) - x_{ref}\| \leq \max\{\|x(0)\| \phi(k), \epsilon\}.$$

For the sake of simplicity, take $x_{ref} = 0$. Then the above condition implies that $x(k)$ does not deviate more than ϵ from 0 in the long run. For any $x(0)$ with $\|x(0)\| \leq \epsilon$ the condition simply requires that $\|x(k)\| \leq \epsilon$ for all k. Then, the problem can be restated as shown below.

Problem 19.1 (Stabilizing allocation rule). *For the sequence of games (19.10), find an ϵ-stabilizing allocation rule such that its average allocation equals $\bar{u}$, i.e.,*

$$\lim_{T \longrightarrow \infty} \frac{1}{T} \sum_{k=0}^{T} u(k) = \bar{u}.$$

Note that the requirement $\lim_{T \to \infty} \frac{1}{T} \sum_{k=0}^{T} u(k) = \bar{u}$ simply represents a constraint on the coalitions' excess in the long run.

Also, observe that the ϵ-stabilization of the excess vector $x(k)$ means that at each time k the excess $x(k)$ does not exceed a predefined threshold ϵ of the game $\langle N, v(k) \rangle$. Using the definition of the ϵ-core from Lehrer (2002) [151], the above problem corresponds to finding an allocation rule that at each time k returns a vector in the ϵ-core of the one-shot game $\langle N, v(k) \rangle$.

19.5 ▪ Allocation rule based on feedback control synthesis

In this section, we develop a constructive method to solve Problem 19.1. The allocation rule is obtained based on a feedback control synthesis in an extended state space.

Let A and D be two matrices satisfying (19.15) and (19.16). We can find two matrices C and F that "square" A and D and satisfy

$$\begin{bmatrix} A \\ C \end{bmatrix} \begin{bmatrix} D & F \end{bmatrix} = I. \tag{19.18}$$

This yields the augmented system

$$\begin{aligned} x(k+1) &= x(k) + Au(k) - v(k), \\ y(k+1) &= y(k) + Cu(k), \end{aligned} \tag{19.19}$$

where $v(k)$ is as in (19.9). Note that the new variable $y(k)$ accounts for the difference between the instantaneous and the average allocations of each player. Define the augmented state variable $z \in \mathbb{R}^{n+m-1}$ as

$$z(k) = \begin{bmatrix} D & F \end{bmatrix} \begin{bmatrix} x(k) \\ y(k) \end{bmatrix}, \quad \begin{bmatrix} x(k) \\ y(k) \end{bmatrix} = \begin{bmatrix} A \\ C \end{bmatrix} z(k).$$

This variable satisfies the equation

$$z(k+1) = \begin{bmatrix} D & F \end{bmatrix} \begin{bmatrix} x(k+1) \\ y(k+1) \end{bmatrix}$$

$$= \begin{bmatrix} D & F \end{bmatrix} \begin{bmatrix} x(k) \\ y(k) \end{bmatrix} + \begin{bmatrix} D & F \end{bmatrix} \begin{bmatrix} A \\ C \end{bmatrix} u(k)$$

$$- \begin{bmatrix} D & F \end{bmatrix} \begin{bmatrix} v(k) \\ 0 \end{bmatrix} \tag{19.20}$$

$$= z(k) + u(k) - Dv(k). \tag{19.21}$$

Then we have that the allocation rule $u(k) = -z(k)$ is a possible allocation rule for the problem under study.

Theorem 19.7. *Consider system (19.21) with $v(k)$ as in (19.9). The allocation rule in feedback form*

$$u(k) = -z(k) \tag{19.22}$$

satisfies

$$\|z(k)\| \leq \|Dv(k)\|. \tag{19.23}$$

Furthermore, if the average coalitions' value is equal to $\bar{v}$, then the average allocation vector converges to $\bar{u}$.

Proof. Let us first prove (19.23). To do this, let us substitute (19.22) in the dynamics (19.21). This yields $z(k+1) = -Dv(k)$ for all k, which in turn implies (19.23).

For the second part of the proof, let us sum both sides of (19.21) over $k = 1, 2, \ldots$. Then we obtain

$$\frac{1}{T}\sum_{k=0}^{T-1} u(k) - \frac{1}{T}\sum_{k=0}^{T-1} Dv(k) = \frac{z(T) - z(0)}{T} \to 0 \quad \text{as } T \to \infty.$$

Actually, note that by taking the limit the numerator remains finite, whereas the denominator goes to infinity. As a consequence, $\bar{u} = D\bar{v}$, and this concludes the proof. □

Let us now find the maximum time period Θ^* such that $\|Dv(k)\| \leq \epsilon$ for given ϵ. It turns out that such a value can be obtained as $\Theta^* = \frac{\epsilon}{\delta}$, where $\delta = \max_{v \in \mathcal{V}_b} |Dv|$. Then we have the following corollary.

Corollary 19.8. *Consider system* (19.21) *with* $v(k)$ *as in* (19.9). *For any* ϵ *and corresponding* Θ^*, *if* $\Theta \leq \min\{\Theta^*, 1\}$, *then the allocation rule in feedback form,*

$$u(k) = -z(k), \tag{19.24}$$

is ϵ-*stabilizing.*

Proof. The thesis follows from

$$\|z(k)\| \leq \|D\mathbf{v}(t)\Theta\| \leq \|D\mathbf{v}(t)\Theta^*\| \leq \max_{v \in \mathcal{V}_b} |Dv\Theta^*| \leq \epsilon. □$$

Note that as $\|z\| \leq \epsilon$, it also holds that $\|u\| \leq \epsilon$ as $u = -z$. That is to say that the smaller the ϵ, the smaller the maximum allocation in magnitude. Furthermore, note that the above results can be extended to the case where $\bar{v}$ is averaged online, with the difference that matrix D must be updated iteratively according to (19.15)–(19.16).

19.6 ▪ The Shapley value as a linear allocation rule

This section shows that the Shapley value can be obtained from the allocation rule (19.17).

Recall that the *Shapley value* ϕ is defined by $\phi = \frac{1}{n!}\sum_{\sigma \in \Pi(N)} m^\sigma$, where $\Pi(N)$ is the set of all permutations of N and m^σ is the marginal vector corresponding to the permutation $\sigma : N \to N$ [219]. A marginal vector m^σ corresponds to a situation in which the players enter a room one by one in the order $\sigma(1), \sigma(2), \ldots, \sigma(n)$ and where each player receives the marginal contribution he creates upon entering. Hence, m^σ is the vector in $\mathbb{R}^n$ with elements

$$m^\sigma_{\sigma(1)} = v(\{\sigma(1)\}),$$
$$m^\sigma_{\sigma(2)} = v(\{\sigma(1), \sigma(2)\}) - v(\{\sigma(1)\}),$$
$$\vdots$$
$$m^\sigma_{\sigma(k)} = v(\{\sigma(1), \sigma(2), \ldots, \sigma(k)\}) - v(\{\sigma(1), \sigma(2), \ldots, \sigma(k-1)\}).$$

Theorem 19.9. *The Shapley value ϕ is linear in v, i.e., $\phi = Lv$, where the matrix $L \in \mathbb{R}^{n \times m}$ is defined by*

$$L_{ij} = \frac{1}{n!} \cdot \begin{cases} -\mu!(n-(\mu+1))! & \text{if } i \notin S, \\ (\mu-1)!(n-\mu)! & \text{if } i \in S, \end{cases} \tag{19.25}$$

if column j corresponds to coalition S with $\mu = |S|$.

Proof. The proof follows immediately from the definition of the Shapley value in Shapley (1953) [219]. □

To emphasize the dependence of ϕ on v we henceforth write $\phi(v)$ instead of ϕ. Let $s(\phi(v))$ be the vector of surplus variables when revenues are allocated according to the Shapley value $\phi(v)$. The idea is now to express $s(\phi(v))$ linearly in v.

Theorem 19.10. *The vector of surplus variables is linear in v, i.e.,*

$$s(\phi(v)) = Qv, \tag{19.26}$$

where $Q \in \mathbb{R}^{(m-1) \times m}$ has row i associated to a surplus variable (a coalition $S \subset N$), column j associated to a coalition $M \subseteq N$, and generic ijth element

$$Q_{ij} = \begin{cases} \sum_{p \in S} L_{pj} & \text{if } i \neq j, \\ \sum_{p \in S} L_{pj} - 1 & \text{if } i = j. \end{cases} \tag{19.27}$$

Proof. First, consider the coalition containing just player 1 and let $L_{i\bullet}$ be the generic ith row of L. The associated surplus variable is

$$s_1(\phi(v)) = \phi_1 - v(\{1\}) = L_{1\bullet}v - v(\{1\}) = (L_{11}-1)v(\{1\}) + L_{12}v(\{2\}) + \cdots + L_{1m}v(N).$$

The latter equation yields $Q_{1\bullet} = [(L_{11}-1) L_{12} \ldots L_{1m}]$, in accordance with (19.27).

If we repeat the same reasoning for a generic coalition $M \subset N$, the surplus variable is

$$s_M(\phi(v)) = \sum_{i \in M} \phi_i - v(M) = \sum_{i \in M} L_{i\bullet}v - v(M).$$

Recall that j is the column associated to coalition M. Then, the latter equation yields $Q_{jk} = \sum_{i \in M} L_{ik}$ if $k \neq j$ and $Q_{jj} = \sum_{i \in M} L_{ij} - 1$, in accordance with (19.27). □

Using the fact that $\phi(v)$ and $s(\phi(v))$ are linear in v, we define the allocation vector associated to the Shapley value by $u(\phi(v)) = [\phi(v)^T \quad s(\phi(v))^T]^T$.

Corollary 19.11. *There exists a matrix $\Phi \in \mathbb{R}^{(n+m-1) \times m}$, defined by $\Phi = [L^T \quad Q^T]^T$, such that $u(\phi(v)) = \Phi v$. Furthermore, Φ is a right inverse of A, i.e., $A\Phi = I$.*

Proof. From Theorems 19.9 and 19.10 we have $[\phi(v)^T \quad s(\phi(v))^T]^T = [L^T \quad Q^T]^T v$. This concludes the first part of the proof.

To prove that $A\Phi = I$, it is sufficient to show that $A_{i\bullet}\Phi_{\bullet j} = 1$ if $i = j$ and zero otherwise. Observe that row i of A, denoted by $A_{i\bullet} \in \mathbb{R}^{1 \times (n+m-1)}$, corresponds to coalition $M \subseteq N$, whereas column j of Φ, denoted by $\Phi_{\bullet j} \in \mathbb{R}^{(n+m-1) \times 1}$, corresponds to coalition $S \subseteq N$. Hence, the condition $i = j$ is equivalent to $M = S$.

Now let the row vector $A_{i\bullet}$ be given. The first n elements of this vector correspond to players $p = 1, \ldots, n$, and the last $m-1$ elements correspond to all coalitions $R \subset N$ (recall the structure of A as described in (19.6)). Now the structure of row $A_{i\bullet}$ may be formulated as

$$A_{i\bullet} = [\cdots \underbrace{1}_{\forall p \in M} \cdots\cdots \underbrace{0}_{\forall p \notin M} \cdots\cdots \underbrace{-1}_{R=M} \cdots\cdots \underbrace{0}_{\forall R \neq M} \cdots]. \tag{19.28}$$

Analogously, the first n elements of $\Phi_{\bullet j}$ correspond to players $p = 1 \ldots n$, and the last $m-1$ elements correspond to all coalitions $R \subset N$ (see (19.25) and (19.27)).

In conclusion, if $i = j$, or $M = S$, then $A_{i\bullet} \Phi_{\bullet j} = \sum_{p \in S} L_{pj} - (\sum_{p \in S} L_{pj} - 1) = 1$. Conversely, if $i \neq j$, or $M \neq S$, then $A_{i\bullet} \Phi_{\bullet j} = \frac{1}{n!}[\sum_{p \in M} L_{pj} - \sum_{p \in M} L_{pj}] = 0$. □

19.7 ▪ Numerical example

Let a three-player TU game be given where the characteristic function satisfies

$$v(\{1\}) = 0, \qquad v(\{2\}) = 0, \qquad v(\{3\}) = 0,$$
$$v(\{1,2\}) \in [0,5], \quad v(\{1,3\}) \in [0,5], \quad v(\{2,3\}) \in [0,7], \quad v(N) \in [0,12].$$

Set the average values

$$\bar{v} = [0, 0, 0, 2, 3, 4, 10]^T, \quad \bar{u} = [3, 5, 2, 3, 5, 2, 6, 2, 3]^T.$$

Note that $A\bar{u} = \bar{v}$. We translate the origin of the u-v space to $\bar{u}$-$\bar{v}$.

First, we calculate D by formulating a linear programming problem. Then we compute the matrices C and F that square B and D. The method is explained in detail in the appendix of [44]. For the maximum sample time we get $\Theta^* > 0.1$ and choose $\Theta = 0.1$.

The nature of the problem does not change even if we consider the bounding polyhedron $\mathcal{U} := \{u \in \mathbb{R}^{10} : -0.5 \cdot \mathbf{1} \leq u \leq 0.5 \cdot \mathbf{1}\}$, where $\mathbf{1}$ is the 10-dimensional vector of ones. Actually, the resulting games in the sequence are balanced.

The algorithm used for the simulations is illustrated below.

ALGORITHM 19.1. Simulation algorithm for the supply-chain example.

Input: Game $\langle N, \{v(t)\}\rangle$, average allocation $\bar{u}$, average values $\bar{v}$, matrices C, D, F
Output: Augmented state $z(k)$, error $\bar{u}(k) - \bar{u}$
  1 : **Initialize.** Set the initial state $z(0)$
  2 : **for** time $t = 0, 1, \ldots, T-1$ **do**
  3 :   run dynamics (19.21) where $u(k)$ is as in (19.22)
  4 : **end for**
  5 : **STOP**

The evolution of the system is displayed in Figs. 19.2 and 19.3. In particular, Fig. 19.2 shows the time plot of the variable $z(.)$. The variable is ϵ-stabilized with $\epsilon = 0.5$. For the same simulation scenario, Fig. 19.3 shows the time plot of $\bar{u}(k) - \bar{u}$, where $\bar{u}(k)$ is the average of $u(k)$ up to time k. All plots tend to zero for increasing time, which means that the average $\bar{u}(k)$ tends to $\bar{u}$.

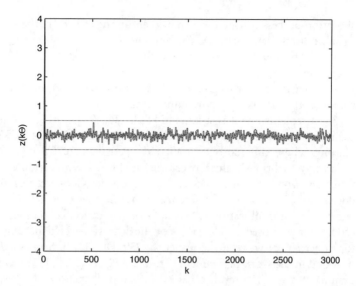

Figure 19.2. *Time plot of $z(.)$. The variable is ϵ-stabilized with $\epsilon = 0.5$. Reprinted with permission from Elsevier* [45].

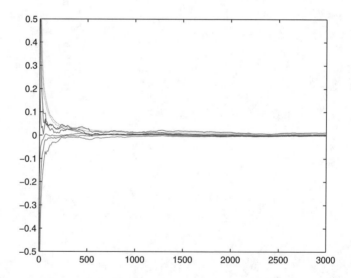

Figure 19.3. *Time plot of $\bar{u}(k) - \bar{u}$. The average tends to $\bar{u}$ for increasing time. Reprinted with permission from Elsevier* [45].

19.8 ▪ Notes and references

This chapter is based on [44] and [45]. We have modeled a supply chain with multiple retailers and uncertain demand as a dynamic TU game. As a main result, we have developed a constructive method to design robust allocation rules based on feedback control synthesis. The rule uses a measure of the extra benefit that a coalition has received up to the current time to redistribute the budget among the players.

A similar idea under the assumption that the values of the coalitions are known and time invariant is also in [75, 151, 216]. Budget distribution occurs iteratively until the allocation process converges to an element in the core or in the ϵ-core if the game is not balanced (in the latter case the core is empty).

It must be noted that to compute the matrix D, used by the allocation rule, the number of constraints of type (19.13) to consider grows exponentially in the number of players n. We refer the reader to [26, Sect. 5] for a method based on *constraints generation* that computes the matrix D in polynomial time.

Joint replenishment is studied in [112, 176, 177]. Other works where the values of coalitions are not known with certainty are [232, 233, 240, 241]. We consider a sequence of games where, differently from Filar and Petrosjan (2000) [95] and Haurie (1975) [114], the *average* coalitions' values (over time) are known with certainty but the instantaneous values are *unknown but bounded* by a polyhedron. Cooperative interval games are introduced in [103]. A classical reference in machine learning is [74]. The definition of the feasible dynamic allocation rule is in [26]. Allocation rules as iterative algorithmic procedures are proposed in [75, 151]. For the definition of balanced map and balanced game for games $\langle N, v \rangle$ we refer the reader to [239, Def. 11.5]. Sets of balanced games can also be found in the work of Kranich, Perea, and Peters (2005) [139] and Lehrer (2002) [151]. Lemma 19.5 recalls a result obtained in Bauso, Blanchini, and Pesenti (2006) [26]. The notion of excess introduced in this chapter is different from the coalitional excess that appears, e.g., in the definition of the nucleolus [214].

Chapter 20

Population of Producers

20.1 ▪ Introduction

We specialize mean-field games to production of exhaustible resources. The novelty in this model lies in the presence of an additional uncertainty capturing the influence of taxation or inflation on the amount of production.

The main contribution of this chapter is three-fold. First, we show that the problem of a population of producers can be formulated as a robust mean-field game. Second, we shed light on the strategic behavior of the producers by establishing mean-field equilibrium production policies. Third, we investigate stability of the microscopic and macroscopic dynamics.

More specifically, we provide an explicit expression of the connection between the optimal production law and the worst-case disturbance law. Exploiting the specific structure of the application involving oil production, we show that the dimension of the mean-field system can be reduced. The chapter also establishes a connection with risk-sensitive mean-field games with a modified *Hamilton–Jacobi–Bellman equation*.

This chapter is structured as follows. Section 20.2 formulates the problem. Section 20.3 turns the problem into a robust mean-field game and shows a mean-field equilibrium. Sections 20.4 and 20.5 examine stability of the microscopic and macroscopic dynamics, respectively. Section 20.6 provides a numerical example. Finally, Section 20.7 provides conclusions, notes, and references.

20.2 ▪ Production of an exhaustible resource

Let a continuum of market producers be given. Producers possess an initial reserve of raw material. For given constant parameters $\alpha, \beta, \sigma \in \mathbb{R}$, let us describe the reserve evolution over the interval $[0, T]$ via the following stochastic differential equation:

$$dx(t) = [\alpha x(t) + \beta u(t)]dt + \sigma[x(t)d\mathscr{B}(t) + \zeta(t)dt], \quad t \in [0, T], \qquad (20.1)$$

where

- $\mathscr{B}(t)$, $t \geq 0$, is a standard Brownian motion, which is independent of the initial state and independent across players; we will occasionally use the Brownian motion notation $\mathscr{B}$ to denote the process over the interval $[0, T]$;

- x_0 is the random initial reserve with distribution m_0;

231

- $u(t):[0,T]\to\mathbb{R}$ is the production rate of the player at time t;

- $\zeta(t):[0,T]\to\mathbb{R}$ is an unknown parameter function, representing taxation or inflation at time t.

Equation (20.1) is commonly referred to as *geometric Brownian motion*.

In addition, let us introduce a probability density function $m(x,t):\mathbb{R}_+\times[0,T]\to[0,+\infty[$, which satisfies $\int_{\mathbb{R}_+}m(x,t)dx=1$ for every t. Furthermore, let $\bar{m}(t)$ be the mean of the process $m(.)$.

A game designer assigns the producers a running cost involving the production cost to which we subtract the total income. The running cost takes the form

$$c(x(t),u(t),m(.))=-h(\bar{m}(t),\zeta(t))u(t)+\left[\frac{a}{2}u(t)^2+bu(t)\right],$$

where $h(\bar{m}(t),\zeta(t))$ is the sale price of oil and thus $h(\bar{m}(t),\zeta(t))u(t)$ is the income collected from producing and selling the quantity $u(t)$; $\frac{a}{2}u(t)^2$ accounts for a production energy consumed, $a>0$, and $bu(t)$ is a known linear taxation on production.

Let us denote the terminal penalty by $g(x_T)$ and set $g(x_T)=\phi|x_T|^2$, where $\phi>0$. The rationale here is to penalize the producers for unexploited reserve at the end of the horizon. In the spirit of H^∞-optimal control, the cost functional over the interval $[0,T]$ is given by

$$J(x(0),u,m,\zeta)=\mathbb{E}\left(g(x(T))+\int_0^T c(x(t),u(t),m(.),t)dt-\gamma^2\int_0^T|\zeta(t)|^2dt\right).$$

The term $h(\bar{m}(t),\zeta(t))$ in the running cost expresses the sale price as function of the distribution. An explicit expression for it is as below:

$$h(\bar{m}(t),\zeta(t))=ke^{r\frac{1}{\beta}\frac{d}{dt}\bar{m}(t)-\frac{\alpha r}{\beta}\bar{m}(t)-\frac{\sigma r}{\beta}\zeta(t)},\quad r<0,k\in\mathbb{R}. \tag{20.2}$$

To obtain (20.2), note that the mean of the state is generated by

$$\frac{d}{dt}[\mathbb{E}x(t)]=\alpha[\mathbb{E}x(t)]+\beta[\mathbb{E}u(t)]+\sigma[\mathbb{E}\zeta(t)].$$

From indistinguishability [149], the mean of the total production is given by

$$[\mathbb{E}u(t)]=\frac{1}{\beta}\left(\frac{d}{dt}\int xm(dx,t)\right)-\frac{\alpha}{\beta}\left(\int xm(dx,t)\right)-\frac{\sigma}{\beta}\zeta(t) \tag{20.3}$$

for $\beta<0$. Thus, by using a standard supply-demand law, we can assume that the price is decreasing in the produced quantity according to the following exponential law:

$$h(\bar{m}(t),\zeta(t))=ke^{r\bar{u}(t)}. \tag{20.4}$$

Equation (20.2) is obtained by substituting the right-hand side in (20.3) in (20.4).

In essence, the above expression describes the price as a function of the mean reserve distribution and taxation, and represents the coupling term between an individual player and the population behavior. Later we will see that as $u(t)$ and $\zeta(t)$ are given by bounded state-feedback closed-loop policies, $\bar{m}(t)$ is differentiable and bounded, which implies that $h(.)$ is uniformly continuous in $\bar{m}(t)$. This in turn implies that J is uniformly continuous in $\bar{m}(t)$.

Problem 20.1 (Population of producers). *Let $\mathcal{B}$ be a one-dimensional Brownian motion process defined on $(\Omega, \mathcal{F}, \mathbb{P})$, where $\mathcal{F}$ is the natural filtration generated by $\mathcal{B}$. Let $x(0)$ be any random variable independent of $\mathcal{B}$ having distribution $m_0(x)$. Introduce the following robust optimization problem:*

$$\inf_{\{u(t)\}_t} \sup_{\{\zeta(t)\}_t} J^\infty(x, u, m^*, \zeta),$$

where the dynamics of $x(t)$ are given by

$$dx(t) = [\alpha x(t) + \beta u(t) + \sigma \zeta(t)]dt + \sigma x(t) d\mathcal{B}(t), \ t \in (0, T], \ x_0 \in \mathbb{R}, \quad (20.5)$$

and $m(.)^*$ is the equilibrium mean-field trajectory obtained when any player at state x implements the control

$$u^*(x) = \arg\inf_{\{u(t)\}_t} \sup_{\{\zeta(t)\}_t} J(x, u, m^*, \zeta).$$

20.3 ▪ Robust mean-field equilibrium production policies

We shall now turn the problem into a robust mean-field game and examine mean-field equilibrium policies.

Set $v(x, t)$ as the (upper) value of the robust optimization problem under worst-case disturbance starting from time t at state x.

Theorem 20.1. *The production problem with exhaustible resources for a population of producers formulated in Problem 20.1 can be modeled via the following robust mean-field game:*

$$
\begin{cases}
\partial_t v(.) + \left[-\frac{1}{2a}\beta^2 + \left(\frac{\sigma}{2\gamma}\right)^2\right]|\partial_x v(.)|^2 + \left[-\frac{1}{2a}(-2h(\bar{m}(t), \zeta^*(t))\beta + 2b\beta)\right.\\
\left. + \alpha x(t)\right]\partial_x v(.) - \frac{1}{2a}\left(h(\bar{m}(t), \zeta^*(t))^2 + b^2 - 2h(\bar{m}(t), \zeta^*(t))b\right) + \frac{1}{2}\sigma^2 x(t)^2 \partial_{xx}^2 v(.) = 0,\\[2mm]
v(x, T) = \phi|x|^2,\\[2mm]
\partial_t m(.) + \partial_x\left[m(.)\left(\alpha x(t) + \beta \frac{h(\bar{m}(t), \zeta^*(t)) - b - \partial_x v(.)\beta}{a} + \frac{\sigma^2}{2\gamma^2}\partial_x v(.)\right)\right]\\[2mm]
+ \frac{\sigma^2}{2\gamma^2}\partial_x(m(.)\partial_x v(.)) - \frac{1}{2}\sigma^2\partial_{xx}^2\left[x(t)^2 m(.)\right] = 0,\\[2mm]
m(x, 0) = m_0(x),
\end{cases}
$$

$$(20.6)$$

where $m_0(x)$ is the initial distribution.

Furthermore, the optimal production is

$$
\begin{cases}
u^*(t) = \dfrac{h(\bar{m}(t), \zeta^*(t)) - b - \partial_x v(.)\beta}{a},\\[4mm]
\zeta^*(t) = \dfrac{\sigma}{2\gamma^2}\partial_x v(.).
\end{cases}
$$

$$(20.7)$$

Proof. To prove condition (20.7), consider the Hamiltonian

$$H(x(t), \partial_x v(.), \bar{m}(t), t) = \inf_u \left\{-h(\bar{m}(t), \zeta^*(t))u\right.$$
$$\left. + \left[\frac{a}{2}u^2 + bu\right] + \partial_x v(.)(\alpha x(t) + \beta u)\right\} = 0.$$

After differentiation we obtain

$$au - h(\bar{m}(t), \zeta^*(t)) + b + \partial_x v(.)\beta = 0,$$

which in turn yields

$$u^*(t) = \frac{h(\bar{m}(t), \zeta^*(t)) - b - \partial_x v(.)\beta}{a},$$

and the first equation in (20.7) is proved. Furthermore, let p be the co-state, and consider the robust Hamiltonian

$$\tilde{H}(x, p, m, t) = \inf_u \sup_\zeta \{ c(x, u, m) - \gamma^2 \zeta^2 + p(\alpha x + \beta u + \sigma \zeta) \}.$$

Assume that the function $\zeta \longmapsto -\gamma^2 \zeta^2 + p\sigma\zeta$ is strictly concave and has a global maximizer given by

$$\zeta^*(t) = \frac{\sigma}{2\gamma^2} p.$$

By setting the co-state equal to $\partial_x v(.)$, we obtain the following expression of the worst-case disturbance:

$$\zeta^*(t) = \frac{\sigma}{2\gamma^2} \partial_x v(.),$$

and the second equation in (20.7) is proved.

We now derive (20.6). Initially, notice that the second and fourth equations are the boundary conditions and derive straightforwardly from the *Hamilton–Jacobi–Isaacs equation* and the evolution of the distribution of states.

The first equation in (20.6) is the *Hamilton–Jacobi–Isaacs equation*. To derive this equation, let us rewrite the robust Hamiltonian as

$$\tilde{H}(x, p, m, t) = \inf_u \{ c(x, u, m) - \gamma^2 \zeta^*(t)^2 + p(\alpha x + \beta u + \sigma \zeta^*(t)) \} \qquad (20.8)$$

$$= \inf_u \{ c(x, u, m) + p(\alpha x + \beta u) \} + \left(\frac{\sigma p}{2\gamma} \right)^2. \qquad (20.9)$$

In (20.9) we have introduced the maximum value of the function $-\gamma^2 \zeta^2 + p\sigma\zeta$, which is $\left(\frac{\sigma p}{2\gamma} \right)^2$. Then for the *Hamilton–Jacobi–Isaacs equation* we have

$$\begin{cases} \partial_t v(.) + H(x, \partial_x v(.), m(.), t) + \left(\frac{\sigma}{2\gamma} \right)^2 |\partial_x v(.)|^2 + \frac{1}{2}\sigma^2 x(t)^2 \partial^2_{xx} v(.) = 0, \\ v(x, T) = g(x). \end{cases} \qquad (20.10)$$

Now, using for u the expression in (20.7), the Hamiltonian in (20.10) can be rewritten as

$$H(x(t), \partial_x v(.), \bar{m}(t), t) = u^*(t)[-h(\bar{m}(t), \zeta^*(t)) + b + \partial_x v(.)\beta] + \frac{a}{2}u^*(t)^2 + \partial_x v(.)\alpha x(t)$$

$$= -\frac{1}{2a}(h(\bar{m}(t), \zeta^*(t)) - b - \partial_x v(.)\beta)^2 + \partial_x v(.)\alpha x(t)$$

$$= -\frac{1}{2a}\Big(h(\bar{m}(t), \zeta^*(t))^2 + b^2 + (\partial_x v(.)\beta)^2 - 2h(\bar{m}(t), \zeta^*(t))b$$

$$\quad - 2h(\bar{m}(t), \zeta^*(t))\partial_x v(.)\beta + 2b\partial_x v(.)\beta \Big) + \partial_x v(.)\alpha x(t)$$

$$= -\frac{1}{2a}\beta^2 |\partial_x v(.)|^2 + \Big[-\frac{1}{2a}(-2h(\bar{m}(t), \zeta^*(t))\beta + 2b\beta)$$

$$\quad + \alpha x(t) \Big]\partial_x v(.) - \frac{1}{2a}\Big(h(\bar{m}(t), \zeta^*(t))^2 + b^2 - 2h(\bar{m}(t), \zeta^*(t))b \Big).$$

Substituting the above expression of the Hamiltonian in the *Hamilton–Jacobi–Isaacs equation* (20.10), we obtain (20.6).

The third equation in (20.6) is the *Kolmogorov–Fokker–Planck equation*. A general expression for this equation is

$$\begin{cases} \partial_t m(x,t) + \partial_x\Big(m(x,t)\partial_p H(x,\partial_x v(.),m(.),t)\Big) + \frac{\sigma^2}{2\gamma^2}\partial_x(m(x,t)\partial_x v(.)) \\ \quad - \frac{1}{2}\sigma^2\partial^2_{xx}\big[x(t)^2 m(x,t)\big] = 0, \\ m(x,0) = m_0(x), \end{cases} \tag{20.11}$$

where $m_0(x)$ is the initial distribution. By substituting (20.7) into (20.11), we obtain the third equation in (20.6), and this concludes the proof. □

The significance of the above result is that to find the optimal production we need to solve the the two coupled partial differential equations in (20.6) in v and m with given boundary conditions. This is usually done by iteratively solving the *Hamilton–Jacobi–Isaacs equation* in (20.6) for fixed m and by substituting the optimal u obtained from (20.7) in the *Kolmogorov–Fokker–Planck* equation until a fixed point in v and m is reached.

20.4 ▪ Stability of the microscopic dynamics

We shall now discuss a sufficient condition under which the microscopic dynamics is exponentially asymptotically stable. To this purpose, from

$$\alpha x(t) + \beta u^*(t) + \sigma \zeta^*(t) = \partial_p H(.) + \sigma \zeta^*(t) \tag{20.12}$$

and from (20.7), we can rewrite (20.5) as

$$\begin{aligned} dx(t) &= \big[\alpha x(t) + \beta u^*(t) + \sigma\zeta^*(t)\big]dt + \sigma x(t)d\mathscr{B}(t) \\ &= \Big[\partial_p H(x(t),\partial_x v(.),m(.),t) + \sigma\zeta^*(t)\Big]dt + \sigma x(t)d\mathscr{B}(t) \\ &= \Big[\partial_p H(x(t),\partial_x v(.),m(.),t) + \frac{\sigma^2}{2\gamma^2}\partial_x v(.)\Big]dt + \sigma x(t)d\mathscr{B}(t), \\ & \quad t \in (0,T],\ x_0 \in \mathbb{R}. \end{aligned} \tag{20.13}$$

Assumption 20.1. *There exists $\hat{k} > 0$ such that*

$$-\hat{k}x(t) \ge \partial_p H(x(t),\partial_x v(.),m(.),t) + \frac{\sigma^2}{2\gamma^2}\partial_x v(.). \tag{20.14}$$

Let us take as Lyapunov function the quadratic function $V(x) = x^2$; then the stochastic derivative of $V(x)$ is obtained by applying the infinitesimal generator to $V(x)$, which yields

$$\mathscr{L}V(x(t)) = \lim_{dt \to 0}\frac{\mathbb{E}[V(x(t+dt)) - V(x(t))|x(t)]}{dt} = [\sigma^2 - 2\hat{k}]x(t)^2.$$

Theorem 20.2 (see [162]). *Let Assumption 20.1 hold true. If $V(x) \ge 0$, $V(0) = 0$, and $\mathscr{L}V(x) \le -\eta V(x)$ on $Q_\epsilon := \{x : V(x) \le \epsilon\}$ for some $\eta > 0$ and for arbitrarily large ϵ, then the origin is asymptotically stable "with probability 1" and*

$$P_{x(0)}\left\{\sup_{T \le t < +\infty} x(t)^2 \ge \lambda\right\} \le \frac{V(x(0))e^{-\psi T}}{\lambda}$$

for some $\psi > 0$.

From the above theorem, we have the following result, which establishes exponential stochastic stability of the mean-field equilibrium provided above (see the Appendix, Chapter E, on stochastic stability).

Corollary 20.3. *Let Assumption 20.1 hold true. If $[\sigma^2 - 2\hat{k}] < 0$, then $\lim_{t \to \infty} x(t) = 0$ almost surely, and*

$$P_{x(0)} \left\{ \sup_{T \leq t < +\infty} x(t)^2 \geq \lambda \right\} \leq \frac{V(x(0))e^{-\psi T}}{\lambda}$$

for some $\psi > 0$.

The above results can be specialized to the case where the value function can be approximated by a quadratic expression of the form $v(.) = \frac{1}{2}\phi(t)x(t)^2$. In this case it holds that $\partial_x v(.) = \phi(t)x(t)$, which means that the gradient is linear in the state. Then, we obtain $h(\bar{m}(t), \zeta^*(t)) - b - \partial_x v(.)\beta > 0$. Furthermore, we can rewrite the dynamics as

$$
\begin{aligned}
dx(t) &= [\alpha x(t) + \beta u^*(t) + \sigma \zeta^*(t)]dt + \sigma x(t)d\mathcal{B}(t) \\
&= \left[\alpha x(t) + \frac{\beta}{a}[h(\bar{m}(t), \zeta^*(t)) - b - \partial_x v(.)\beta] + \frac{\sigma^2}{2\gamma^2}\partial_x v(.)\right]dt + \sigma x(t)d\mathcal{B}(t) \\
&= \left[\alpha + \frac{\beta}{a}\phi(t) + \frac{\sigma^2}{2\gamma^2}\phi(t)\right]x(t)dt + \frac{1}{a}(h(\bar{m}(t), \zeta^*(t)) - b)dt + \sigma x(t)d\mathcal{B}(t), \\
&\quad t \in (0, T], \; x_0 \in \mathbb{R}.
\end{aligned}
$$
(20.15)

Consequently, condition (20.14) takes the form

$$-\hat{k}x(t) \geq \left[\alpha + \frac{\beta}{a}\phi(t) + \frac{\sigma^2}{2\gamma^2}\phi(t)\right]x(t) + \frac{h(\bar{m}(t), \zeta^*(t)) - b}{a}.$$
(20.16)

We can conclude that the existence of a $\hat{k}$ satisfying the above expression implies stability of the microscopic dynamics.

20.5 ▪ Stability of the macroscopic dynamics

Let Assumption 20.1 hold. Then, we can approximate the macroscopic dynamics describing the evolution of $\bar{m}(t)$ over the horizon $(0, T]$ as follows:

$$\frac{d}{dt}\bar{m}(t) \leq -\hat{k}\bar{m}(t), \quad t \in (0, T], \; \bar{m}_0 \in \mathbb{R},$$
(20.17)

which yields the following upper bound for $\bar{m}(t)$:

$$\bar{m}(t) \leq \bar{m}_0 e^{-\hat{k}t}, \quad t \in (0, T], \; \bar{m}_0 \in \mathbb{R}.$$

The inequality above describes a converging linear dynamics which upper bounds the time evolution of $\bar{m}(t)$ for all $t \in (0, T]$.

Furthermore, from (20.15), the macroscopic dynamics describing the evolution of $\bar{m}(t)$ is given by

$$
\frac{d}{dt}\bar{m}(t) = \left[\alpha + \frac{\beta}{a}\phi(t) + \frac{\sigma^2}{2\gamma^2}\phi(t)\right]\bar{m}(t) + \frac{1}{a}(h(\bar{m}(t), \zeta^*(t)) - b),
$$
$$
t \in (0, T], \; \bar{m}_0 \in \mathbb{R}.
$$
(20.18)

Table 20.1. *Simulation parameters for the population of producers.*

α	β	T	$\bar{m}_0$	h_{min}	h_{max}	b	σ	$std(m_0)$	$a[10^1]$	Q
0	-1	40	70	1	10^2	0	0.09	7	$\{1,5,50\}$	$\{1,5,50\}$

The interpretation of the above equation is straightforward if we take $b = 0$ and $\tilde{h}\bar{m}(t) = h(\bar{m}(t), \zeta^*(t))$. The latter means that the sale price $h(\bar{m}(t), \zeta^*(t))$ is linear in $\bar{m}(t)$. In this case, we can derive the following tight bound for the microscopic dynamics:

$$\begin{cases} \bar{m}(t) = \bar{m}_0 e^{\rho t}, \\ \rho = \alpha + \dfrac{\beta}{a}\phi(t) + \dfrac{\sigma^2}{2\gamma^2}\phi(t) + \dfrac{\tilde{h}}{a}. \end{cases} \tag{20.19}$$

That is to say that if ρ is strictly negative, the mean stock converges exponentially to zero, and therefore the macroscopic dynamics is stable.

20.6 ▪ Numerical example

We provide next a numerical example showing a feasible macroscopic evolution pattern. The game involves $n = 10^3$ players and a discretized set of states $\mathcal{X} = \{x_{min}, x_{min} + 1, \ldots, x_{max}\}$, where $x_{min} = 0$ (no reserve) and $x_{max} = 100$ (maximal reserve). The simulation parameters are listed in Table 20.1. The reserve changes only with the production quantity and the stochastic disturbance $\mathcal{B}(t)$, and therefore let us set $\alpha = 0$ and $\beta = -1$. We also set $v(.) = Qx^2$ and assume that the deterministic disturbance $\zeta(t)$ enters into the picture indirectly through the coefficient Q. Indeed, Q is monotonically decreasing in γ as a higher γ leads to a lower ζ^*, and this in turn implies a lower optimal cost $v(.)$. Parameter σ is set equal to 0.09 to guarantee that, given the initial distribution m_0, $x \in \mathcal{X}$ at each time. The horizon length is $T = 40$. The dynamic equation (20.5) is then given by

$$\begin{cases} dx(t) = -u(t)dt + \sigma x(t)d\mathcal{B}(t), \quad t = 0, 1, \ldots, T, \\ x_0 \in \{x_{min}, x_{min} + 1, \ldots, x_{max}\}. \end{cases} \tag{20.20}$$

Furthermore, m_0 is Gaussian with mean $\bar{m}_0 = 70$ and standard deviation $std(m_0) = 7$. For the sale price $h(\bar{m}, .)$ we consider a linear approximation between the minimum $h_{min} = 1$ when $\bar{m} = x_{max}$ and the maximum $h_{max} = 10^2$ when $\bar{m} = 0$:

$$\hat{h}(\bar{m}(t)) = \left(\frac{10^2 - \bar{m}(t)}{10^2}\right)h_{max} + \left(1 - \frac{10^2 - \bar{m}(t)}{10^2}\right)h_{min}. \tag{20.21}$$

Furthermore, we approximate $\partial_x v(.) = Qx$, which fits the case of quadratic value function, and replace the optimal production in (20.7) by

$$u^*(t) = \frac{\hat{h}(\bar{m}(t)) + Qx}{a}. \tag{20.22}$$

We have obtained the simulations via the algorithm displayed below.

ALGORITHM 20.1. **Simulation algorithm for a population of producers.**

Input: Set of parameters as in Table 20.1.
Output: Distribution function $m(.)$, mean $\bar{m}(t)$, and standard deviation $std(m(.))$.
  1 : **Initialize.** Generate $x(0)$ as n random samples from Gaussian distribution
        with mean $\bar{m}_0$ and standard deviation $std(m_0)$,
  2 : **for** time $t = 0, 1, \ldots, T - 1$ **do**
  3 :   **if** $t > 0$, **then** compute distribution $m(.)$, mean distribution $\bar{m}(t)$,
               standard deviation $std(m(.))$,
  4 :   **end if**
  5 :   compute sale price $\hat{h}(\bar{m}(t))$,
  6 :   **for** player $i = 1, 2, \ldots, n$ **do**
  7 :   compute $u(t)$ from (20.22),
  8 :   generate Brownian motion $d\mathcal{B}(t)$,
  9 :   compute new state $x(t + 1)$ by executing (20.20),
10 :   **end for**
11 : **end for**
12 : **STOP**

The macroscopic evolution pattern highlights the impact of the disturbance $\zeta(t)$. Here, the mean distribution $\bar{m}(t)$ decreases monotonically, and the standard deviation $std(m(.))$ first increases due to the influence of the Brownian motion and then decreases to zero.

This is shown in Fig. 20.1(left). From top to bottom, the figure depicts the time plot of the microscopic evolution $x(t)$. In the horizontal axis we plot the time t. The linear term $\frac{Qx}{a}$ dominates more and more in comparison with the constant term $\frac{\hat{h}(\bar{m}(t))}{a}$ from top to bottom. In particular we have set $Q = 1$, $a = 10$ (top); $Q = 5$, $a = 50$ (middle); and $Q = 50$, $a = 500$ (bottom). Note that the ratio $\frac{Q}{a}$ is kept constant, whereas $\frac{\hat{h}(.)}{a}$ is strongly decreasing from top to bottom. Apparently, the speed of convergence reduces as well. This is clear from observing the graphs on the right column which display the time plot $\bar{m}(t)$ (solid line and y-axis labeling on the left) and the evolution of the standard deviation $std(m(.))$ (dashed line and y-axis labeling on the right).

20.7 ▪ Notes and references

This chapter is based on [42, 43]. This chapter has provided insights on how to specialize mean-field games to uncertain production applications. We have first applied the methodology to production of an exhaustible resource. Then, we have established a mean-field system for the resulting robust games.

We highlight three key directions for current and future research. A first key direction examines the connection with *risk-sensitive* optimal control problems.

A second key direction aims at extending the results to the vector state case and to an infinite horizon problem involving both a discounted cost functional and a time-average cost functional. A third direction is concerned with the study of a market where some producers are leaders and others are followers. The model takes the form of a Stackelberg game.

Mean-field games applied to production of exhaustible resources were first introduced in [105].

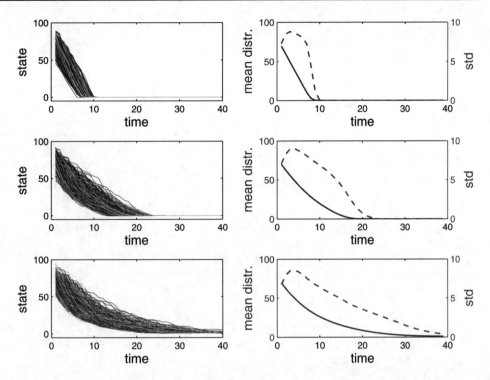

Figure 20.1. *Macroscopic evolution pattern: showing the effects of a higher control coefficient Q (associated with a stronger disturbance $\zeta(t)$): both the mean distribution $\bar{m}(t)$ and the standard deviation $std(m(.))$ decrease monotonically. Reprinted with kind permission of Springer Science+Business Media* [43].

Chapter 21

Cyber-Physical Systems

21.1 ▪ Introduction

This chapter provides a robust mean-field game modeling cyber-attacks and concurrency in cyber-physical systems (CPSs).

CPSs include (i) computation processes, (ii) physical processes, and (iii) humans in the loop. Let us consider the paradigmatic example depicted in Fig. 21.1. There, we have a physical plant, a sensor network, distributed controllers, and actuators. A communication network connects sensors, controllers, and actuators.

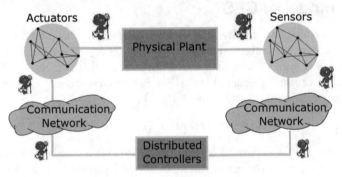

Figure 21.1. *Example of cyber-attacks.*

In the design of CPSs, a main goal is to guarantee a reasonable good performance under the following circumstances:

- *Cyber-attack*: A cyber-attack is any malicious behavior that disturbs the measurement and control process, such as measurement distortions, communication noise or disruptions, actuator failures, etc.

- *Concurrency*: Physical processes involve several parallel dynamics, in contrast to software processes, which are compositions of sequential steps. Thus, there is the *need to bridge an inherently sequential semantics with an intrinsically concurrent physical world* [85]. In hybrid systems, a similar aspect yields to *minimum attention control*.

Reframing CPSs within the theory of mean-field games is motivated by the following reasons. First, the game is large and highly distributed, in that we decentralize decisions, information, and objectives. Second, worst-case adversarial disturbances may well describe the effects of cyber-attacks. Third, a mean-field term in the cost functional keeps in consideration the level of congestion in the communication network. Last but not least, sticking to heuristics rather than sophisticated strategies may accurately simulate bounded rationality and limited computation capabilities of the humans in the loop.

In this chapter, we adopt the following model. We have a large number of physical processes. The players are the controllers of the physical processes. The players' state dynamics are given by linear stochastic differential equations. Players minimize a cost functional which includes a mean-field term. Such a penalty term deters the use of the communication network when this is congested. Adversarial disturbances with bounded energy disturb the players and interfere with the optimization of their performances.

Our study provides fundamental insights on *consistency*, *scalability*, and *stability* in CPSs.

In particular, the provided mean-field game is consistent, in that the macroscopic part provides an accurate and reliable description of what happens when all players are rational. The players are shown to use best-response policies which are scalable, in that they build on simple information structures. The best-response policies lead to stable microscopic and macroscopic behavior.

In Section 21.2, we formulate the problem. In Section 21.3, we turn the problem into a mean-field game. In Section 21.4, we study equilibria. In Section 21.5, we study stability. In Section 21.6, we provide numerical studies. Finally, in Section 21.7, we provide conclusions, notes, and references for this chapter.

21.2 • A model of CPS

Let a continuum of physical processes be given. Let us model such physical processes as homogeneous players' dynamics. Let $x(0) \in \mathbb{R}_+$ be the initial state of a generic player, which is realized according to the probability distribution m_0. The state of the generic player at time t, denoted by $x(t) \in \mathbb{R}_+$ evolves, according to the following controlled stochastic process in the interval $[0, T]$:

$$dx(t) = [\alpha x(t) + \beta u(t)]dt + \sigma[x(t)d\mathcal{B}(t) + \zeta(t)dt], \qquad (21.1)$$

where $u(t) \in \mathbb{R}_+$ is the control input, $\mathcal{B}(t)$ is a standard Brownian motion, which is independent of the initial state $x(0)$ and independent across players and time, $\alpha, \sigma \in \mathbb{R}$ and $\beta < 0$ are parameters, and $\zeta(t)$ is an adversarial disturbance.

In addition, let us introduce a probability density function $m(x, t) : \mathbb{R}_+ \times [0, T] \to [0, +\infty[$, which satisfies $\int_{\mathbb{R}_+} m(x, t)dx = 1$ for every t. Similarly, let a probability density function for the control be given as $z(u, t) : \mathbb{R} \times [0, T] \to [0, +\infty[$, which satisfies $\int_{\mathbb{R}} z(u, t)du = 1$ for every t. Furthermore, let $\bar{z}(t)$ and $\bar{m}(t)$ be the mean of the processes $z(.)$ and $m(.)$, respectively. The players are assigned the following cost functional:

$$J(x(0), u, \bar{z}, \zeta) = \mathbb{E}\Big(g(x(T)) + \int_0^T c(x(t), u(t), \bar{z}(t))dt - \gamma^2 \int_0^T |\zeta(t)|^2 dt\Big).$$

Players wish to stabilize their states to zero, and therefore we can take for the stage cost

$$c(x(t), u(t), \bar{z}_t, \zeta(t)) = h(\bar{z}(t))u(t) + \Big[\frac{a}{2}x(t)^2 + \frac{b}{2}u(t)^2\Big],$$

where $h(\bar{z}(t))$ is a measure of the congestion. Thus $h(\bar{z}(t))u(t)$ is a penalty on the control of the single player which is proportional to the congestion in the control loop. The term $\frac{a}{2}x(t)^2$, where $a > 0$, is the cost of being in a nonnull state, and $\frac{b}{2}u(t)^2$, where $b > 0$, accounts for the control energy. The terminal penalty on final state is given by $g(x(T)) = \phi x(T)^2$ for a given scalar $\phi > 0$. This term penalizes nonnull states at the end of the horizon. The congestion term depends on the magnitude of the average control and is given by

$$h(\bar{z}(t)) = k|\bar{z}(t)| = k\left|\frac{1}{n}\sum_{j=1}^{n} u_{j,t}\right| = k\left|\frac{1}{\beta}\frac{d}{dt}\bar{m}(t) - \frac{\alpha}{\beta}\bar{m}(t) - \frac{\sigma}{\beta}\zeta(t)\right|, \quad k \in \mathbb{R}_+. \tag{21.2}$$

Actually, to obtain the last equation we have introduced expectations in (21.1) and considered deterministic disturbance $\zeta(t)$. By using indistinguishability we obtain

$$\begin{aligned}[\mathbb{E}u(t)] &= \frac{1}{\beta}\left(\frac{d}{dt}[\mathbb{E}x(t)]\right) - \frac{\alpha}{\beta}([\mathbb{E}x(t)]) - \frac{\sigma}{\beta}\zeta(t)\\ &= \frac{1}{\beta}\left(\frac{d}{dt}\int xm(dx,t)\right) - \frac{\alpha}{\beta}\left(\int xm(dx,t)\right) - \frac{\sigma}{\beta}\zeta(t).\end{aligned}$$

We are in a position to give a precise statement of the problem under study.

Problem 21.1 (Continuum of physical processes). *Let $\mathcal{B}$ be a one-dimensional Brownian motion defined on $(\Omega, \mathcal{F}, \mathbb{P})$, where $\mathcal{F}$ is the natural filtration generated by $\mathcal{B}$. Let $x(0)$ be independent of $\mathcal{B}$ and with density $m_0(x)$. Consider the problem in $\mathbb{R}$ and $(0, T]$:*

$$\begin{cases} \inf_{\{u(t)\}_t} \sup_{\{\zeta(t)\}_t} J(x, u, \bar{z}, \zeta), \\ dx(t) = [\alpha x(t) + \beta u(t) + \sigma\zeta(t)]dt + \sigma x(t)d\mathcal{B}(t). \end{cases}$$

In the context of CPSs, dynamics (21.1) is suitable to describe a multi-tank system, where the tank level is the state variable. Here the control input tries to stabilize the tank level to zero while at the same time an adversarial disturbance tries to impede stabilization. Dynamics (21.1) may represent a power grid, where the angle of the rotor of each generator is the state. Here the control acts to guarantee transient stability despite the volatility of the renewable power sources (wind or solar power). As a further example, we can think of cyber-physical economic systems; here (21.1) shares similarity with the Black and Scholes model derived in the context of portfolio selection.

21.3 ▪ Turning a CPS into a mean-field game

This section shows that the above problem can be modeled as a robust mean-field game.

To this purpose, let $v(.)$ be the (upper) value of the robust optimization problem under worst-case disturbance starting from time t at state x. Consider the Hamiltonian function

$$H(x, p, \bar{z}) = \inf_u\{c(x, u, \bar{z}) + p(\alpha x + \beta u)\},$$

where p is the co-state.

The problem of a continuum of physical processes can be turned into a robust mean-field game as established in the following.

Theorem 21.1. *The closed-loop robust mean-field game for the crowd-averse CPSs takes on the form*

$$
\begin{cases}
\partial_t v(.) + \left[-\frac{1}{2b}\beta^2 + \left(\frac{\sigma}{2\gamma}\right)^2 \right] |\partial_x v(.)|^2 + \left[-\frac{1}{2b}(2h(\bar{z}(t))\beta) + \alpha x(t) \right] \partial_x v(.) \\
\quad - \frac{1}{2b}h(\bar{z}(t))^2 + \frac{a}{2}x(t)^2 + \frac{1}{2}\sigma^2 x(t)^2 \partial_{xx}^2 v(.) = 0 \ in \ \mathbb{R}_+ \times [0,T), \\
v(x,T) = \phi|x|^2 \ in \ \mathbb{R}_+, \\[2mm]
\partial_t m(.) + \partial_x \left[m(.) \left(\alpha x(t) + \beta \frac{-h(\bar{z}(t)) - \partial_x v(.)\beta}{b} + \frac{\sigma^2}{2\gamma^2}\partial_x v(.) \right) \right] + \frac{\sigma^2}{2\gamma^2}\partial_x(m(.)\partial_x v(.)) \\
\quad - \frac{1}{2}\sigma^2 \partial_{xx}^2 \left[x(t)^2 m(.) \right] = 0 \ in \ \mathbb{R}_+ \times [0,T), \\
m_0(x) \ given \ in \ \mathbb{R}_+, \\
\dot{\bar{m}}(t) = \alpha \bar{m}(t) + \beta \bar{z}(t)^* + \sigma \bar{\zeta}^*(t), \ \bar{m}_0 \ given,
\end{cases}
$$
(21.3)

where $\bar{z}(t)^* := \int_{\mathbb{R}_+} u^*(x,t)m(x,t)dx$, $\bar{\zeta}^*(t) := \int_{\mathbb{R}_+} \zeta^*(t)(x)m(x,t)dx$, *and the optimal closed-loop control and disturbance are*

$$
\begin{cases}
u^*(x) = \frac{-h(\bar{z}(t)) - \partial_x v(.)\beta}{b}, \\[2mm]
\zeta^*(x) = \frac{\sigma}{2\gamma^2}\partial_x v(.).
\end{cases}
$$
(21.4)

Proof. Let us start by proving condition (21.4). To this purpose, let the Hamiltonian be given by

$$
H(x(t), \partial_x v(.), \bar{z}(t)) = \inf_u \left\{ h(\bar{z}(t))u(t) + \left[\frac{a}{2}x(t)^2 + \frac{b}{2}u(t)^2 \right] \right. \\
\left. + \partial_x v(.)(\alpha x(t) + \beta u(t)) \right\} = 0.
$$
(21.5)

By differentiating with respect to $u(t)$ we obtain $bu(t) + h(\bar{z}(t)) + \partial_x v(.)\beta = 0$. From the latter we get $u^*(x)$ as in (21.4). The worst-case disturbance $\zeta^*(x,t)$ is obtained by solving

$$
\sup_{\zeta(t)} \left\{ -\gamma^2 \zeta(t)^2 + \partial_x v(.)\sigma \zeta(t) \right\}.
$$

Assuming concavity in $\zeta(t)$, and after differentiation, we get $-2\gamma^2\zeta^*(t) + \partial_x v(.)\sigma = 0$. From the latter, we obtain $\zeta^*(x,t) = \frac{\sigma}{2\gamma^2}\partial_x v(.)$. This concludes the first part of the proof related to condition (21.4).

Let us now focus on the set of (21.3). From Chapter 12 the *Hamilton–Jacobi–Isaacs equation* of the robust mean-field game for Problem 21.1 is given by

$$
\begin{cases}
\partial_t v(.) + H(x, \partial_x v(.), \bar{z}(t)) + \left(\frac{\sigma}{2\gamma}\right)^2 |\partial_x v(.)|^2 + \frac{1}{2}\sigma^2 x(t)^2 \partial_{xx}^2 v(.) = 0 \\
\quad in \ \mathbb{R}_+ \times [0,T), \\
v(x,T) = \phi x^2 \ in \ \mathbb{R}_+.
\end{cases}
$$
(21.6)

Now, let us substitute the value of u^* from (21.4) in the Hamiltonian (21.5):

$$
H(x(t), \partial_x v(.), \bar{z}(t)) = u^*(t)[h(\bar{z}(t)) + \partial_x v(.)\beta] + \frac{a}{2}x(t)^2 + \frac{b}{2}(u^*(t))^2 + \partial_x v(.)\alpha x(t) \\
= -\frac{1}{2b}\beta^2 |\partial_x v(.)|^2 + \left[-\frac{1}{2b}(2h(\bar{z}(t))\beta) + \alpha x(t) \right] \partial_x v(.) - \frac{1}{2b}h(\bar{z}(t))^2 + \frac{a}{2}x(t)^2.
$$

Using the above expression of the Hamiltonian in the *Hamilton–Jacobi–Isaacs equation* in (21.6), we obtain the first equation in (21.3).

To obtain the third equation in (21.3), we know from Chapter 12 that the *Kolmogorov–Fokker–Planck equation* is given by

$$
\begin{cases}
\partial_t m(x,t) + \partial_x \Big(m(x,t)\partial_p H(x,\partial_x v(.),\bar{z}(t)) \Big) + \frac{\sigma^2}{2\gamma^2}\partial_x(m(x,t)\partial_x v(.)) \\
\quad - \frac{1}{2}\sigma^2 \partial^2_{xx}\big[x(t)^2 m(x,t) \big] = 0 \text{ in } \mathbb{R}_+ \times [0,T), \\
m_0(x) \text{ given in } \mathbb{R}_+.
\end{cases}
\tag{21.7}
$$

By introducing u^* from (21.4) into the above equation we obtain the third equation in (21.3).

Finally, the last equation, which is the ordinary differential equation describing the evolution of the mean $\bar{m}(t)$, is obtained from (21.1) by averaging over the state space, and this concludes the proof. □

The interpretation of the above result is that the solution of the above game must be obtained by solving the two coupled partial differential equations in (21.3) in the value function v and density function m with given boundary conditions.

Recall from Chapter 12 that such a solution is called *worst-disturbance feedback mean-field equilibrium*.

21.4 ▪ Humans in the loop and heuristic policies

After introducing the robust mean-field game, we now develop a heuristic method to compute an approximation of the mean-field equilibrium. Such a method builds on a state space extension involving the microscopic state and the average state distribution.

In the following, we assume that there exists a lower bound on the rate of change of the mean $\bar{m}(t)$.

Assumption 21.1. *There exist a $\theta > 0$ and an $\tilde{m}(t)$ satisfying*

$$
\begin{cases}
\frac{d}{dt}\bar{m}(t) \geq \frac{d}{dt}\tilde{m}(t) = -\theta\tilde{m}(t) \quad \forall t \in [0,T], \\
\bar{m}_0 = \tilde{m}_0.
\end{cases}
\tag{21.8}
$$

In addition to this, let us also assume that $\bar{\zeta}(t) = \delta\tilde{m}(t)$.

From (21.2) and using the approximate dynamics $\frac{d}{dt}\tilde{m}(t) = -\theta\tilde{m}(t)$, $\bar{m}_0 = \tilde{m}_0$, we obtain

$$
h(\bar{z}(t)) = k\left| \frac{-\theta - \alpha - \sigma\delta}{\beta}\tilde{m}(t) \right| := 2s\tilde{m}(t).
\tag{21.9}
$$

The problem under study is then given by

$$
\inf_{\{u(t)\}_t} \sup_{\{\zeta(t)\}_t} \int_0^T \left[2s\tilde{m}(t)u(t) + \frac{q}{2}\tilde{m}(t)^2 + \left(\frac{a}{2}x(t)^2 + \frac{b}{2}u(t)^2 - \gamma^2\zeta(t)^2 \right) \right]dt + g(x_T)
$$

$$
\text{s.t. } \begin{bmatrix} dx(t) \\ d\tilde{m}(t) \end{bmatrix} = \left(\begin{bmatrix} \alpha & 0 \\ 0 & -\theta \end{bmatrix}\begin{bmatrix} x(t) \\ \tilde{m}(t) \end{bmatrix} + \begin{bmatrix} \beta \\ 0 \end{bmatrix}u(t) \right.
$$

$$
\left. + \begin{bmatrix} \sigma \\ 0 \end{bmatrix}\zeta(t) \right)dt + \begin{bmatrix} \sigma x(t)d\mathcal{B}(t) \\ 0 \end{bmatrix},
$$

where the term $\frac{q}{2}\tilde{m}_t^2$ is here introduced to guarantee convexity of the cost as formalized later in Assumption 21.2.

After introducing the extended state and control

$$X(t) = \left[\begin{array}{c} x(t) \\ \tilde{m}(t) \end{array} \right], \quad \tilde{u}(t) = u(t) + \frac{2}{b}s\tilde{m}, \tag{21.10}$$

and by completing the square in the objective function, we get the linear-quadratic problem

$$\left\{ \begin{array}{l} \displaystyle \inf_{\{\tilde{u}(t)\}_t} \sup_{\{\zeta(t)\}_t} \int_0^T \left[\frac{1}{2}(X(t)^T \tilde{Q}X(t) + R\tilde{u}(t)^2 - \Gamma\zeta(t)^2) \right] dt + g(x_T), \\[2mm] dX(t) = (\tilde{A}X(t) + B\tilde{u}(t) + C\zeta(t))dt + Cx(t)d\mathscr{B}(t), \end{array} \right.$$

where

$$\tilde{Q} = \left[\begin{array}{cc} a & 0 \\ 0 & q - \frac{4}{b}s^2 \end{array} \right], \qquad R = b, \qquad \Gamma = 2\gamma^2,$$

$$\tilde{A} = \left[\begin{array}{cc} \alpha & -\beta\frac{2}{b}s \\ 0 & -\theta \end{array} \right], \quad B = \left[\begin{array}{c} \beta \\ 0 \end{array} \right], \quad C = \left[\begin{array}{c} \sigma \\ 0 \end{array} \right].$$

Consider a value function $\mathscr{V}(x, \tilde{m}, t)$ (in compact form $\mathscr{V}(X, t)$) in the extended state space, for which it holds that

$$\left\{ \begin{array}{l} \partial_t \mathscr{V}(X,t) + H(X, \partial_X \mathscr{V}(X,t)) + \left(\frac{\sigma}{2\gamma}\right)^2 |\partial_x \mathscr{V}(X,t)|^2 + \frac{1}{2}\sigma^2 x(t)^2 \partial_{xx}^2 \mathscr{V}(X,t) = 0, \\[2mm] \mathscr{V}(X,T) = g(x). \end{array} \right.$$

$$\tag{21.11}$$

Let us assume that the above value function has a quadratic structure as given below:

$$\mathscr{V}(x, \tilde{m}, t) = [x(t) \quad \tilde{m}(t)] \underbrace{\left[\begin{array}{cc} P_{11}(t) & P_{12}(t) \\ P_{21}(t) & P_{22}(t) \end{array} \right]}_{P(t)} \left[\begin{array}{c} x(t) \\ \tilde{m}(t) \end{array} \right].$$

The matrix $P(t)$ appearing in the above equation must be solution of the differential Riccati equation

$$\dot{P}(t) + P(t)\tilde{A} + \tilde{A}^T P(t) - 2P(t)(BR^{-1}B^T - C\Gamma^{-1}C^T)P(t) + \frac{\tilde{Q}}{2} + W = 0, \tag{21.12}$$

where

$$BR^{-1}B^T - C\Gamma^{-1}C^T = \left[\begin{array}{cc} \frac{1}{b}\beta^2 + \frac{1}{2\gamma^2}\sigma^2 & 0 \\ 0 & 0 \end{array} \right],$$

$$W = \left[\begin{array}{cc} \sigma^2 P_{11} & 0 \\ 0 & 0 \end{array} \right]. \tag{21.13}$$

Assumption 21.2. *For the parameters q and s it holds that*

$$\tilde{Q} = \left[\begin{array}{cc} a & 0 \\ 0 & q - \frac{4}{b}s^2 \end{array} \right] \geq 0. \tag{21.14}$$

Now, given a solution P of the Riccati equation, the best response μ is given by

$$
\begin{aligned}
\mu(X(t), t) &= -2R^{-1}B^T P(t)X(t) \\
&= -\frac{2}{b}[\beta \ 0]\begin{bmatrix} P_{11}(t) & P_{12}(t) \\ P_{21}(t) & P_{22}(t) \end{bmatrix}\begin{bmatrix} x(t) \\ \tilde{m}(t) \end{bmatrix} \\
&= -\frac{2\beta}{b}(P_{11}(t)x(t) + P_{12}(t)\tilde{m}(t)).
\end{aligned}
\tag{21.15}
$$

From the above expression and from (21.10) we get

$$
\tilde{u}^*(X(t), t) = -\frac{2}{b}[\beta P_{11}(t)x(t) + (\beta P_{12}(t) + s)\tilde{m}(t)].
\tag{21.16}
$$

Analogously, for the worst-case disturbance we obtain

$$
\begin{aligned}
\tilde{\zeta}^*(X(t), t) &= 2\Gamma^{-1}C^T P(t)X(t) \\
&= \frac{1}{\gamma^2}[\sigma \ 0]\begin{bmatrix} P_{11}(t) & P_{12}(t) \\ P_{21}(t) & P_{22}(t) \end{bmatrix}\begin{bmatrix} x(t) \\ \tilde{m}(t) \end{bmatrix} \\
&= \frac{1}{\gamma^2}\sigma(P_{11}(t)x(t) + P_{12}(t)\tilde{m}(t)).
\end{aligned}
\tag{21.17}
$$

It is worth noting that by taking the average in (21.17), the condition $\bar{\zeta}(t) = \delta \tilde{m}(t)$ in Assumption 21.1 is satisfied.

21.5 • Asymptotic stability

In this section we provide an analysis of the stability of the mean-field equilibrium established in the previous section. Introducing the best-response and worst-case disturbance (21.16)–(21.17) in the stochastic differential equation (21.1) we get

$$
\begin{aligned}
dx(t) &= \alpha x(t) + (-\frac{2\beta^2}{b} + \frac{\sigma^2}{\gamma^2})P_{11}(t)x(t) \\
&\quad + [(-\frac{2\beta^2}{b} + \frac{\sigma^2}{\gamma^2})P_{12}(t) - \beta \frac{2}{b}s]\tilde{m}(t) + \sigma x(t)d\mathscr{B}(t), \\
&\quad t \in (0, T], \ x_0 \in \mathbb{R}.
\end{aligned}
$$

The above stochastic differential equation can be studied in the framework of stochastic stability theory [162]. To do this, consider the candidate Lyapunov function $V(x) = \Phi x^2$. The stochastic derivative of $V(x)$ yields

$$
\mathscr{L}V(x(t)) = [\sigma^2 + 2(\alpha - \frac{2\beta^2}{b} + \frac{\sigma^2}{\gamma^2})]\Phi x(t)^2.
$$

Theorem 21.2 (see [162]). *If $V(x) \geq 0$, $V(0) = 0$, and $\mathscr{L}V(x) \leq -\eta V(x)$ on $Q_\epsilon := \{x : V(x) \leq \epsilon\}$ for some $\eta > 0$ and for arbitrarily large ϵ, then the origin is asymptotically stable "with probability 1" and*

$$
P_{x(0)}\left\{ \sup_{T \leq t < +\infty} x(t)^2 \geq \lambda \right\} \leq \frac{V(x(0))e^{-\psi T}}{\lambda}
$$

for some $\psi > 0$.

From the above theorem, we get the result below, which states exponential stochastic stability of the mean-field equilibrium.

Table 21.1. *Simulation parameters for a CPS.*

$\sigma\,[10^{-1}]$	$std(m_0)$	b	Q
1	5	$\{20, 25, 100\}$	$\{4, 5, 20\}$

Corollary 21.3. *If* $[\sigma^2 + 2(\alpha - \frac{2\beta^2}{b} + \frac{\sigma^2}{\gamma^2})]\Phi < 0$, *then* $lim_{t \to \infty} x(t) = 0$ *almost surely and*

$$P_{x(0)}\left\{\sup_{T \le t < +\infty} x(t)^2 \ge \lambda\right\} \le \frac{V(x(0))e^{-\psi T}}{\lambda}$$

for some $\psi > 0$.

From the above result we conclude that the players stabilize their states to zero asymptotically.

We can approximate the mean-field equilibrium, which is captured by the evolution of $\bar{m}(t)$ over the horizon $(0, T]$, as

$$\frac{d}{dt}\bar{m}(t) = \left[\alpha + (-\frac{2\beta^2}{b} + \frac{\sigma^2}{\gamma^2})(P_{11}(t) + P_{12}(t)) - \beta\frac{2}{b}s\right]\bar{m}_t, \quad t \in (0, T], \; x_0 \in \mathbb{R}.$$

Actually, we can derive a differential equation describing the evolution of the mean distribution which represents a bound, namely

$$\begin{cases} \bar{m}(t) = \bar{m}_0 e^{\rho t}, \\ \rho = \alpha + (-\frac{2\beta^2}{b} + \frac{\sigma^2}{\gamma^2})(P_{11}(t) + P_{12}(t)) - \beta\frac{2}{b}s. \end{cases}$$

The equation above corresponds to saying that the mean distribution converges exponentially to zero in absence of the stochastic disturbances (the Brownian motion), under the assumption that ρ is strictly negative.

21.6 • Numerical example

This section highlights a stereotypical evolution pattern for a CPS made by a continuum of physical processes.

Let a number of processes $n = 10^3$ and a discretized set of states $\mathcal{X} = \{x_{min}, x_{min} + 1, \ldots, x_{max}\}$ be given, where $x_{min} = 0$ and $x_{max} = 100$ (see parameters in Table 21.1).

Let us set the parameters $\alpha = 0$ and $\beta = -1$, and let us examine the influence of $\zeta(t)$ implicitly by increasing the coefficient Q used in the quadratic approximation of the value function $v(.) = Qx^2$. The horizon length is $T = 40$.

Let the initial density m_0 be Gaussian with mean $\bar{m}_0 = 70$ and standard deviation $std(m_0) = 5$. Consider a linear function of type

$$\hat{h}(\bar{m}) = \left(\frac{10^2 - \bar{m}}{10^2}\right)h_{min} + \left(1 - \frac{10^2 - \bar{m}}{10^2}\right)h_{max}. \tag{21.18}$$

The above expression is a linear approximation of $h(\bar{m}, .)$ in the interval from $h_{min} = 0$ to $h_{max} = 10^2$. The minimum $h_{min} = 0$ is obtained for $\bar{m} = x_{min}$, and the maximum $h_{max} = 10^2$ is obtained for $\bar{m} = x_{max}$. In addition, let us take $\partial_x v(.) = 2Qx$. As a consequence, from (21.4) we get

$$u^*(t) = \frac{-\hat{h}(\bar{m}(t)) + 2Qx}{b}. \tag{21.19}$$

Simulations are performed using the following algorithm.

ALGORITHM 21.1. **Simulation algorithm for the CPS example.**

Input: Set of parameters as in Table 21.1.
Output: Distribution function $m(.)$, mean $\bar{m}(t)$,
   and standard deviation $std(m(.))$.
 1 : **Initialize.** Generate $x(0)$ given $\bar{m}_0$ and $std(m_0)$
 2 : **for** time $t = 0, 1, \ldots, T-1$ **do**
 3 :   **if** $t > 0$, **then** compute $m(.)$, $\bar{m}(t)$, and $std(m(.))$
 4 :   **end if**
 5 :   compute congestion term $\hat{h}(\bar{m}(t))$,
 6 :   **for** player $i = 1, 2, \ldots, n$ **do**
 7 :   compute new state $x(t+1)$ by executing (21.1)
 8 :   **end for**
 9 : **end for**
12 : **STOP**

The evolution pattern highlights the effects of a higher linear term $\frac{Qx}{b}$ in comparison with the constant term $\frac{\hat{h}(\bar{m}(t))}{b}$ in the control input expression (21.19). A higher value for Q can be linked back to the effects of the disturbance $\zeta(t)$. Agents far from zero show a faster converging dynamics. As a consequence, the standard deviation $std(m(.))$ and the sparsity decrease with time, while the mean distribution $\bar{m}(t)$ decreases to zero. This is shown in Fig. 21.2(left). From top to bottom, the figure shows the time plot of the microscopic dynamics $x(t)$. The linear term $\frac{Qx}{b}$ dominates the constant term $\frac{\hat{h}(\bar{m}(t))}{b}$ more and more in the graphs from top to bottom. Actually, $Q = 4$, $b = 20$ (top); $Q = 5$, $a = 25$ (middle); and $Q = 20$, $a = 100$ (bottom). The ratio $\frac{Q}{b}$ is constant, whereas $\frac{\hat{h}(.)}{b}$ is decreasing from top to bottom. Apparently, the speed of convergence increases. The graphs on the right column plot the time evolution of $\bar{m}(t)$ and the evolution of the standard deviation $std(m(.))$. The average $\bar{m}(t)$ is plotted using a solid line and the corresponding y-axis labels are indicated on the left, while the standard deviation $std(m(.))$ is plotted using a dashed line and the corresponding y-axis labels are indicated on the right. It is worth noting that both the mean distribution $\bar{m}(t)$ and the standard deviation $std(m(.))$ decrease to zero.

21.7 ▪ Notes and references

This chapter is based on [41] and discusses robust mean-field games as paradigmatic models for CPSs. The players simulate a continuum of physical processes, adversarial disturbances account for cyber-attacks, and a mean-field term in the individual cost functionals models congestion. After introducing the robust mean-field game we examine mean-field equilibrium policies and study stability.

Future directions include the study of (i) the connection with *risk-sensitive* optimal control problems, (ii) the vector state case and infinite horizon (with discounted payoff and time-average payoff), and (iii) a cyber-physical economic market with some big players and many other small players.

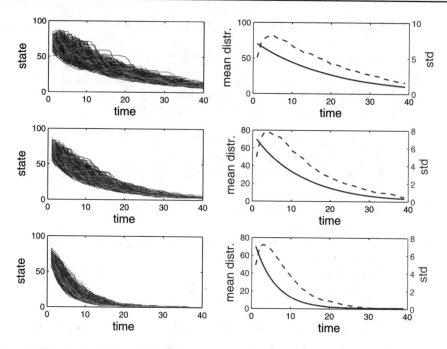

Figure 21.2. *Macroscopic evolution pattern: showing the effects of a higher control coefficient Q (associated with a stronger disturbance $\zeta(t)$): both the mean distribution $\bar{m}(t)$ and the standard deviation $std(m(.))$ decrease monotonically. Reprinted with permission from IEEE [41].*

An analysis of the main issue arising when combining sequential semantics and intrinsically concurrent physical world is available in [85]. Minimum attention control is discussed in [87]. Adversarial disturbances with bounded energy in CPSs are studied in [235]. A multi-tank system is studied in [129]. Transient stability and the volatility of the renewable power sources (wind or solar power) are developed in [227]. Cyber-physical economic systems and the Black and Scholes model are discussed in [56].

Appendix A

Mathematical Review

This text makes use of some fundamentals of real analysis which we survey in this appendix. This chapter of the appendix is largely subsumed by [23, Appx. I]. We refer the reader to the classical book by Luenberger [164] for more details.

A.1 ▪ Sets and vector spaces

A set S is defined as a collection of elements. The notation $s \in S$ indicates that s is an element of the set S. Conversely, the notation $s \notin S$ means that s is not an element of the set S. We call S a *finite set* if the number of elements is finite; otherwise, it is called an *infinite set*. If there is a one-to-one correspondence between the elements of a set and the positive integers, we say that set is *countable*; otherwise, it is a *uncountable*. A *space* is a set S enjoying a specific structure. For a linear (*vector*) *space*, the structure is of algebraic nature and enjoys well-known properties which we assume that the reader is familiar with. Given a set S, if S is a vector space, we call *subspace* a subset of S that is also a vector space. As an example, the *n-dimensional Euclidean space* is a vector space. Here each element is determined by n real numbers. We denote such a space by $\mathbb{R}^n$. An element $x \in \mathbb{R}^n$ is written as a *column vector* $x = (x_1, \ldots, x_n)^T$, where $x_1, \ldots, x_n$ are real numbers and denote the components of x.

A.1.1 ▪ Linear independence and basis

Given a vector space S and a finite number of vectors $s_1, \ldots, s_n$ in S, these vectors are *linearly independent* if $\sum_{i=1}^n \alpha_i s_i = 0$ implies that $\alpha_i = 0$ for all $i = 1, \ldots, n$. In addition, if any element in S can be expressed as a linear combination of these vectors, this set of vectors, denoted by X, is said to be a *basis* of the space S. As a consequence, S is said to be *finite dimensional*, and the "dimension" is the number of elements of X; otherwise, S is said to be *infinite dimensional*.

A.2 ▪ Normed linear vector spaces

Given a linear (vector) space S, we say that S is *normed* if there is a structure induced on S by a real-valued function mapping every element $u \in S$ into a real number, denoted by $\|u\|$ and referred to as the *norm* of u. The norm satisfies the following three axioms:

(1) $\|u\| \geq 0$ for all $u \in S$; $\|u\| = 0$ if and only if $u = 0$.

(2) $\|u + v\| \leq \|u\| + \|v\|$ for each $u, v \in S$.

(3) $\|\alpha u\| = |\alpha| \cdot \|u\|$ for all $\alpha \in \mathbb{R}$ and for each $u \in S$.

A.2.1 • Convergent sequences, limit points, and Cauchy sequence

Given a normed vector space S and an infinite sequence of vectors $\{s_1, s_2, \ldots, s_i, \ldots\}$ in S, this sequence is said to converge to a vector s if, for a given $\epsilon > 0$, there exists an N for which it holds that $\|s - s_i\| < \epsilon$ for all $i \geq N$. To indicate that the sequence converges we use the notation $s_i \to s$, or $\lim_{i \to \infty} s_i \to s$. The vector s is referred to as the *limit point* of the sequence $\{s_i\}$. More generally, a vector s is a *limit point* of an infinite sequence $\{s_i\}$ if there is an infinite subsequence $\{s_{i_k}\}$ that converges to s.

A *Cauchy sequence* is an infinite sequence $\{s_i\}$ in a normed vector space such that, for a given $\epsilon > 0$, there exists an N such that $\|s_n - s_m\| < \epsilon$ for all $n, m \geq N$. Letting S be a normed vector space, S is *complete*, or a *Banach space*, if every Cauchy sequence in S converges to an element of S.

A.2.2 • Open, closed, and compact sets

Given a normed vector space S, an element $s \in S$, and an $\epsilon > 0$, we call the set $N_\epsilon(s) = \{x \in S : \|x - s\| < \epsilon\}$ an ϵ-*neighborhood* of s. Letting a subset X of S be given, X is said to be *open* if, for every $x \in X$, there exists an $\epsilon > 0$ such that $N_\epsilon(x) \subset X$. Letting a subset X of S be given, X is said to be *closed* if its complement in S is open. This is equivalent to saying that every convergent sequence in X has its limit point in X. Letting a set $X \in S$ be given, the largest open subset of X is referred to as the *interior* of X and denoted as $int\{X\}$.

Let a normed vector space S be given. A subset X of S is *compact* if every infinite sequence in X has a convergent subsequence whose limit point is in X. If the subset X is finite dimensional, then the property of compactness is equivalent to the properties of being closed and bounded.

A.2.3 • Functions, functionals, and continuity

Given two vector spaces S and T, a *function* is a mapping f of S into T and is denoted by $f : S \to T$ or $y = f(x)$ for $x \in S$, $y \in T$. If $T = \mathbb{R}$, then f is a *functional*.

Let S and T be two normed linear spaces, and let $f : S \to T$ be given. f is *continuous* at $x_0 \in S$ if, for every $\epsilon > 0$, there exists a $\delta > 0$ such that $f(x) \in N_\epsilon(f(x^o))$ for every $x \in N_\delta(x_0)$. f is *continuous everywhere* or, simply, *continuous* if it is continuous at every point of S.

A.3 • Matrices

A rectangular array A of numbers arranged in m rows and n columns is called an $(m \times n)$ *matrix*. The numbers are the *elements* or *entries* of the matrix. We use a subscript ij to denote the element in the ith row and jth column of A, namely a_{ij} or $[A]_{ij}$ (occasionally also a_j^i). We also write $A = \{a_{ij}\}$. A matrix with the same number of rows and columns is a *square* matrix. An *identity matrix* is an $(n \times n)$ square matrix A such that $a_{ii} = 1$, $i = 1, \ldots, n$, and $a_{ij} = 0$, $i \neq j$, $i, j = 1, \ldots, n$. We symbolically write I_n or, simply, I for such a matrix.

Given an $(m \times n)$ matrix A, the *transpose* of A is the $(n \times m)$ matrix A^T with entries $a_{ij}^T = a_{ji}$. A square matrix A is said to be symmetric if $A = A^T$. A square matrix A is nonsingular if there is an $(n \times n)$ matrix called the inverse of A, which is symbolically written as A^{-1}, that satisfies the equations $A^{-1}A = I = AA^{-1}$.

A.3.1 ▪ Eigenvalues and quadratic forms

Given a square matrix A, a scalar λ and a nonzero vector x such that $Ax = \lambda x$ are an *eigenvalue* and an *eigenvector* of A, respectively.

Let a square symmetric matrix A be given, and let the eigenvalues be all positive (respectively, nonnegative). A is a *positive definite* (respectively, *non·egative definite* or *positive semidefinite*). Equivalently, a symmetric $(n \times n)$ matrix A is positive definite (respectively, nonnegative definite) if $x^T A x > 0$ (respectively, $x^T A x \geq 0$) for all nonzero vectors $x \in \mathbb{R}^n$. The matrix A is *negative definite* (respectively, *positive definite*) if the matrix $(-A)$ is positive (respectively, nonnegative) definite. We use the notation $A > 0$ (respectively, $A \geq 0$) to mean that A is positive (respectively, nonnegative) definite. An $n \times n$ matrix A is row-stochastic if the matrix has nonnegative entries a_j^i and $\sum_{j=1}^{n} a_j^i = 1$ for all $i = 1, \ldots, n$. A matrix A is doubly stochastic if both A and its transpose A^T are row-stochastic.

A.4 ▪ Convex sets and convex functionals

Given a vector space S, a subset C of S is *convex* if for any pair $u, v \in C$ and every scalar $\alpha \in [0, 1]$, it holds that $\alpha u + (1 - \alpha)v \in C$. Let a functional $f : C \to \mathbb{R}$ be given, which is defined over a convex subset C of a vector space S. The functional f is *convex* if, for any pair $u, v \in C$ and any scalar $\alpha \in [0, 1]$, it holds that $f(\alpha u + (1 - \alpha)v) \leq \alpha f(u) + (1 - \alpha)f(v)$. If the latter is a strict inequality for all $\alpha \in (0, 1)$, the functional f is *strictly convex*. Conversely, f is *concave* if $(-f)$ is convex, and *strictly concave* if $(-f)$ is strictly convex.

Given a functional $f : \mathbb{R}^n \to \mathbb{R}$, if the partial derivatives of f with respect to the components of $x = (x_1, \ldots, x_n)^T \in \mathbb{R}^n$ exist, then f is *differentiable*. In this case we use the symbol

$$\nabla f(x) = [\partial f(x)/\partial x_1, \ldots, \partial f(x)/\partial x_n].$$

We refer to $\nabla f(x)$ as the *gradient* of f at x. The gradient is also symbolically expressed as $f_x(x)$ or $df(x)/dx$. Consider a partition of x involving two vectors y and z of dimensions n_1 and $n - n_1$, respectively. We use the notation $\nabla f(y, z)$ or $\partial f(y, z)/\partial y$ to denote the partial gradient of f with respect to y. Consider a vector-valued function $g : \mathbb{R}^n \to \mathbb{R}^m$, and let its components be differentiable with respect to $x \in \mathbb{R}^n$. Then, $g(x)$ is said to be differentiable, and the derivative $dg(x)/dx$ is an $(m \times n)$ matrix determined by an ijth entry of the form $\partial g_i(x)/\partial x_j$. Here we write g_i to mean the ith component of g.

Consider the gradient $\nabla f(x)$ which is a vector. Its derivative corresponds to the second derivative of $f : \mathbb{R}^n \to \mathbb{R}$ and is an $(n \times n)$ matrix. If $f(x)$ is twice continuously differentiable with respect to x, then such a matrix, denoted by $\nabla^2 f(x)$, is symmetric and is referred to as the *Hessian matrix* of f at x. A necessary and sufficient condition for the Hessian matrix to be nonnegative definite for all $x \in \mathbb{R}^n$ is that f be convex.

Let a subset $C \subseteq \mathbb{R}^n$ be given, and consider a point x_0 in its boundary. Furthermore, let $a \neq 0$ be given such that $a^T x \leq a^T x_0$ for all $x \in C$. Then, the hyperplane $\{x : a^T x = a^T x_0\}$ is referred to as the *supporting hyperplane* to C at the point x_0. That is to say that the hyperplane $\{x : a^T x = a^T x_0\}$ *separates* the vector x_0 and the set C. Geometrically this corresponds to saying that the hyperplane $\{x : a^T x = a^T x_0\}$ is tangent to C at x_0, and the half-space $\{x : a^T x \leq a^T x_0\}$ contains C.

Appendix B

Optimization

In this appendix we give some basic notions of mathematical optimization which we use throughout the text.

B.1 ▪ Optimizing functionals

This section is based on [23, Appx. I.5]. Consider a vector space S, a functional $f : S \to \mathbb{R}$, and a subset $X \subseteq S$. The optimization problem

$$\text{minimize } f(x) \text{ subject to } x \in X$$

consists in finding an element $x^* \in X$, which we call a *minimizing element* or an *optimal solution*, for which it holds that

$$f(x^*) \leq f(x) \quad \forall x \in X.$$

This is occasionally called a *globally minimizing solution,* in contrast with a *locally minimizing solution,* as defined next. An element $x^o \in X$ is a locally minimizing solution if there is an $\epsilon > 0$ for which it holds that

$$f(x^o) \leq f(x) \quad \forall x \in N_\epsilon(x^o) \cap X.$$

An optimal solution may not exist. An optimal solution exists if the set of real numbers $\{f(x) : x \in X\}$ is bounded below and there exists an $x^* \in X$ such that $\inf\{f(x) : x \in X\} = f(x*)$, in which case we write

$$f(x^*) = \inf f(x) = \min f(x).$$

We say that an optimal solution does not exist if such an x^* cannot be computed, even though $\inf\{f(x) : x \in X\}$ is finite. In this case the notation

$$\inf\{f(x) : x \in X\} \quad \text{or} \quad \inf_{x \in X} f(x)$$

stands for the *optimal value* of the optimization problem. In the case where $\{f(x) : x \in X\}$ is not bounded below, namely $\inf_{x \in X} f(x) = -\infty$, then neither an optimal solution nor an optimal value exists.

We can always convert a maximization problem into a minimization problem by replacing f by $-f$. Any solution of one problem is also solution to the other problem. The optimal value of the maximization problem is symbolically expressed as $\sup_{x \in X} f(x)$, which is equal to minus the optimal value of the minimization problem. For a given *maximizing element* $x^* \in X$, we have $\sup_{x \in X} f(x) = \max_{x \in X} f(x) = f(x^*)$.

B.1.1 ▪ Existence of optimal solutions

Keeping in mind the aforementioned minimization problem, if X is a finite set, there is only a finite number of comparisons, and therefore an optimal solution exists.

If X is not finite, an optimal solution exists if f is continuous and X is compact. This is a result known as the *Weierstrass theorem*. When X is finite dimensional, recall that compactness is equivalent to being closed and bounded.

B.1.2 ▪ Necessary and sufficient conditions for optimality

Consider $S = \mathbb{R}^n$ and $f : \mathbb{R}^n \to \mathbb{R}$, which is differentiable. If X is an open set, an optimal solution must satisfy the first-order necessary condition

$$\nabla f(x^*) = 0.$$

Furthermore, if f is also twice continuously differentiable on $\mathbb{R}^n$, an optimal solution must satisfy the second-order necessary condition

$$\nabla^2 f(x^*) \geq 0.$$

The two conditions mentioned above are necessary for $x^* \in X$ to be a locally minimizing solution. Furthermore, if X is also a convex set and f is a convex functional on X, then the above conditions are also sufficient for global optimality.

B.2 ▪ Mathematical optimization

This section is based on [64, Chap. 1] and [117, Chap. 13]. A *mathematical optimization problem* can be symbolically expressed by

$$\begin{aligned} \text{minimize} \quad & f_0(x) \\ \text{subject to} \quad & f_i(x) \leq b_i, \quad i = 1, \ldots, m. \end{aligned} \tag{B.1}$$

In the above problem, the vector $x = (x_1, \ldots, x_n)^T$ is the *optimization variable* of the problem, the function $f_0 : \mathbb{R}^n \to \mathbb{R}$ represents the objective function, the functions $f_i : \mathbb{R}^n \to \mathbb{R}$, $i = 1, \ldots, m$, are the (inequality) constraint functions, and the constants $b_1, \ldots, b_m$ are the bounds for the constraints. A vector x^* is the *optimal solution* to the problem (B.1) if it provides the smallest value of the objective functions among all vectors that satisfy the constraints. In other words we have that for any z with $f_1(z) \leq b_1, \ldots, f_m(z) \leq b_m$, we have $f_0(z) \geq f_0(x^*)$.

The optimization problem (B.1) is said to be a *linear program* if the objective and constraint functions $f_0, \ldots, f_m$ are linear. That is to say that they satisfy

$$f_i(\alpha x + \beta y) = \alpha f_i(x) + \beta f_i(y)$$

for all $x, y \in \mathbb{R}^n$ and all $\alpha, \beta \in \mathbb{R}$. Conversely, a *nonlinear program* is an optimization problem which is not linear. An optimization problem is said to be *convex* if the objective and the constraint functions are convex, i.e., if it holds that

$$f_i(\alpha x + \beta y) \leq \alpha f_i(x) + \beta f_i(y), \quad \alpha + \beta = 1, \alpha, \beta \geq 0.$$

Note that convex optimization is a generalization of linear programming, as any linear program is a convex optimization problem.

B.2.1 ▪ Linear programming

Linear programming involves optimization problems where the objective and all constraint functions are linear. A linear program is given by

$$\begin{aligned}
\text{minimize} \quad & c^T x \\
\text{subject to} \quad & a_i^T x \le b_i, \quad i = 1, \ldots, m,
\end{aligned} \tag{B.2}$$

where the vectors $c, a_1, \ldots, a_m \in \mathbb{R}^n$ and the scalars $b_1, \ldots, b_m \in \mathbb{R}$ represent the parameters of the problem.

The solution of a linear program has no simple analytical formula. However, there exist effective solution algorithms, such as Dantzig's simplex algorithm, or interior point methods described in [117, Chap. 4] and [64, Chap. 11].

B.2.2 ▪ The complementarity problem

Consider the variables $w_1, w_2, \ldots, w_p$ and $z_1, z_2, \ldots, z_p$. The *complementarity problem* consists in finding a solution of the constraints

$$w = F(z), \quad w \ge 0, \quad z \ge 0$$

that verifies also the following complementarity constraint:

$$w^T z = 0.$$

In the above problem, w and z represent column vectors, and F is a given vector-valued function. Note that the problem has no objective function. Such a problem is referred to as the complementarity problem because of the complementary relationships that either $w_i = 0$ or $z_i = 0$ (or both) for each $i = 1, 2, \ldots, p$.

A special complementarity problem is the *linear complementarity problem*, which is symbolically expressed as

$$F(z) = q + Mz.$$

In the above, q is a column vector and M is a $p \times p$ matrix.

B.2.3 ▪ Quadratic programming

The convex optimization problem (B.1) is said to be a *quadratic program* if the objective function is (convex) quadratic and the constraint functions are affine. A quadratic program is in general given by

$$\begin{aligned}
\text{minimize} \quad & (1/2)x^T P x + q^T x + r \\
\text{subject to} \quad & Gx \le h, \\
& Ax = b,
\end{aligned} \tag{B.3}$$

where $P \in \mathbf{S}_+^n$, $G \in \mathbb{R}^{m \times n}$, and $A \in \mathbb{R}^{p \times n}$. The symbol $\mathbf{S}_+^n$ indicates the set of symmetric positive semidefinite matrices. A quadratic program involves the minimization of a convex quadratic function over a polyhedron. If the objective in (B.1) as well as the inequality constraint functions are (convex) quadratic, as in

$$\begin{aligned}
\text{minimize} \quad & (1/2)x^T P x + q^T x + r \\
\text{subject to} \quad & (1/2)x^T P_i x + q_i^T x + r_i \le 0, \quad i = 1, \ldots, m, \\
& Ax = b,
\end{aligned} \tag{B.4}$$

where $P_i \in \mathbf{S}_+^n$, $i = 0, 1, \ldots, m$, the problem is called a *quadratically constrained quadratic program*. A quadratically constrained quadratic program involves the minimization of a convex quadratic function over a feasible region that is the intersection of ellipsoids (when $P_i > 0$). Linear programs are special quadratic programs, which are obtained by taking $P = 0$ in (B.3). Quadratically constrained quadratic programs include quadratic programs (and therefore also linear programs), in which case we have $P_i = 0$ in (B.4) for $i = 1, \ldots, m$.

Appendix C

Lyapunov Stability

This appendix introduces Lyapunov stability and related concepts. It is based on [162]. These concepts are extensively used throughout the text.

Consider the trajectory of a dynamic system starting from x_0 at time t_0, and denote such a trajectory by $x(t; x_0, t_0)$. In the following we give three different definitions of Lyapunov stability. Let the equilibrium solution be 0 unless stated otherwise.

Definition C.1 (Lyapunov stability). *The equilibrium solution is* stable *if, given $\epsilon > 0$, there exists a $\delta(\epsilon, t_0) > 0$ such that, for all $\|x_0\| < \delta$,*

$$\sup_{t \geq t_0} \|x(t; x_0, t_0)\| < \epsilon.$$

Definition C.2 (Asymptotic Lyapunov stability). *The equilibrium solution is* asymptotically stable *if it is stable and if there exists a $\delta' > 0$ such that $\|x_0\| < \delta'$ guarantees that*

$$\lim_{t \to \infty} \|x(t; x_0, t_0)\| = 0.$$

If the convergence condition is satisfied for all initial times, t_0, then the equilibrium solution is uniformly asymptotically stable.

Definition C.3 (Exponential Lyapunov stability). *The equilibrium solution is* exponentially stable *if it is asymptotically stable and if there exist a $\delta > 0$ and $\alpha > 0$ and a $\beta > 0$ such that $\|x_0\| \leq \delta$ guarantees that*

$$\|x(t; x_0, t_0)\| \leq \beta \|x_0\| e^{-\alpha(t - t_0)}.$$

If the convergence condition is satisfied for all initial times, t_0, then the equilibrium solution is uniformly exponentially stable.

Let a nonnegative continuous function $V(x)$ on $\mathbb{R}^n$ be given, where $V(0) = 0$ and $V(x) > 0$ for $x \neq 0$. Assume that for some $m \in \mathbb{R}$ the set $Q_m = \{x \in \mathbb{R}^n : V(x) < m\}$ is bounded and $V(x)$ has continuous first partial derivatives in Q_m. Given the initial time $t_0 = 0$, let $x(t) = x(t, x_0)$ be the unique solution of the initial value problem

$$\begin{cases} \dot{x}(t) = f[x(t)], & t \geq 0, \\ x(0) = x_0 \in \mathbb{R}^n, & f(0) = 0, \end{cases} \tag{C.1}$$

for $x_0 \in Q_m$. From the continuity of $V(x)$, the open set Q_r for $r \in (0, m]$ given by $Q_r = \{x \in \mathbb{R}^n : V(x) < r\}$ includes the origin and decreases monotonically to the singleton set $\{0\}$ as $r \to 0^+$. Consider the total derivative $\dot{V}(x)$ of $V(x)$ along the solution trajectory $x(t, x_0)$, which is given by

$$\dot{V}(x) = \frac{dV(x)}{dt} = f^T(x)\frac{\partial V}{\partial x} := -k(x). \tag{C.2}$$

If such a derivative satisfies $-k(x) \leq 0$ for all $x \in Q_m$, where $k(x)$ is continuous, then $V(x(t))$ is a nonincreasing function of t. That is to say that $V(x_0) < m$ implies $V(x(t)) < m$ for all $t \geq 0$. This corresponds to the stability of the zero solution of (C.1) in the sense of Lyapunov, and $V(x)$ is called a *Lyapunov function* for (C.1). Furthermore, if $k(x) > 0$ for $x \in Q_m \setminus \{0\}$, then $V(x(t))$, as a function of time t, is strictly monotone decreasing. This implies that $V(x(t)) \to 0$, as $t \to +\infty$, from (C.2). From this we also have that $x(t) \to 0$ as $t \to +\infty$. We can arrive at the same result from a different perspective. Actually, after integration of (C.2) we have

$$0 < V(x_0) - V(x(t)) = \int_0^t k(x(s))ds < +\infty \quad \text{for } t \in [0, +\infty). \tag{C.3}$$

From (C.2) we then get $x(t) \to \{0\} = \{x \in Q_m : k(x) = 0\}$ as $t \to +\infty$. This proves the asymptotic stability for system (C.1).

The Lyapunov function $V(x)$ may be regarded as a generalized energy function of the system (C.1). The above argument illustrates the physical intuition that if the energy of a physical system is always decreasing near an equilibrium state, then the equilibrium state is stable.

Appendix D

Some Notions of Probability Theory

This appendix reviews some basics of probability theory used throughout the text. It is largely subsumed by [23, Appx. II].

D.1 ▪ Basics of probability theory

Consider a set Ω whose elements are the outcomes of a random experiment. As an example, if the experiment is the toss of a coin, then the set Ω has two elements. Another experiment could be the selection of an integer from the set $[0, \infty)$, in which case Ω is countably infinite. Furthermore, if the experiment is the continuous roulette wheel, then we have a nondenumerable set Ω. We call *event* any subset of Ω on which a probability measure can be defined. In particular, let us denote by $\mathbf{F}$ the class of all such events, namely subsets of Ω. Then $\mathbf{F}$ enjoys the following properties:

(1) $\Omega \in \mathbf{F}$.

(2) If $A_1, A_2 \in \mathbf{F}$, then its complement $A^c = \{\omega \in \Omega : \omega \notin A\}$ belongs to $\mathbf{F}$. Note that the empty set, $\emptyset$, is the complement of Ω and also belongs to $\mathbf{F}$.

(3) If $A_1, A_2 \in \mathbf{F}$, then $A_1 \cap A_2$ and $A_1 \cup A_2$ also belong to $\mathbf{F}$.

(4) If $A_1, A_2, \ldots, A_i, \ldots$ denote a countable number of events, then the countable intersection $\cap_{i=1}^{\infty} A_i$ and the countable union $\cup_{i=1}^{\infty} A_i$ are also events (i.e., $\in \mathbf{F}$).

The class $\mathbf{F}$ defined as above is a *sigma algebra (σ-algebra)*. We call *probability measure* P a nonnegative functional defined on the elements of this σ-algebra. Such a probability measure P satisfies the following axioms:

(1) For every event $A \in \mathbf{F}$, $0 \leq P(A) \leq 1$, and $P(\Omega) = 1$.

(2) If $A_1, A_2 \in \mathbf{F}$ and $A_1 \cap A_2 = \phi$ (i.e., A_1 and A_2 are disjoint events), then $P(A_1 \cup A_2) = P(A_1) + P(A_2)$.

(3) Let $\{A_i\}$ denote a (countably) infinite sequence in $\mathbf{F}$, with the properties $A_{i+1} \subset A_i$ and $\cap_{i=1}^{\infty} A_i = \emptyset$. Then, the limit of the sequence of real numbers $\{P(A_i)\}$ is zero (i.e., $\lim_{i \to \infty} P(A_i) \to 0$).

The aforementioned triple $(\Omega, \mathbf{F}, \mathsf{P})$ is referred to as a *probability space*. The pair $(\Omega, \mathbf{F})$ is a measurable space. If $\Omega = \mathbb{R}^n$, then the subsets are n-dimensional rectangles, and the smallest σ-algebra generated by these rectangles is called the n-dimensional Borel σ-algebra and is denoted by $\mathbb{B}^n$. We call *Borel sets* the elements of $\mathbb{B}^n$. We refer to the pair $(\mathbb{R}^n, \mathbb{B}^n)$ as a *Borel (measurable) space*. A probability measure defined on this space is known as a *Borel probability measure*.

D.1.1 • Finite and countable probability spaces

Let Ω be a finite set, and denote it by $\Omega = \{\omega_1, \omega_2, \ldots, \omega_n\}$. Then, we can assign probability weights on individual elements of Ω rather than on subsets of Ω. In doing this, we use the symbol p_i to indicate the probability of the event ω_i. We refer to the n-tuple $(p_1, p_2, \ldots, p_n)$ as a *probability distribution* over Ω. Evidently, it holds that $0 \leq p_i \leq 1$ for all $i = 1, \ldots, n$. Furthermore, if the elements of Ω are all independent events, we get $\sum_{i=1}^{n} p_i = 1$. The same convention applies when Ω is a countable set (i.e., $\Omega = \{\omega_1, \omega_2, \ldots\}$), in which case we simply replace n by ∞.

D.2 • Random vectors

Consider two measurable spaces $(\Omega_1, \mathbf{F}_1)$ and $(\Omega_2, \mathbf{F}_2)$, and let f be a function defined from Ω_1 into Ω_2. f is a *measurable function* if for every $A \in \mathbf{F}_2$ we have $f^{-1}(A) := \{\omega \in \Omega_1 : f(\omega) \in A\} \in \mathbf{F}_1$. Equivalently we can say that f is a *measurable transformation of* $(\Omega_1, \mathbf{F}_1)$ into $(\Omega_2, \mathbf{F}_2)$. Furthermore, if $(\Omega_2, \mathbf{F}_2)$ is a Borel space, then f is said to be a *Borel function*, in which case we denote it by x. In addition to this, if the Borel space is $(\Omega_2, \mathbf{F}_2) = (\mathbb{R}, \mathbb{B})$, then the Borel function x is referred to as a *random variable*. In the special case when $(\Omega_2, \mathbf{F}_2) = (\mathbb{R}^n, \mathbb{B}^n)$, x is an *n-dimensional random vector*.

If there is a probability measure P defined on $(\Omega_1, \mathbf{F}_1)$—which we henceforth write simply as $(\Omega, \mathbf{F})$—then the random vector x will induce a probability measure P_x on the Borel space $(\mathbb{R}^n, \mathbb{B}^n)$, so that for every $B \in \mathbb{B}^n$ we have $\mathsf{P}_x(B) = \mathsf{P}(x^{-1}(B))$. Since every element of $\mathbb{B}^n$ is an n-dimensional rectangle, the arguments of P_x are in general infinite sets; however, considering the collection of sets $\{\xi \in \mathbb{R}^n : \xi_i < a_i, i = 1, \ldots, n\}$ in $\mathbb{B}^n$, where $a_i (i = 1, \ldots, n)$ are real numbers, restriction of P_x to this class is also a probability measure whose argument is now a finite set. This probability measure is denoted by $P_x = P_x(a_1, a_2, \ldots, a_n)$ and is referred to as *probability distribution function* of the random vector x. It is worth noting that

$$P_x(a_1, a_2, \ldots, a_n) = \mathsf{P}(\{\omega \in \Omega : x_1(\omega) < a_1, x_2(\omega) < a_2, \ldots, x_n(\omega) < a_n\}).$$

In the above x_i is a random variable which denotes the ith component of x. If $n > 1$, P_x is occasionally called the *cumulative (joint) probability distribution function*. It is a well-established fact that there is a one-to-one correspondence between P_x and P_x, and the subspace on which P_x is defined can generate the whole $\mathbb{B}^n$.

D.2.1 • Independence

Let the probability distribution function of a random vector $x = (x_1, \ldots, x_n)$ be given. The (*marginal*) distribution function of each random variable x_i is given by

$$P_{x_i}(a_i) = \lim_{a_j \to \infty, j \neq i} P_x(a_1, \ldots, a_n).$$

The random variables $x_1, \ldots, x_n$ are *independent* if

$$P_x(a_1, \ldots, a_n) = P_{x_1}(a_1) P_{x_2}(a_2) \ldots P_{x_n}(a_n)$$

for all scalars $a_1, \ldots, a_n$.

D.2.2 ▪ Probability density function

A *Lebesgue measure* is a measure defined on subintervals of the real line and is equal to the length of the corresponding subinterval(s). Such a measure assigns zero weight to countable subsets of the real line, and its definition can be extended to n-dimensional rectangles in $\mathbb{R}^n$. Consider a Borel probability measure P defined on $(\mathbb{R}^n, \mathbb{B}^n)$. Such a measure P is said to be absolutely continuous with respect to the Lebesgue measure if any element of $\mathbb{B}^n$ which has a Lebesgue measure of zero has also a P-measure of zero. With this in mind, recall a result of probability theory which says that if $x : (\Omega_1, \mathbf{F}, \mathbf{P}) \to (\mathbb{R}^n, \mathbb{B}^n, \mathbf{P}_x)$ is a random vector and if $\mathbf{P}_x$ is absolutely continuous with respect to the Lebesgue measure, there exists a nonnegative Borel function $p_x(.)$ such that, for every $A \in \mathbb{B}^n$,

$$\mathbf{P}_x(A) = \int_A p_x(\xi) d\xi.$$

We refer to $p_x(.)$ as the *probability density function* of the random vector x. Equivalently, the preceding relation can be written as

$$P_x(a_1, \dots, a_n) = \int_{-\infty}^{a_1} \dots \int_{-\infty}^{a_n} p_x(\xi_1, \dots, \xi_n) d\xi_1 d\xi_n$$

for every scalar $a_1, \dots, a_n$.

D.3 ▪ Integrals and expectation

Consider a random vector $x : (\Omega, \mathbf{F}, \mathbf{P}) \to (\mathbb{R}^n, \mathbb{B}^n, \mathbf{P}_x)$ and a nonnegative Borel function $f : (\mathbb{R}^n, \mathbb{B}^n) \to (\mathbb{R}^m, \mathbb{B}^m)$. As a consequence, f can also be considered as a random vector from $(\Omega, \mathbf{F})$ into $(\mathbb{R}^m, \mathbb{B}^m)$. Its *average value (expected value)* is determined by $\int_\Omega f(x(\omega))\mathbf{P}(d\omega)$ or equivalently by $\int_{\mathbf{R}^n} f(\xi)\mathbf{P}_x(d\xi)$. Both integrals are well defined and are uniquely equal in value. If f changes signs, then we take $f = f^+ - f^-$, where both f^+ and f^- are nonnegative, and write the expected value of f as

$$\mathbb{E}[f(x)] = \int_{\mathbb{R}^n} f^+(\xi)\mathbf{P}_x(d\xi) - \int_{\mathbb{R}^n} f^-(\xi)\mathbf{P}_x(d\xi) := \int_{\mathbb{R}^n} f(\xi)\mathbf{P}_x(d\xi),$$

provided that at least one of the pairs $\mathbb{E}[f^+(x)]$ and $\mathbb{E}[f^-(x)]$ is finite. Since, by definition, $\mathbf{P}_x(d\xi) = P_x(\xi + d\xi) - P_x(\xi) := dP_x(\xi)$, this integral can further be written as

$$\mathbb{E}[f(x)] = \int_{\mathbb{R}^n} f(\xi) dP_x(\xi),$$

which is a Lebesgue–Stieltjes integral and which is the convention that we shall adopt. When $f(x) = x$ we have

$$\mathbb{E}[x] := \int_{\mathbb{R}^n} \xi dP_x(\xi) := \bar{x},$$

which is the *mean (expected) value* of x. We call the *covariance* of the n-dimensional random vector x the quantity given by

$$\mathbb{E}[(x - \bar{x})(x - \bar{x})^T] = \int_{\mathbb{R}^n} (\xi - \bar{x})(\xi - \bar{x})^T dP_x(\xi) := cov(x).$$

This is a nonnegative definite matrix of dimension $(n \times n)$. Under the assumption that P_x is absolutely continuous with respect to the Lebesgue measure, we can equivalently write

$$\mathbb{E}[f(x)] = \int_{\mathbb{R}^n} f(\xi) p_x(\xi) d\xi,$$

where $p_x(.)$ is the corresponding probability density function. In the case where Ω involves a finite number of independent events $\omega_1, \omega_2, \ldots, \omega_n$, the integrals are replaced by the summation

$$\mathbb{E}[f(x(\omega))] = \sum_{i=1}^{n} f(x(\omega_i)) p_i.$$

In the above, the symbol p_i defines the probability of occurrence of event ω_i. For a countable set Ω, we have the counterpart

$$\mathbb{E}[f(x(\omega))] = \lim_{n \to \infty} \sum_{i=1}^{n} f(x(\omega_i)) p_i.$$

Appendix E

Stochastic Stability

This appendix provides some of the foundations of stochastic stability theory. It is based on [162]. We first extend to stochastic systems the definitions of Lyapunov stability given in Appendix C for deterministic systems. Then, we introduce some basic results on stability.

E.1 ▪ Different definitions of stochastic stability

We shall note that there are at least three times as many definitions for the stability of stochastic systems as there are for deterministic systems. This is due to the fact that in a stochastic setting we have three different types of convergence, such as convergence in probability, convergence in mean (or moment), and convergence in an almost sure (sample path, probability 1) sense.

Definition E.1 (Lyapunov stability in probability). *The equilibrium solution is stable in probability if, given $\epsilon, \epsilon' > 0$, there exists a $\delta(\epsilon, \epsilon', t_0)$ such that, for all $\|x_0\| < \delta$,*

$$\mathbb{P}\left\{\sup_{t \geq t_0} \|x(t; x_0, t_0, \omega)\| > \epsilon'\right\} < \epsilon.$$

Here, $\mathbb{P}$ denotes probability.

Definition E.2 (Lyapunov stability in the pth moment). *The equilibrium solution is stable in the pth moment, $p > 0$, if, given $\epsilon > 0$, there exists a $\delta(\epsilon, t_0) > 0$ such that $\|x_0\| < \delta$ guarantees that[1]*

$$\mathbb{E}\left\{\sup_{t \geq t_0} \|x(t; x_0, t_0, \omega)\|^p\right\} < \epsilon.$$

Definition E.3 (Almost sure Lyapunov stability). *The equilibrium solution is almost sure stable if*

$$\mathbb{P}\left\{\lim_{\|x_0\| \to 0} \sup_{t \geq t_0} \|x(t; x_0, t_0, \omega)\| = 0\right\} = 1.$$

It is worth noting that almost sure stability corresponds to saying that, with probability 1, all sample solutions are Lyapunov stable.

---

[1] Here, $\mathbb{E}$ denotes expectation.

We provide next definitions of asymptotic stability for stochastic systems.

Definition E.4 (Asymptotic stability in probability). *The equilibrium solution is* asymptotically stable in probability *if it is stable in probability and if there exists a $\delta' > 0$ such that $\|x_0\| < \delta'$ guarantees that*

$$\lim_{\delta \to \infty} \mathbb{P}\left\{\sup_{t \geq \delta} \|x(t; x_0, t_0, \omega)\| > \epsilon\right\} = 0.$$

If the convergence condition is verified for all initial times, t_0, then the equilibrium solution is *uniform asymptotic stability in probability*.

Definition E.5 (Asymptotic stability in the pth moment). *The equilibrium solution is* asymptotically pth moment stable *if it is stable in the pth moment and if there exists a $\delta' > 0$ such that $\|x_0\| < \delta'$ guarantees that*

$$\lim_{\delta \to \infty} \mathbb{E}\left\{\sup_{t \geq \delta} \|x(t; x_0, t_0, \omega)\|\right\} = 0.$$

Definition E.6 (Almost sure asymptotic Lyapunov stability). *The equilibrium solution is* almost surely asymptotically stable *if it is surely stable and if there exists a $\delta' > 0$ such that $\|x_0\| < \delta'$ guarantees that*

$$\lim_{\delta \to \infty} \left\{\sup_{t \geq \delta} \|x(t; x_0, t_0, \omega)\| > \epsilon\right\} = 0.$$

We introduce next the definitions of exponential stability for stochastic systems.

Definition E.7 (pth moment exponential Lyapunov stability). *The equilibrium solution is* pth moment exponentially stable *if there exist $\delta > 0$ and $\alpha > 0$ and a $\beta > 0$ such that $\|x_0\| < \delta$ guarantees that*

$$\mathbb{E}\{\|x(t; x_0, t_0, \omega)\|\} \leq \beta \|x_0\| e^{-\alpha(t - t_0)}.$$

Definition E.8 (Almost sure exponential Lyapunov stability). *The equilibrium solution is* almost surely exponentially stable *if there exist $\delta > 0$ and $\alpha > 0$ and a $\beta > 0$ such that $\|x_0\| < \delta$ guarantees that*

$$\mathbb{P}\{\|x(t; x_0, t_0, \omega)\|\} \leq \beta \|x_0\| e^{-\alpha(t - t_0)} = 1.$$

E.2 ▪ Some fundamental theorems

We now examine the key ideas behind the Lyapunov function approach to the stability analysis of stochastic systems.

Consider the stochastic system (E.1), which is defined on a probability space $(\Omega, \mathcal{F}, P)$, where Ω is the set of elementary events (sample space), $\mathcal{F}$ is the σ field which consists of all subsets of Ω that are measurable, and P is a probability measure:

$$\begin{cases} \dot{x}(t) = f[x(t), \omega], & t \geq 0, \\ x(0) = x_0. \end{cases} \tag{E.1}$$

To study stability let us focus on the time derivative of the expectation of $V(x(t))$, which is denoted by $\mathscr{L}V(x(t))$, where $\mathscr{L}$ is the infinitesimal generator of the process $x(t)$. If the system is such that the conditional probability distribution of future states of the process depends only upon the present state, not on the sequence of events that preceded it, then the system is said to be *Markovian*.

If the system is Markovian, the solution process is a strong, time homogeneous *Markov process*. Then, $\mathscr{L}$ is defined by

$$\mathscr{L}V(x_0) = \lim_{\Delta t \to 0} \frac{E_{x_0}(V(x(\Delta t))) - V(x_0)}{\Delta t}, \tag{E.2}$$

where the domain of $\mathscr{L}$ is the space of functions $V(x)$ for which (E.2) is well defined. Note that by doing this we construct a natural analogue of the total derivative of $V(x)$ along the solution $x(t)$ which we use in the deterministic case. With this in mind, assume that there exists a Lyapunov function $V(x)$ which satisfies the conditions stated above and for which it also holds that $\mathscr{L}V(x) \leq -k(x) \leq 0$. From this we obtain

$$\begin{aligned} 0 \leq V(x_0) - E_{x_0} V(x(t)) &= E_{x_0} \int_0^t k(x(s)) ds \\ &= -E_{x_0} \int_0^t \mathscr{L}V(x(s)) ds < +\infty \end{aligned} \tag{E.3}$$

and for $t, s > 0$

$$E_{x(s)}(V(x(t+s))) - V(x(s)) \leq 0 \quad \text{almost surely.} \tag{E.4}$$

Equation (E.4) essentially states that the Lyapunov function $V(x(t))$ is a supermartingale. Then, from the martingale convergence theorem, we have that $V(x(t)) \to 0$ almost surely as $t \to +\infty$. That is to say that $x(t) \to 0$ almost surely as $t \to +\infty$. From (E.3) we have that $x(t) \to \{x \in \mathbb{R}^n : k(x) = 0\}$ almost surely.

From (E.3) for given x_0, we obtain the following supermartingale inequality:

$$\mathbb{P}_{x_0} \left\{ \sup_{0 \leq t < +\infty} V(x(t)) \geq \epsilon \right\} \leq \frac{V(x_0)}{\epsilon}. \tag{E.5}$$

The above yields some of the key results by Kushner [142, 143, 144]. To keep the analysis reasonably simple, suppose that $x(t) \in Q_m$ almost surely for some $m > 0$.

Theorem E.9.

1. *Stability with probability* 1:
   If $\mathscr{L}V(x) \leq 0$ with $V(x) > 0$ for $x \in Q_m \setminus \{0\}$, then the origin is stable with probability 1.

2. *Asymptotic stability with probability* 1:
   If $\mathscr{L}V(x) = -k(x) \leq 0$ with $k(x) > 0$ for $x \in Q_m \setminus \{0\}$ and $k(0) = 0$, and if for any $d > 0$ small, $\epsilon_d > 0$ exists so that $k(x) \geq d$ for $x \in \{Q_m : \|x\| \geq \epsilon_d\}$, then the origin is stable with probability 1 with

$$\mathbb{P}_{x_0} \{x(t) \to 0 \text{ as } t \to +\infty\} \geq 1 - \frac{V(x_0)}{m}.$$

In particular if the conditions are satisfied for arbitrarily large m, then the origin is asymptotically stable with probability 1.

3. *Exponential asymptotic stability with probability 1:*
   If $V(x) \geq 0$, $V(0) = 0$, and $\mathscr{L}V(x) \leq -\alpha V(x)$ on Q_m for some $\alpha > 0$, then the origin is stable with probability 1, and

$$\mathbb{P}_{x_0}\{\sup_{T \leq t < +\infty} V(x(t)) \geq \lambda\} \leq \frac{V(x_0)}{m} + \frac{V(x_0)e^{-\alpha T}}{\lambda} \quad \forall T \geq 0.$$

In particular, if the conditions are satisfied for arbitrarily large m, then the origin is asymptotical stable with probability 1, and

$$\mathbb{P}_{x_0}\{\sup_{T \leq t < +\infty} V(x(t)) \geq \lambda\} \leq \frac{V(x_0)e^{-\alpha T}}{\lambda}.$$

We borrow from Kushner's work a few examples that show the application of the stability theorems.

Example E.10. Let the following scalar Itô equation be given:

$$dx = ax\,dt + \sigma x\,dw, \tag{E.6}$$

where w is a standard Wiener process. For this system, the infinitesimal generator takes the form

$$\mathscr{L} = \frac{1}{2}\sigma^2 x^2 \frac{d^2}{dx^2} + ax\frac{d}{dx}.$$

Let us consider the Lyapunov function $V(x) = x^2$. Then we have

$$\mathscr{L}V(x) = (\sigma^2 + 2a)x^2.$$

Now, if $\sigma^2 + 2a < 0$, then with $Q_m = \{x : x^2 < m^2\}$, from item 1 of the previous theorem, the zero solution is stable with probability 1. Furthermore, set $m \to +\infty$. By item 2 of the theorem,

$$\lim_{t \to \infty} x(t) = 0 \quad \text{almost surely},$$

where $x(t)$ is solution of (E.6). By item 3 of the theorem,

$$\mathbb{P}_{x_0}\{\sup_{T \leq t < +\infty} x(t)^2 \geq \lambda\} \leq \frac{V(x_0)e^{-\alpha T}}{\lambda}$$

for some $\alpha > 0$. ∎

In Has'minskii's work [113] and references therein, we find a detailed analysis of diffusion processes which are solutions of a stochastic system described by the following Itô differential equation:

$$\begin{cases} dx(t) = b(t,x)dt + \sum_{r=1}^{k} \sigma_r(t,x)d\xi_r(t), & t \geq s, \\ x(s) = x_s, \end{cases} \tag{E.7}$$

where $\xi_r(t)$ are independent standard Wiener processes and the coefficients $b(t,x)$ and $\sigma_r(t,x)$ satisfy Lipschitz and growth conditions. The infinitesimal generator $\mathcal{L}$ is given by

$$\mathcal{L}V(t,x) = \frac{\partial V}{\partial t} + \sum_{i=1}^{n} b_i(t,x)\frac{\partial V}{\partial x_i} + \frac{1}{2}\sum_{i,j=1}^{n} a_{ij}(t,x)\frac{\partial^2 V}{\partial x_i \partial x_j}. \tag{E.8}$$

Note that $\mathcal{L}$ is a second-order partial differential operator on $V(t,x)$, which are twice continuously differentiable with respect to x and continuously differentiable with respect to t.

In the following we provide some typical results obtained by Has'minskii's work. The key idea is to derive an inequality like (E.5).

Consider a neighborhood of 0, say U, and let $U_1 = \{t > 0\} \times U$. Let us denote by $C_2^0(U_1)$ the set of functions $V(t,x)$ defined in U_1, which are twice continuously differentiable in x except at the point $x = 0$ and continuously differentiable in t. A function $V(t,x)$ is said to be positive definite in the Lyapunov sense if $V(t,0) = 0$ for all $t \geq 0$ and $V(t,x) \geq \omega(x) > 0$ for $x \neq 0$ and some continuous function $\omega(x)$.

Theorem E.11 (Has'minskii).

1. *The trivial solution of (E.7) is stable in probability (same as our definition) if there exists $V(t,X) \in C_2^0(U_1)$, positive definite in the Lyapunov sense, so that $\mathcal{L}V(t,x) \leq 0$ for $x \neq 0$.*

2. *If the system (E.7) is time homogeneous, i.e., $b(t,x) = b(x)$ and $\sigma_r(t,x) = \sigma_r(x)$, and if the nondegeneracy condition*

$$\sum_{i,j=1}^{n} a_{ij}(x)\lambda_i\lambda_j > m(x)\sum_{i=1}^{n} \lambda_i^2 \quad \text{for } \lambda = (\lambda_1 \ldots \lambda_n)^T \in \mathbb{R}^n$$

is satisfied with continuous $m(x) > 0$ for $x \neq 0$, then a necessary and sufficient condition for the trivial solution to be stable in probability is that a twice continuously differentiable function $V(x)$ exists, except perhaps at $x = 0$, so that

$$\mathcal{L}_0 V(x) = \sum_{i=1}^{n} b_i(x)\frac{\partial V}{\partial x_i} + \sum_{i,j=1}^{n} a_{ij}\frac{\partial^2 V(x)}{\partial x_i \partial x_j} \leq 0,$$

where $\mathcal{L}_0$ is the generator of the time homogeneous system.

3. *If the system (E.7) is linear, i.e., $b(t,x) = b(t)x$ and $\sigma_r(t,x) = \sigma_r(t)x$, then the system is exponentially p-stable (the pth moment is exponentially stable), i.e.,*

$$\mathbb{E}_{x_0}\{\|x(t,x_0,s)\|^p\} \leq A\|x\|^p e^{-\alpha(t-s)}, \quad p > 0,$$

for some constant $\alpha > 0$ if and only if a function $V(t,x)$ exists, homogeneous of degree p in x, so that for some constants $k_i > 0$, $i = 1,2,3,4$,

$$\begin{aligned} k_1\|x\|^p &\leq V(t,x) \leq k_2\|x\|^p, \\ \mathcal{L}V(t,x) &\leq -k_3\|x\|^p \end{aligned} \tag{E.9}$$

and

$$\left\|\frac{\partial V}{\partial x}\right\| \leq k_4\|x\|^{p-1}, \quad \left\|\frac{\partial^2 V}{\partial x^2}\right\| \leq k_4\|x\|^{p-2}.$$

Example E.12. Let the following system be given:

$$\begin{cases} dx(t) = Ax(t)dt + \sum_{i=1}^{m} B_i x(t) d\xi_i(t), & t \geq 0, \\ x(0) = x_0, \end{cases}$$ (E.10)

where $\xi_i(t)$ are independent standard Wiener processes. The process $x(t)$ is then a Markov diffusion process with the infinitesimal generator

$$\mathscr{L}u = (Ax)^T \frac{\partial u}{\partial x} + \frac{1}{2} \sum_{i,j=1}^{d} \sigma_{ij}(x) \frac{\partial^2 u}{\partial x_i \partial x_j},$$

where u is a real-valued twice continuously differentiable function and $\Sigma(x) = (\sigma_{ij}(x))_{d \times d} = \sum_{i=1}^{m} B_i x x^T B_i$. ∎

Appendix F

Indistinguishability and Mean-Field Convergence

This appendix presents the notion of indistinguishability (or exchangeability) and discusses the existence of a limiting measure and mean-field convergence in the framework of de Finetti, Hewitt, Savage (see, e.g., [83, 116]).

Definition F.1 (Indistinguishability). *A family of processes* $(x_1, x_2, \ldots, x_n)$ *is said to be indistinguishable (or exchangeable) if the joint distribution is invariant by permutation over the index set* $\{1, \ldots, n\}$. *That is to say that for any permutation* π *over* $\{1, 2, \ldots, n\}$ *we have*

$$\mathscr{L}(x_1, x_2, \ldots, x_n) = \mathscr{L}(x_{\pi(1)}, \ldots, x_{\pi(n)}), \tag{F.1}$$

where $\mathscr{L}(X)$ *is the distribution of the random variable* X.

The *de Finetti–Hewitt–Savage* theorem establishes the mean-field convergence with speed $O(\frac{1}{\sqrt{n}})$. The theorem makes use of the Monge–Kantorovich distance given by

$$d(\mu, v) = \sup_{\phi} \left\{ \int_{\mathscr{X}} \phi(x) \, \mathrm{d}(\mu - v)(x) \right\},$$

where μ and v are two measures and the supremum is over all $\phi : \mathscr{X} \to \mathbb{R}$, continuous, and with Lipschitz constant $\mathrm{Lip}(\phi) \leq 1$.

Theorem F.2 (de Finetti–Hewitt–Savage). *Let* $x_1, x_2, \ldots$ *be a sequence of* $\mathscr{X}$ *–valued random variables, where* $\mathscr{X}$ *is a Polish space, and for each* n, $\{x_1, x_2, \ldots, x_n\}$ *is indistinguishable. Let* $\mathscr{P}(\mathscr{X})$ *denote the space of probability measures on* $\mathscr{X}$. *Then, there is a* $\mathscr{P}(\mathscr{X})$ *–valued random variable* m *such that*

$$m = \lim_{n \to \infty} \frac{1}{n} \sum_{k=1}^{n} \delta_{x_k} \text{ almost surely.}$$

In addition, if the moments of x_k *are finite, then*

$$d\left(m, \frac{1}{n} \sum_{k=1}^{n} \delta_{x_k} \right) \leq O\left(\frac{1}{\sqrt{n}} \right) \quad \text{almost surely.}$$

This result has been proved by [83] for infinite binary sequences and has been extended by [116] to continuous and compact state spaces.

271

Bibliography

[1] D. ACEMOĞLU, G. COMO, F. FAGNANI, AND A. OZDAGLAR, *Opinion fluctuations and disagreement in social networks*, Mathematics of Operations Research, 38 (2013), pp. 1–27. (Cited on p. 193)

[2] D. ACEMOĞLU AND A. OZDAGLAR, *Opinion dynamics and learning in social networks*, International Review of Economics, 1 (2011), pp. 3–49. (Cited on pp. 192, 193)

[3] Y. ACHDOU, F. CAMILLI, AND I. CAPUZZO-DOLCETTA, *Mean field games: Numerical methods for the planning problem*, SIAM Journal on Control and Optimization, 50 (2012), pp. 77–109. (Cited on pp. 139, 216)

[4] Y. ACHDOU AND I. CAPUZZO-DOLCETTA, *Mean field games: Numerical methods*, SIAM Journal on Numerical Analysis, 48 (2010), pp. 1136–1162. (Cited on p. 139)

[5] S. ADLAKHA AND R. JOHARI, *Mean field equilibrium in dynamic games with strategic complementarities*, Operations Research, 61 (2013), pp. 971–989. (Cited on p. 139)

[6] D. AEYELS AND F. DE SMET, *A mathematical model for the dynamics of clustering*, Physica D: Nonlinear Phenomena, 237 (2008), pp. 2517–2530. (Cited on p. 192)

[7] E. ALTMAN, *Applications of dynamic games in queues*, in Advances in Dynamic Games, Annals of the International Society of Dynamic Games 7, Birkhäuser, Boston, MA, 2005, pp. 309–342. (Cited on p. 105)

[8] R. AMIR, *Continuous stochastic games of capital accumulation with convex transitions*, Games & Economic Behaviors, 15 (1996), pp. 111–131. (Cited on p. 105)

[9] D. ANGELI AND P.-A. KOUNTOURIOTIS, *A stochastic approach to dynamic-demand refrigerator control*, IEEE Transactions on Control Systems Technology, 20 (2012), pp. 581–592. (Cited on pp. 156, 164)

[10] M. ARCAK, *Passivity as a design tool for group coordination*, IEEE Transactions on Automatic Control, 52 (2007), pp. 1380–1390. (Cited on pp. 165, 180)

[11] J. P. AUBIN, *Viability Theory*, Birkhäuser, Boston, MA, 1991. (Cited on p. 120)

[12] J. P. AUBIN AND A. CELLINA, *Differential Inclusions: Set-Valued Maps and Viability Theory*, Springer, Berlin, Heidelberg, 1984. (Cited on p. 120)

[13] J. P. AUBIN AND H. FRANKOWSKA, *Set-Valued Analysis*, Birkhäuser, Boston, MA, 1990. (Cited on p. 120)

[14] R. J. AUMANN, *Markets with a continuum of players*, Econometrica, 32 (1964), pp. 39–50. (Cited on p. 139)

[15] ———, *Game theory*, in The New Palgrave, Vol. 2, J. Eatwell, M. Milgate, and P. Newman, editors, Macmillan, London, 1987. (Cited on p. 16)

[16] F. BAGAGIOLO AND D. BAUSO, *Objective function design for robust optimality of linear control under state-constraints and uncertainty*, ESAIM: Control, Optimisation and Calculus of Variations, 17 (2011), pp. 155–177. (Cited on pp. 120, 164)

[17] ——, *Mean-field games and dynamic demand management in power grids*, Dynamic Games and Applications, 4 (2014), pp. 155–176. (Cited on pp. 139, 156, 164)

[18] S. BALSEIRO, O. BESBES, AND G. Y. WEINTRAUB, *Repeated auctions with budgets in ad exchanges: Approximations and design*, Management Science, 61 (2015). (Cited on p. 139)

[19] A. V. BANERJEE, *A simple model of herd behavior*, Quarterly Journal of Economics, 107 (1992), pp. 797–817. (Cited on p. 192)

[20] M. BARDI, *Explicit solutions of some linear-quadratic mean field games*, Network and Heterogeneous Media, 7 (2012), pp. 243–261. (Cited on pp. 127, 139)

[21] T. BAŞAR, *Nash equilibria of risk-sensitive nonlinear stochastic differential games*, Journal of Optimization Theory and Applications, 100 (1999), pp. 479–498. (Cited on p. 133)

[22] T. BAŞAR AND P. BERNHARD, H^∞ *Optimal Control and Related Minimax Design Problems: A Dynamic Game Approach*, 2nd ed., Birkhäuser, Boston, MA, 1995. (Cited on pp. xiii, xv, 23, 26, 94, 96, 130, 137, 139)

[23] T. BAŞAR AND G. J. OLSDER, *Dynamic Noncooperative Game Theory*, 2nd ed., Classics in Applied Mathematics 23, SIAM, Philadelphia, 1999. (Cited on pp. 17, 40, 49, 132, 137, 251, 255, 261)

[24] D. BAUSO, *Game theory: Models, numerical methods and applications*, Foundations and Trends in Systems and Control, 1 (2014), pp. 379–522. (Cited on p. xxiv)

[25] D. BAUSO, F. BAGAGIOLO, AND R. PESENTI, *Mean-field game modeling the bandwagon effect with activation costs*, Dynamic Games and Applications, Special Issue on Game Theoretic Analysis of Novel Network Technologies and Complex Systems. (Cited on p. 182)

[26] D. BAUSO, F. BLANCHINI, AND R. PESENTI, *Robust control strategies for multi inventory systems with average flow constraints*, Automatica, 42 (2006), pp. 1255–1266. (Cited on pp. 111, 230)

[27] ——, *Optimization of long run average-flow cost in networks with time-varying unknown demand*, IEEE Transactions on Automatic Control, 55 (2010), pp. 20–31. (Cited on pp. xv, 111, 120, 217)

[28] D. BAUSO AND M. CANNON, *Consensus in cooperative games, bargaining, and opinion dynamics*, submitted. (Cited on p. 192)

[29] D. BAUSO, M. CANNON, AND J. FLEMING, *Robust consensus in social networks and coalitional games*, in Proceedings of 2014 IFAC World Congress, 2014, pp. 1537–1542. (Cited on p. 192)

[30] D. BAUSO, L. GIARRÉ, AND R. PESENTI, *Nonlinear protocols for the optimal distributed consensus in networks of dynamic agents*, Systems & Control Letters, 55 (2006), pp. 918–928. (Cited on pp. xvi, 153, 154)

[31] ——, *Consensus in noncooperative dynamic games: A multi-retailer inventory application*, IEEE Transactions on Automatic Control, 53 (2008), pp. 998–1003. (Cited on pp. 59, 154)

[32] ——, *Distributed consensus in noncooperative inventory games*, European Journal of Operational Research, 192 (2009), pp. 866–878. (Cited on pp. 59, 154)

[33] ——, *Robust control of uncertain multi-inventory systems via linear matrix inequality*, International Journal of Control, 83 (2010), pp. 1723–1740. (Cited on p. 120)

[34] D. BAUSO, T. MYLVAGANAM, AND A. ASTOLFI, *Approximate solutions for crowd-averse robust mean-field games*, in Proceedings of the 2014 European Control Conference (ECC), Strasbourg, France, 2014, pp. 1217–1222. (Cited on pp. 139, 217)

[35] ——, *Crowd-averse robust mean-field games: Approximation via state space*, IEEE Transactions on Automatic Control, in press; first online September 17, 2015. (Cited on pp. 139, 217)

[36] ——, *Multi-population robust mean-field games in power-grids*, submitted. (Not cited)

[37] D. BAUSO, A. PAPACHRISTODOULOU, AND X. ZHANG, *Density flow in dynamical networks via mean-field games*, submitted. (Not cited)

[38] D. BAUSO AND R. PESENTI, *Team theory and person-by-person optimization with binary decisions*, SIAM Journal on Control and Optimization, 50 (2012), pp. 3011–3028. (Cited on p. 17)

[39] ——, *Mean field linear quadratic games with set up costs*, Dynamic Games and Applications, 3 (2013), pp. 89–104. (Cited on p. 139)

[40] D. BAUSO, E. SOLAN, E. LEHRER, AND X. VENEL, *Attainability in repeated games with vector payoffs*, INFORMS Mathematics of Operations Research, 40 (2015), pp. 739–755. (Cited on pp. 114, 119, 120, 211, 213, 217)

[41] D. BAUSO AND H. TEMBINE, *Crowd-averse cyber-physical systems: The paradigm of robust mean field games*, IEEE Transactions on Automatic Control, in press; first online October 19, 2015. (Cited on pp. 249, 250)

[42] D. BAUSO, H. TEMBINE, AND T. BAŞAR, *Robust mean field games with application to production of an exhaustible resource*, in Proceedings of 7th IFAC Symposium on Robust Control Design, Aalborg, Denmark, 2012. (Cited on pp. 139, 238)

[43] ——, *Robust mean field games*, Dynamic Games and Applications, in press; first online June 6, 2015. (Cited on pp. 139, 238, 239)

[44] D. BAUSO AND J. TIMMER, *Robust dynamic cooperative games*, International Journal of Game Theory, 38 (2009), pp. 23–36. (Cited on pp. 59, 228, 229)

[45] ——, *On robustness and dynamics in (un)balanced coalitional games*, Automatica, 48 (2012), pp. 2592–2596. (Cited on pp. xvii, 59, 229)

[46] R. W. BEARD, T. W. MCLAIN, M. A. GOODRICH, AND E. P. ANDERSON, *Coordinated target assignment and intercept for unmanned air vehicles*, IEEE Transactions on Robotics and Automation, 18 (2002), pp. 911–922. (Cited on p. 154)

[47] R. E. BELLMAN, *Dynamic Programming*, Princeton University Press, Princeton, NJ, 1957; reprinted, Dover, New York, 2003. (Cited on pp. 89, 96)

[48] M. BENAÏM, J. HOFBAUER, AND S. SORIN, *Stochastic approximations and differential inclusions*, SIAM Journal on Control and Optimization, 44 (2005), pp. 328–348. (Cited on pp. 186, 214)

[49] ——, *Stochastic approximations and differential inclusions, Part II: Applications*, INFORMS Mathematics of Operations Research, 31 (2006), pp. 673–695. (Cited on pp. 186, 214)

[50] J.-D. BENAMOU AND Y. BRENIER, *A computational fluid mechanics solution to the Monge-Kantorovich mass transfer problem*, Numerische Mathematik, 84 (2000), pp. 375–393. (Cited on p. 139)

[51] B. BERGEMANN AND J. VÄLIMÄKI, *The dynamic pivot mechanism*, Econometrica, 78 (2010), pp. 771–789. (Cited on p. 164)

[52] D. P. BERTSEKAS, *Dynamic Programming and Optimal Control*, Athena Scientific, Belmont, MA, 1995. (Cited on p. 149)

[53] D. P. BERTSEKAS AND I. B. RHODES, *On the minimax reachability of target set and target tubes*, Automatica, 7 (1971), pp. 233–247. (Cited on pp. 111, 112)

[54] J. BEWERSDORFF, *Luck, Logic, and White Lies: The Mathematics of Games*, A. K. Peters/CRC Press, Boca Raton, FL, 2004. (Cited on p. 17)

[55] D. T. BISHOP AND C. CANNINGS, *A generalized war of attrition*, Journal of Theoretical Biology, 70 (1978), pp. 85–124. (Cited on p. 78)

[56] F. BLACK AND M. SCHOLES, *The pricing of options and corporate liabilities*, Journal of Political Economy, 81 (1973), pp. 637–654. (Cited on p. 250)

[57] D. BLACKWELL, *An analog of the minmax theorem for vector payoffs*, Pacific Journal of Mathematics, 6 (1956), pp. 1–8. (Cited on pp. 120, 187)

[58] F. BLANCHINI, *Set invariance in control: A survey*, Automatica, 35 (1999), pp. 1747–1768. (Cited on p. 120)

[59] F. BLANCHINI, S. MIANI, AND W. UKOVICH, *Control of production-distribution systems with unknown inputs and system failures*, IEEE Transactions on Automatic Control, 45 (2000), pp. 1072–1081. (Cited on pp. 120, 217)

[60] F. BLANCHINI, F. RINALDI, AND W. UKOVICH, *Least inventory control of multi-storage systems with non-stochastic unknown input*, IEEE Transactions on Robotics and Automation, 13 (1997), pp. 633–645. (Cited on pp. 120, 217)

[61] V. D. BLONDEL, J. M. HENDRICKX, AND J. N. TSITSIKLIS, *Continuous-time average-preserving opinion dynamics with opinion-dependent communications*, SIAM Journal on Control and Optimization, 48 (2010), pp. 5214–5240. (Cited on p. 192)

[62] L. BLUME, *The statistical mechanics of strategic interaction*, Games and Economic Behavior, 5 (1993), pp. 387–424. (Cited on p. 85)

[63] O. N. BONDAREVA, *Some applications of linear programming methods to the theory of cooperative game*, Problemi Kibernetiki, 10 (1963), pp. 119–139. (Cited on pp. 67, 222)

[64] S. BOYD AND L. VANDENBERGHE, *Convex Optimization*, Cambridge University Press, Cambridge, UK, 2004. (Cited on pp. 150, 256, 257)

[65] A. BRESSAN, *Noncooperative Differential Games. A Tutorial*, 2010; available online at http://www.math.psu.edu/bressan/PSPDF/game-lnew.pdf, 2010. (Cited on pp. xiv, 12, 17, 47, 49, 91, 96)

[66] A. BRESSAN, *Noncooperative differential games*, Milan Journal of Mathematics, 79 (2011), pp. 357–427. (Cited on pp. xiv, 12, 47)

[67] A. BRESSAN AND B. PICCOLI, *Introduction to Mathematical Theory of Control*, AIMS 2007. (Cited on pp. xv, 88, 90)

[68] D. S. CALLAWAY AND I. A. HISKENS, *Achieving controllability of electric loads*, Proceedings of the IEEE, 99 (2011), pp. 184–199. (Cited on p. 164)

[69] C. CANUTO, F. FAGNANI, AND P. TILLI, *An Eulerian approach to the analysis of Krause's consensus models*, SIAM Journal on Control and Optimization, 50 (2012), pp. 243–265. (Cited on p. 193)

[70] P. CARDALIAGUET, *Notes on Mean Field Games* (P.-L. Lions' lectures), Collège de France; available online at https://www.ceremade.dauphine.fr/~cardalia/MFG100629.pdf, 2012. (Cited on pp. 126, 136)

[71] P. CARDALIAGUET, M. QUINCAMPOIX, AND P. SAINT-PIERRE, *Differential games through viability theory: Old and recent results*, in Advances in Dynamic Games Theory, Annals of International Society of Dynamic Games, Birkhäuser, Boston, MA, 2007. (Cited on p. 120)

[72] C. CASTELLANO, S. FORTUNATO, AND V. LORETO, *Statistical physics of social dynamics*, Reviews of Modern Physics, 81 (2009), pp. 591–646. (Cited on p. 192)

[73] N. CESA-BIANCHI AND G. LUGOSI, *Prediction, Learning and Games*, Cambridge University Press, Cambridge, UK, 2006. (Cited on pp. 120, 187)

[74] N. CESA-BIANCHI, G. LUGOSI, AND G. STOLTZ, *Regret minimization under partial monitoring*, Mathematics of Operations Research, 31 (2006), pp. 562–580. (Cited on p. 230)

[75] J. C. CESCO, *A convergent transfer scheme to the core of a TU-game*, Revista de Matemáticas Aplicadas, 19 (1998), pp. 23–35. (Cited on pp. 229, 230)

[76] V. CHARI AND P. KEHOE, *Sustainable plans*, Journal of Political Economics, 98 (1990), pp. 783–802. (Cited on p. 105)

[77] G. C. CHASPARIS, A. ARAPOSTATHIS, AND J. S. SHAMMA, *Aspiration learning in coordination games*, SIAM Journal on Control and Optimization, 51 (2013), pp. 465–490. (Cited on p. 85)

[78] G. C. CHASPARIS, J. S. SHAMMA, AND A. RANTZER, *Nonconvergence to saddle boundary points under perturbed reinforcement learning*, International Journal of Game Theory, 44 (2015), pp. 667–699. (Cited on p. 85)

[79] G. COMO AND F. FAGNANI, *Scaling limits for continuous opinion dynamics systems*, The Annals of Applied Probability, 21 (2011), pp. 1537–1567. (Cited on p. 193)

[80] G. COMO, K. SAVLA, D. ACEMOGLU, M. DAHLEH, AND E. FRAZZOLI, *Distributed robust routing in dynamical networks—Part I: Locally responsive policies and weak resilience*, IEEE Transactions on Automatic Control, 58 (2013), pp. 317–332. (Cited on p. 216)

[81] ——, *Distributed robust routing in dynamical networks—Part II: Strong resilience, equilibrium selection and cascaded failures*, IEEE Transactions on Automatic Control, 58 (2013), pp. 333–348. (Cited on p. 216)

[82] R. COUILLET, S. M. PERLAZA, H. TEMBINE, AND M. DEBBAH, *Electrical vehicles in the smart grid: A mean field game analysis*, IEEE Journal on Selected Areas in Communications, 30 (2012), pp. 1086–1096. (Cited on p. 164)

[83] B. DE FINETTI, *Funzione caratteristica di un fenomeno aleatorio*, Atti della R. Academia Nazionale dei Lincei, Serie 6. Memorie, Classe di Scienze Fisiche, Mathematice e Naturale, 4 (1931), pp. 251–299. (Cited on p. 271)

[84] G. DEFFUANT, D. NEAU, F. AMBLARD, AND G. WEISBUCH, *Mixing beliefs among interacting agents*, Advances in Complex Systems, 3 (2000), pp. 87–98. (Cited on p. 193)

[85] P. DERLER, E. A. LEE, AND A. SANGIOVANNI-VINCENTELLI, *Modeling cyber-physical systems*, Proceedings of the IEEE (special issue on Cyber-Physical Systems), 100 (2012), pp. 13–28. (Cited on pp. 241, 250)

[86] A. DI MARE AND V. LATORA, *Opinion formation models based on game theory*, International Journal of Modern Physics C, Computational Physics and Physical Computation, 18 (2007). (Cited on pp. 17, 193)

[87] M. C. F. DONKERS, P. TABUADA, AND W. P. M. H. HEEMELS, *Minimum attention control for linear systems: A linear programming approach*, Discrete Event Dynamic Systems: Theory and Applications (special issue on Event-Based Control and Optimization), 24 (2014), pp. 199–218. (Cited on p. 250)

[88] F. DÖRFLER AND F. BULLO, *Synchronization and transient stability in power networks and nonuniform Kuramoto oscillators*, SIAM Journal on Control Optimization, 50 (2012), pp. 1616–1642. (Cited on pp. 165, 180)

[89] W. B. DUNBAR AND R. M. MURRAY, *Distributed receding horizon control with application to multi-vehicle formation stabilization*, Automatica, 4 (2006), pp. 549–558. (Cited on p. 154)

[90] P. DUTTA AND R. K. SUNDARAM, *The tragedy of the commons?*, Economic Theory, 3 (1993), pp. 413–426. (Cited on p. 105)

[91] N. J. ELLIOT AND N. J. KALTON, *The existence of value in differential games of pursuit and evasion*, Journal of Differential Equations, 12 (1972), pp. 504–523. (Cited on p. 120)

[92] J. ENGWERDA, *LQ Dynamic Optimization and Differential Games*, John Wiley & Sons, New York, 2005. (Cited on p. 59)

[93] F. FACCHINEI AND J-S. PANG, *Finite-Dimensional Variational Inequalities and Complementarity Problems*, Springer-Verlag, New York, 2003. (Cited on p. 204)

[94] A. FAX AND R. M. MURRAY, *Information flow and cooperative control of vehicle formations*, IEEE Transactions on Automatic Control, 49 (2004), pp. 1465–1476. (Cited on p. 154)

[95] J. A. FILAR AND L. A. PETROSJAN, *Dynamic cooperative games*, International Game Theory Review, 2 (2000), pp. 47–65. (Cited on p. 230)

[96] J. A. FILAR AND K. VRIEZE, *Competitive Markov decision processes*, Springer, New York, 1996. (Cited on p. 105)

[97] A. M. FINK, *Equilibrium in a stochastic n-person game*, Journal of Science of the Hiroshima University, Series A-I (Mathematics), 28 (1964), pp. 89–93. (Cited on pp. 101, 105)

[98] D. FOSTER AND R. VOHRA, *Regret in the on-line decision problem*, Games and Economic Behavior, 29 (1999), pp. 7–35. (Cited on p. 120)

[99] D. FUDENBERG AND D. K. LEVINE, *The Theory of Learning in Games*, MIT Press, Cambridge, MA, 1998. (Cited on p. 85)

[100] V. GAZI AND K. PASSINO, *Stability analysis of social foraging swarms*, IEEE Transactions on Systems, Man, and Cybernetics, 34 (2004), pp. 539–557. (Cited on p. 154)

[101] R. GIBBONS, *Game Theory for Applied Economists*, Princeton University Press, Princeton, NJ, 1992. (Cited on p. 17)

[102] F. GIULIETTI, L. POLLINI, AND M. INNOCENTI, *Autonomous formation flight*, IEEE Control Systems Magazine, 20 (2000), pp. 34–44. (Cited on p. 154)

[103] S. Z. ÃLPARSLAN-GÖK, S. MIQUEL, AND S. TIJS, *Cooperation under interval uncertainty*, Mathematical Methods of Operations Research, 69 (2009), pp. 99–109. (Cited on p. 230)

[104] D. A. GOMES AND J. SAÚDE, *Mean field games models—a brief survey*, Dynamic Games and Applications, 4 (2014), pp. 110–154. (Cited on pp. 136, 139)

[105] O. GUÉANT, J.-M. LASRY, AND P.-L. LIONS, *Mean field games and applications*, in Paris-Princeton Lectures on Mathematical Finance 2010, Springer, Berlin, 2010, pp. 205–266. (Cited on pp. 121, 127, 139, 238)

[106] T. HAMILTON AND R. MESIC, *A Simple Game-Theoretic Approach to Suppression of Enemy Defenses and Other Time Critical Target Analyses*, RAND Project Air Force; available online at http://www.rand.org/pubs/documented_briefings/DB385.html, 2004. (Cited on p. 17)

[107] J. C. HARSANYI AND R. SELTEN, *A General Theory of Equilibrium Selection in Games*, MIT Press, Cambridge, MA, 1988. (Cited on p. 49)

[108] S. HART, *Shapley value*, in The New Palgrave: Game Theory, J. Eatwell, M. Milgate, and P. Newman, editors, Norton, New York, 1989, pp. 210–216. (Cited on p. 67)

[109] ———, *Adaptive heuristics*, Econometrica, 73 (2005), pp. 1401–1430. (Cited on pp. 85, 111, 120)

[110] S. HART AND A. MAS-COLELL, *A general class of adaptive strategies*, Journal of Economic Theory, 98 (2001), pp. 26–54. (Cited on pp. 85, 111, 120)

[111] ———, *Regret-based continuous-time dynamics*, Games and Economic Behavior, 45 (2003), pp. 375–394. (Cited on pp. 85, 111, 120)

[112] B. C. HARTMAN, M. DROR, AND M. SHAKED, *Cores of inventory centralization games*, Games and Economic Behavior, 31 (2000), pp. 26–49. (Cited on p. 230)

[113] R. Z. HAS'MINSKII, *Stochastic Stability of Differential Equations*, Sijthoff and Noordhoff, Germantown, MD, 1980. (Cited on p. 268)

[114] A. HAURIE, *On some properties of the characteristic function and the core of a multistage game of coalitions*, IEEE Transactions on Automatic Control, 20 (1975), pp. 238–241. (Cited on p. 230)

[115] R. HEGSELMANN AND U. KRAUSE, *Opinion dynamics and bounded confidence models, analysis, and simulations*, Journal of Artificial Societies and Social Simulation, 5 (2002). (Cited on pp. 192, 193)

[116] E. HEWITT AND L. J. SAVAGE, *Symmetric measures on Cartesian products*, Transactions of the American Mathematical Society, 80 (1955), pp. 470–501. (Cited on p. 271)

[117] F. S. HILLIER AND G. J. LIEBERMAN, *Introduction to Operations Research*, 7th ed., McGraw–Hill, New York, 2001. (Cited on pp. 34, 40, 54, 55, 59, 256, 257)

[118] Y.-C. HO, *Team decision theory and information structures*, Proceedings of the IEEE, 68 (1980), pp. 644–654. (Cited on p. 17)

[119] J. HOFBAUER, *Deterministic Evolutionary Game Dynamics*, Proceedings of Symposia in Applied Mathematics, Vol. 69, 2011, pp. 61–79. (Cited on p. 85)

[120] A. J. HOFFMAN, *On approximate solutions of systems of linear inequalities*, Journal of Research of the National Bureau of Standards, 49 (1952), pp. 263–265. (Cited on p. 204)

[121] T.-F. HOU, *Approachability in a two-person game*, The Annals of Mathematical Statistics, 42 (1971), pp. 735–744. (Cited on p. 120)

[122] M. Y. HUANG, P. E. CAINES, AND R. P. MALHAMÉ, *Individual and mass behaviour in large population stochastic wireless power control problems: Centralized and Nash equilibrium solutions*, in IEEE Conference on Decision and Control, Honolulu, HI, 2003, pp. 98–103. (Cited on p. 139)

[123] ———, *Large population stochastic dynamic games: Closed loop Kean-Vlasov systems and the Nash certainty equivalence principle*, Communications in Information and Systems, 6 (2006), pp. 221–252. (Cited on p. 139)

[124] ———, *Nash certainty equivalence in large population stochastic dynamic games: Connections with the physics of interacting particle systems*, in 45th IEEE Conference on Decision and Control, San Diego, CA, 2006, pp. 4921–4926. (Cited on p. 139)

[125] ———, *Large population cost-coupled LQG problems with non-uniform agents: Individual-mass behaviour and decentralized ϵ-nash equilibria*, IEEE Transactions on Automatic Control, 52 (2007), pp. 1560–1571. (Cited on p. 139)

[126] R. ISAACS, *Differential Games: A Mathematical Theory with Applications to Warfare and Pursuit, Control and Optimization*, Wiley, New York, 1965; reprinted, Dover, New York, 1999. (Cited on p. 96)

[127] K. IYER, R. JOHARI, AND M. SUNDARARAJAN, *Mean field equilibria of dynamic auctions with learning*, Management Science, 60 (2014), pp. 2949–2970. (Cited on p. 139)

[128] A. JADBABAIE, J. LIN, AND A. MORSE, *Coordination of groups of mobile autonomous agents using nearest neighbor rules*, IEEE Transactions on Automatic Control, 48 (2003), pp. 988–1001. (Cited on p. 154)

[129] K. H. JOHANSSON, *The quadruple-tank process: A multivariable laboratory process with an adjustable zero*, IEEE Transactions on Control Systems Technology, 8 (2000), pp. 456–465. (Cited on p. 250)

[130] J. S. JORDAN, *Three problems in learning mixed-strategy Nash equilibria*, Games and Economic Behavior, 5 (1993), pp. 368–386. (Cited on p. 85)

[131] B. JOVANOVIC AND R. W. ROSENTHAL, *Anonymous sequential games*, Journal of Mathematical Economics, 17 (1988), pp. 77–87. (Cited on p. 139)

[132] S. KAKADE, I. LOBEL, AND H. NAZERZADEH, *Optimal dynamic mechanism design and the virtual pivot mechanism*, Operations Research, 61 (2013), pp. 837–854. (Cited on p. 164)

[133] E. KALAI, *Proportional solutions to bargaining situations: Interpersonal utility comparisons*, Econometrica, 45 (1977), pp. 1623–1630. (Cited on p. 59)

[134] ———, *Large robust games*, Econometrica, 72 (2004), pp. 1631–1665. (Cited on p. 139)

[135] E. KALAI AND E. LEHRER, *Rational learning leads to Nash equilibrium*, Econometrica, 61 (1993), pp. 1019–1045. (Cited on p. 85)

[136] E. KALAI AND D. SAMET, *Monotonic solutions to general cooperative games*, Econometrica, 53 (1985), pp. 307–327. (Cited on p. 59)

[137] E. KALAI AND M. SMORODINSKY, *Other solution to Nash's bargaining problem*, Econometrica, 43 (1975), pp. 513–518. (Cited on p. 59)

[138] M. KANDORI, G. J. MAILATH, AND R. ROB, *Learning, mutation, and long-run equilibria in games*, Econometrica, 61 (1993), pp. 29–56. (Cited on pp. 42, 49)

[139] L. KRANICH, A. PEREA, AND H. PETERS, *Core concepts in dynamic TU games*, International Game Theory Review, 7 (2005), pp. 43–61. (Cited on p. 230)

[140] U. KRAUSE, *A discrete nonlinear and non-autonomous model of consensus formation*, in Communications in Difference Equations, S. Elaydi, G. Ladas, J. Popenda, and J. Rakowski, editors, Gordon and Breach, Amsterdam, 2000, pp. 227–236. (Cited on pp. 192, 193)

[141] K. KUGA, *Brouwer's fixed point theorem: An alternative proof*, SIAM Journal on Mathematical Analysis, 5 (1974), pp. 893–897. (Cited on p. 17)

[142] H. J. KUSHNER, *Stochastic Stability and Control*, Academic Press, New York, 1967. (Cited on p. 267)

[143] ——, *Introduction to Stochastic Control Theory*, Holt, Rinehart and Winston, New York, 1971. (Cited on p. 267)

[144] ——, *Stochastic stability*, in Stability of Stochastic Dynamical Systems, Lecture Notes in Mathematics 294, Springer, New York, 1972, pp. 97–124. (Cited on p. 267)

[145] A. LACHAPELLE, J. SALOMON, AND G. TURINICI, *Computation of mean field equilibria in economics*, Mathematical Models and Methods in Applied Sciences, 20 (2010), pp. 1–22. (Cited on p. 139)

[146] A. LACHAPELLE AND M.-T. WOLFRAM, *On a mean field game approach modeling congestion and aversion in pedestrian crowds*, Transportation Research Part B, 15 (2011), pp. 1572–1589. (Cited on p. 127)

[147] J.-M. LASRY AND P.-L. LIONS, *Jeux à champ moyen. I. Le cas stationnaire*, Comptes Rendus Mathematique, 343 (2006), pp. 619–625. (Cited on p. 139)

[148] ——, *Jeux à champ moyen. II. Horizon fini et controle optimal*, Comptes Rendus Mathematique, 343 (2006), pp. 679–684. (Cited on p. 139)

[149] ——, *Mean field games*, Japanese Journal of Mathematics, 2 (2007), pp. 229–260. (Cited on pp. 126, 136, 139, 232)

[150] E. LEHRER, *Allocation processes in cooperative games*, International Journal of Game Theory, 31 (2002), pp. 341–351. (Cited on p. 120)

[151] ——, *Approachability in infinite dimensional spaces*, International Journal of Game Theory, 31 (2002), pp. 253–268. (Cited on pp. 110, 120, 225, 229, 230)

[152] ——, *A wide range no-regret theorem*, Games and Economic Behavior, 42 (2003), pp. 101–115. (Cited on p. 120)

[153] E. LEHRER AND E. SOLAN, *Excludability and bounded computational capacity strategies*, Mathematics of Operations Research, 31 (2006), pp. 637–648. (Cited on p. 120)

[154] ——, *Approachability with bounded memory*, Games and Economic Behavior, 66 (2009), pp. 995–1004. (Cited on p. 120)

[155] E. LEHRER, E. SOLAN, AND D. BAUSO, *Repeated games over networks with vector payoffs: The notion of attainability*, in Proceedings of the International Conference on Network Games, Control and Optimization (NETGCOOP 2011), 2011, pp. 1–5. (Cited on pp. xv, 113, 114, 119, 120, 211, 213, 217)

[156] E. LEHRER AND S. SORIN, *Minmax via differential inclusion*, Convex Analysis, 14 (2007), pp. 271–273. (Cited on pp. 111, 120)

[157] D. LEVHARI AND L. MIRMAN, *The great fish war: An example using a dynamic Cournot-Nash solution*, The Bell Journal of Economics, 11 (1980), pp. 322–334. (Cited on p. 105)

[158] W. LI AND C. G. CASSANDRAS, *A cooperative receding horizon controller for multivehicle uncertain environments*, IEEE Transactions on Automatic Control, 51 (2006), pp. 242–257. (Cited on p. 154)

[159] D. LIBERZON, *Calculus of Variations and Optimal Control Theory: A Concise Introduction*, Princeton University Press, Princeton, NJ, 2012. (Cited on p. 87)

[160] T. M. LIGGETT, *Stochastic interacting systems: Contact, voter, and exclusion processes*, Springer-Verlag, New York, 1999. (Cited on p. 193)

[161] Y. LIU AND K. PASSINO, *Stable social foraging swarms in a noisy environment*, IEEE Transactions on Automatic Control, 49 (2004), pp. 30–44. (Cited on p. 154)

[162] K. A. LOPARO AND X. FENG, *Stability of stochastic systems*, in The Control Handbook, CRC Press, Boca Raton, FL, 1996, pp. 1105–1126. (Cited on pp. 235, 247, 259, 265)

[163] J. LORENZ, *Continuous opinion dynamics under bounded confidence: A survey*, International Journal of Modern Physics C, 18 (2007), pp. 1819–1838. (Cited on p. 193)

[164] D. G. LUENBERGER, *Optimization by Vector Space Methods*, Wiley, New York, 1969. (Cited on p. 251)

[165] Z. MA, D. S. CALLAWAY, AND I. A. HISKENS, *Decentralized charging control of large populations of plug-in electric vehicles*, IEEE Transactions on Control Systems Technology, 21 (2013), pp. 67–78. (Cited on p. 164)

[166] S. MANNOR AND J. S. SHAMMA, *Multi-agent learning for engineers*, Artificial Intelligence (special issue on Foundations of Multi-Agent Learning), 171 (2007), pp. 417–422. (Cited on p. 85)

[167] J. R. MARDEN, G. ARSLAN, AND J. S. SHAMMA, *Cooperative control and potential games*, IEEE Transactions on Systems, Man, and Cybernetics, Part B: Cybernetics, 39 (2009), pp. 1393–1407. (Cited on p. 85)

[168] ———, *Joint strategy fictitious play with inertia for potential games*, IEEE Transactions on Automatic Control, 54 (2009), pp. 208–220. (Cited on p. 85)

[169] J. R. MARDEN AND J. S. SHAMMA, *Revisiting log-linear learning: Asynchrony, completeness, and payoff-based implementation*, Games and Economic Behavior, 75 (2012), pp. 788–808. (Cited on p. 85)

[170] ———, *Game theory and distributed control*, in Handbook of Game Theory, Vol. 4, H. P. Young and S. Zamir, editors, Elsevier, Amsterdam, 2015, pp. 861–899. (Cited on p. 17)

[171] J. R. MARDEN, H. P. YOUNG, G. ARSLAN, AND J. S. SHAMMA, *Payoff-based dynamics for multiplayer weakly acyclic games*, SIAM Journal on Control and Optimization (special issue on Control and Optimization in Cooperative Networks), 48 (2009), pp. 373–396. (Cited on p. 85)

[172] J. MARSCHAK AND R. RADNER, *Economic Theory of Teams*, Yale University Press, New Haven, CT, 1972. (Cited on p. 17)

[173] M. MASCHLER, E. SOLAN, AND S. ZAMIR, *Game Theory*, Cambridge University Press, Cambridge, UK, 2013. (Cited on pp. xv, 16, 67, 109, 120)

[174] J. L. MATHIEU, S. KOCH, AND D. S. CALLAWAY, *State estimation and control of electric loads to manage real-time energy imbalance*, IEEE Transactions on Power Systems, 28 (2013), pp. 430–440. (Cited on p. 164)

[175] D. Q. MAYNE, J. B. RAWLINGS, C. V. RAO, AND P. O. M. SCOKAERT, *Constrained model predictive control: Stability and optimality*, Automatica, 36 (2000), pp. 789–814. (Cited on p. 186)

[176] A. MECA, I. GARCIA-JURADO, AND P. BORM, *Cooperation and competition in inventory games*, Mathematical Methods of Operations Research, 57 (2003), pp. 481–493. (Cited on p. 230)

[177] A. MECA, J. TIMMER, I. GARCIA-JURADO, AND P. BORM, *Inventory games*, European Journal of Operational Research, 156 (2004), pp. 127–139. (Cited on p. 230)

[178] I. MENACHE AND A. OZDAGLAR, *Network Games: Theory, Models, and Dynamics*, Synthesis Lecture, Morgan and Claypool, 2010. (Cited on pp. xiii, 11, 14)

[179] J. F. MERTENS AND A. NEYMAN, *Stochastic games*, International Journal of Game Theory, 10 (1981), pp. 53–66. (Cited on pp. 104, 105)

[180] A. MIRTABATABAEI, P. JIA, AND F. BULLO, *Eulerian opinion dynamics with bounded confidence and exogenous inputs*, SIAM Journal on Applied Dynamical Systems, 13 (2014), pp. 425–446. (Cited on p. 193)

[181] M. MOBILIA, A. PETERSEN, AND S. REDNER, *On the role of zealotry in the voter model*, Journal of Statistical Mechanics: Theory and Experiment, 2007, no. 8. (Cited on p. 193)

[182] J. D. MORROW, *Game Theory for Political Scientists*, Princeton University Press, Princeton, NJ, 1994. (Cited on p. 17)

[183] J. F. NASH JR., *Equilibrium points in n-person games*, Proceedings of the National Academy of Sciences, 36 (1950), pp. 48–49. (Cited on pp. 10, 17)

[184] ——, *Non-cooperative games*, Annals of Mathematics, 54 (1951), pp. 286–295. (Cited on p. 17)

[185] A. NEDIĆ AND D. BAUSO, *Dynamic coalitional TU games: Distributed bargaining among players' neighbors*, IEEE Transactions Automatic Control, 58 (2013), pp. 1363–1376. (Cited on pp. xvii, 59, 186, 197, 198, 200, 202, 203, 204, 220)

[186] A. NEDIĆ, A. OLSHEVSKY, A. OZDAGLAR, AND J. N. TSITSIKLIS, *On distributed averaging algorithms and quantization effects*, IEEE Transactions on Automatic Control, 54 (2009), pp. 2506–2517. (Cited on p. 200)

[187] A. NEDIĆ AND A. OZDAGLAR, *Distributed subgradient methods for multi-agent optimization*, IEEE Transactions on Automatic Control, 54 (2009), pp. 48–61. (Cited on pp. 197, 204)

[188] A. NEDIĆ, A. OZDAGLAR, AND P. A. PARRILO, *Constrained consensus and optimization in multi-agent networks*, IEEE Transactions on Automatic Control, 55 (2010), pp. 922–938. (Cited on pp. 186, 200)

[189] A. NEYMAN AND S. SORIN, *Stochastic Games and Applications*, NATO Science Series C: Mathematical and Physical Sciences, Kluwer, Dordrecht, 2003. (Cited on p. 105)

[190] N. NOAM, T. ROUGHGARDEN, E. TARDOS, AND V. VAZIRANI, *Algorithmic Game Theory*, Cambridge University Press, Cambridge, UK, 2007. (Cited on pp. 17, 40)

[191] A. S. NOWAK, *On a new class of nonzero-sum discounted stochastic games having stationary nash equilibrium points*, International Journal of Game Theory, 32 (2003), pp. 121–132. (Cited on p. 105)

[192] U.S. DEPARTMENT OF ENERGY, *Benefits of Demand Response in Electricity Markets and Recommendations for Achieving Them*, report to the United States Congress, February 2006; available online at http://eetd.lbl.gov, 2006. (Cited on p. 163)

[193] R. OLFATI-SABER, J. A. FAX, AND R. M. MURRAY, *Consensus and cooperation in networked multi-agent systems*, Proceedings of the IEEE, 95 (2007), pp. 215–233. (Cited on pp. 154, 165, 180, 183, 193)

[194] R. OLFATI-SABER AND R. M. MURRAY, *Consensus problems in networks of agents with switching topology and time-delays*, IEEE Transactions on Automatic Control, 49 (2004), pp. 1520–1533. (Cited on pp. 144, 154)

[195] M. J. OSBORNE AND A. RUBINSTEIN, *Bargaining and Markets*, Series in Economic Theory, Econometrics, and Mathematical Economics, Academic Press, New York, 1990. (Cited on p. 17)

[196] ———, *A Course in Game Theory*, MIT Press, Cambridge, MA, 1994. (Cited on pp. 16, 17, 49, 67, 154, 164)

[197] A. OZDAGLAR, *Game Theory with Engineering Applications*, MITOPENCOURSEWARE, available at http://ocw.mit.edu/courses/electrical-engineering-and-computer-science/6-254-game-theory-with-engineering-applications-spring-2010/, 2010. (Cited on pp. xiii, 11, 14, 85)

[198] F. PARISE, M. COLOMBINO, S. GRAMMATICO, AND J. LYGEROS, *Mean field constrained charging control policy for large populations of plug-in electric vehicles*, in Proceedings of the IEEE Conference on Decision and Control, Los Angeles, CA, 2014. (Cited on p. 164)

[199] R. PESENTI AND D. BAUSO, *Mean field linear quadratic games with set up costs*, Dynamic Games and Applications, 3 (2013), pp. 89–104. (Cited on p. 139)

[200] C. PHELAN AND E. STACCHETTI, *Sequential equilibria in a Ramsey tax model*, Econometrica, 69 (2001), pp. 1491–1518. (Cited on p. 105)

[201] A. PLUCHINO, V. LATORA, AND A. RAPISARDA, *Compromise and synchronization in opinion dynamics*, European Physical Journal B—Condensed Matter and Complex Systems, 50 (2006), pp. 169–176. (Cited on pp. 183, 192)

[202] B. POLAK, *Econ 159: Game Theory* (Yale University: Open Yale Courses); available online from http://oyc.yale.edu (Accessed 2007). (Cited on pp. xiv, xv, 70, 72, 73, 74, 75, 76, 77, 78)

[203] L. S. PONTRYAGIN, V. G. BOLTYANSKII, R. V. GAMKRELIDZE, AND E. F. MISHCHENKO, *The Mathematical Theory of Optimal Processes*, Interscience, New York, 1962. (Cited on p. 96)

[204] P. V. REDDY AND J. C. ENGWERDA, *Pareto optimality in infinite horizon linear quadratic differential games*, Automatica, 49 (2013), pp. 1705–1714. (Cited on p. 59)

[205] W. REN AND R. BEARD, *Consensus seeking in multi-agent systems under dynamically changing interaction topologies*, IEEE Transactions on Automatic Control, 50 (2005), pp. 655–661. (Cited on p. 154)

[206] W. REN, R. BEARD, AND E. M. ATKINS, *A survey of consensus problems in multi-agent coordination*, in Proceedings of the American Control Conference, Portland, OR, 2005, pp. 1859–1864. (Cited on pp. 144, 154)

[207] M. ROOZBEHANI, M. A. DAHLEH, AND S. K. MITTER, *Volatility of power grids under real-time pricing*, IEEE Transactions on Power Systems, 27 (2012), pp. 1926–1940. (Cited on pp. 164, 165, 180)

[208] E. ROXIN, *The axiomatic approach in differential games*, Journal of Optimization Theory and Applications, 3 (1969), pp. 153–163. (Cited on p. 120)

[209] W. SAAD, Z. HAN, M. DEBBAH, A. HJÖRUNGNES, AND T. BAŞAR, *Coalitional game theory for communication networks: A tutorial*, IEEE Signal Processing Magazine (special issue on Game Theory), 26 (2009), pp. 77–97. (Cited on pp. 17, 67)

[210] Y. E. SAGDUYU AND A. EPHREMIDES, *Power control and rate adaptation as stochastic games for random access*, in Proceedings of the 42nd IEEE Conference on Decision and Control, Vol. 4, 2003, pp. 4202–4207. (Cited on p. 105)

[211] W. H. SANDHOLM, *Population Games and Evolutionary Dynamics*, MIT Press, Cambridge, MA, 2010. (Cited on pp. 78, 85)

[212] M. SASSANO AND A. ASTOLFI, *Dynamic approximate solutions of the HJ inequality and of the HJB equation for input-affine nonlinear systems*, IEEE Transactions on Automatic Control, 57 (2012), pp. 2490–2503. (Cited on p. 139)

[213] ——, *Approximate finite-horizon optimal control without PDEs*, Systems & Control Letters, 62 (2013), pp. 97–103. (Cited on p. 139)

[214] D. SCHMEIDLER, *The nucleolus of a characteristic function game*, SIAM Journal on Applied Mathematics, 17 (1969), pp. 1163–1170. (Cited on p. 230)

[215] R. SELTEN, *Preispolitik der Mehrproduktenunternehmung in der statischen theorie*, Springer-Verlag, Berlin, 1970. (Cited on p. 139)

[216] A. SENGUPTA AND K. SENGUPTA, *A property of the core*, Games and Economic Behavior, 12 (1996), pp. 266–273. (Cited on p. 229)

[217] J. S. SHAMMA AND G. ARSLAN, *Unified convergence proofs of continuous-time fictitious play*, IEEE Transactions on Automatic Control, 49 (2004), pp. 1137–1142. (Cited on p. 85)

[218] ——, *Dynamic fictitious play, dynamic gradient play, and distributed convergence to Nash equilibria*, IEEE Transactions on Automatic Control, 50 (2005), pp. 312–327. (Cited on p. 85)

[219] L. S. SHAPLEY, *Stochastic games*, in Proceedings of the National Academy of Sciences, 39 (1953), pp. 1095–1100. (Cited on pp. 67, 101, 105, 226, 227)

[220] ——, *Some topics in two-person games*, Annals of Mathematical Studies, 5 (1964), pp. 1–8. (Cited on p. 85)

[221] ——, *On balanced sets and cores*, Naval Research Logistics Quarterly, 14 (1967), pp. 453–460. (Cited on pp. 67, 222)

[222] Y. SHOHAM AND K. LEYTON-BROWN, *Multiagent Systems Algorithmic, Game-Theoretic, and Logical Foundations*, Cambridge University Press, Cambridge, UK, 2009. (Cited on p. 17)

[223] J. M. SMITH, *Game theory and the evolution of fighting*, in On Evolution, Edinburgh University Press, Edinburgh, Scotland, 1972, pp. 8–28. (Cited on p. 78)

[224] ——, *Evolution and the Theory of Games*, Cambridge University Press, Cambridge, UK, 1982. (Cited on p. 78)

[225] J. M. SMITH AND G. R. PRICE, *The logic of animal conflict*, Nature, 246 (1973), pp. 15–18. (Cited on p. 78)

[226] E. SOLAN, *Stochastic games*, in Encyclopedia of Database Systems, Springer, Berlin, 2009; also available at http://www.math.tau.ac.il/~eilons/encyclopedia.pdf, 2009. (Cited on p. 105)

[227] K. C. SOU, H. SANDBERG, AND K. H. JOHANSSON, *Data attack isolation in power networks using secure voltage magnitude measurements*, IEEE Transactions on Smart Grid, 5 (2014), pp. 14–28. (Cited on p. 250)

[228] S. E. Z. SOUDJANI, S. GERWINN, C. ELLEN, M. FRAENZLE, AND A. ABATE, *Formal synthesis and validation of inhomogeneous thermostatically controlled loads*, in Quantitative Evaluation of Systems, Springer-Verlag, Berlin, 2014, pp. 74–89. (Cited on p. 164)

[229] S. A. SOULAIMANI, M. QUINCAMPOIX, AND S. SORIN, *Repeated games and qualitative differential games: Approachability and comparison of strategies*, SIAM Journal of Control and Optimization, 48 (2009), pp. 2461–2479. (Cited on pp. 111, 119, 120)

[230] X. SPINAT, *A necessary and sufficient condition for approachability*, Mathematics of Operations Research, 27 (2002), pp. 31–44. (Cited on p. 120)

[231] D. SUBBARAO, R. UMA, B. SAHA, AND M. V. R. PHANENDRA, *Self-organization on a power system*, IEEE Power Engineering Review, 21 (2001), pp. 59–61. (Cited on pp. 165, 180)

[232] J. SUIJS AND P. BORM, *Stochastic cooperative games: Superadditivity, convexity, and certainty equivalents*, Games and Economic Behavior, 27 (1999), pp. 331–345. (Cited on p. 230)

[233] J. SUIJS, P. BORM, A. DE WAEGENAERE, AND S. TIJS, *Cooperative games with stochastic payoffs*, European Journal of Operational Research, 113 (1999), pp. 193–205. (Cited on p. 230)

[234] A.-S. SZNITMAN, *Topics in propagation of chaos*, in École d'Été de Probabilités de Saint-Flour XIX—1989, Lecture Notes in Mathematics 1464, Springer, Berlin, 1991, pp. 165–251. (Cited on p. 193)

[235] A. TEIXEIRA, D. PEREZ, H. SANDBERG, AND K. H. JOHANSSON, *Attack models and scenarios for networked control systems*, in Conference on High Confidence Networked Systems (HiCoNS), CPSWeek, Beijing, China, 2012. (Cited on p. 250)

[236] H. TEMBINE, J. Y. LE BOUDEC, R. ELAZOUZI, AND E. ALTMAN, *Mean field asymptotic of Markov decision evolutionary games*, in Proceedings of GameNets, 2009. (Cited on p. 139)

[237] H. TEMBINE, Q. ZHU, AND T. BAŞAR, *Risk-sensitive mean-field stochastic differential games*, in Proceedings of the 2011 IFAC World Congress, Milan, Italy, 2011, pp. 3222–3227. (Cited on p. 139)

[238] ———, *Risk-sensitive mean-field games*, IEEE Transactions on Automatic Control, 59 (2014), pp. 835–850. (Cited on p. 139)

[239] S. TIJS, *Introduction to Game Theory*, Hindustan Book Agency, Gurgaon, India, 2003. (Cited on pp. xiii, xiv, 7, 8, 9, 11, 17, 23, 40, 49, 52, 53, 54, 55, 57, 59, 65, 67, 230)

[240] J. TIMMER, P. BORM, AND S. TIJS, *On three Shapley-like solutions for cooperative games with random payoffs*, International Journal of Game Theory, 32 (2003), pp. 595–613. (Cited on p. 230)

[241] ———, *Convexity in stochastic cooperative situations*, International Game Theory Review, 7 (2005), pp. 25–42. (Cited on p. 230)

[242] P. P. VARAIYA, *On the existence of solutions to a differential game*, SIAM Journal on Control and Optimization, 5 (1967), pp. 153–162. (Cited on p. 120)

[243] N. VIEILLE, *Weak approachability*, Mathematics of Operations Research, 17 (1992), pp. 781–791. (Cited on p. 120)

[244] ———, *Equilibrium in 2-person stochastic games* I: *A reduction*, Israel Journal of Mathematics, 119 (2000), pp. 55–91. (Cited on pp. 104, 105)

[245] R. B. VINTER, *Minimax optimal control*, SIAM Journal on Control and Optimization, 44 (2005), pp. 939–968. (Cited on p. 96)

[246] ———, *Optimal Control*, Modern Birkhäuser Classics, Birkhäuser, Boston, MA, 2010. (Cited on p. 96)

[247] J. VON NEUMANN, *Zur theorie der gesellschaftspiele*, Mathematische Annalen, 100 (1928), pp. 295–320. (Cited on pp. 16, 26)

[248] J. VON NEUMANN AND O. MORGENSTERN, *Theory of Games and Economic Behavior*, Princeton University Press, Princeton, NJ, 1944. (Cited on pp. 16, 26)

[249] H. VON STACKELBERG, *Marktform und Gleichgewicht*, Springer-Verlag, Vienna, 1934 (an English translation appeared in 1952 entitled *The Theory of the Market Economy*, published by Oxford University Press, Oxford, UK) (Cited on pp. 16, 49)

[250] J. WEIBULL, *Evolutionary Game Theory*, MIT Press, Cambridge, MA, 1995. (Cited on pp. 78, 85)

[251] G. Y. WEINTRAUB, C. BENKARD, AND B. VAN ROY, *Oblivious equilibrium: A mean field approximation for large-scale dynamic games*, in Advances in Neural Information Processing Systems, MIT Press, Cambridge, MA, 2005. (Cited on p. 139)

[252] L. XIAO AND S. BOYD, *Fast linear iterations for distributed averaging*, Systems & Control Letters, 53 (2004), pp. 65–78. (Cited on pp. 144, 154)

[253] D. W. K. YEUNG AND L. A. PETROSJAN, *Cooperative Stochastic Differential Games*, Springer series in Operations Research and Financial Engineering, Springer, New York, 2006. (Cited on p. 59)

[254] H. P. YOUNG, *The evolution of conventions*, Econometrica, 61 (1993), pp. 57–84. (Cited on pp. 42, 49)

[255] Q. ZHU AND T. BAŞAR, *A multi-resolution large population game framework for smart grid demand response management*, in Proceedings of the International Conference on Network Games, Control and Optimization (NETGCOOP 2011), Paris, France, 2011. (Cited on p. 139)

Index